Ethics of Science and Technology Assessment

Volume 45

Schriftenreihe der EA European Academy
of Technology and Innovation Assessment GmbH
edited by Petra Ahrweiler

More information about this series at http://www.springer.com/series/4094

Kristin Hagen · Margret Engelhard
Georg Toepfer
Editors

Ambivalences of Creating Life

Societal and Philosophical Dimensions of Synthetic Biology

Editors
Kristin Hagen
EA European Academy of Technology and
 Innovation Assessment GmbH
Bad Neuenahr-Ahrweiler
Germany

Margret Engelhard
EA European Academy of Technology and
 Innovation Assessment GmbH
Bad Neuenahr-Ahrweiler
Germany

Georg Toepfer
Center for Literary and Cultural Research
 Berlin
Berlin
Germany

Series editor
Prof. Dr. Petra Ahrweiler
EA European Academy of Technology and
 Innovation Assessment GmbH
Bad Neuenahr-Ahrweiler
Germany

ISSN 1860-4803   ISSN 1860-4811   (electronic)
Ethics of Science and Technology Assessment
ISBN 978-3-319-21087-2   ISBN 978-3-319-21088-9   (eBook)
DOI 10.1007/978-3-319-21088-9

Library of Congress Control Number: 2015946585

Springer Cham Heidelberg New York Dordrecht London

Springer International Publishing AG Switzerland
(www.springer.com)

Kristin Hagen · Margret Engelhard
Georg Toepfer

Editors

Ambivalences of Creating Life

Societal and Philosophical Dimensions
of Synthetic Biology

 Springer

Editors
Kristin Hagen
EA European Academy of Technology and
  Innovation Assessment GmbH
Bad Neuenahr-Ahrweiler
Germany

Margret Engelhard
EA European Academy of Technology and
  Innovation Assessment GmbH
Bad Neuenahr-Ahrweiler
Germany

Georg Toepfer
Center for Literary and Cultural Research
  Berlin
Berlin
Germany

Series editor
Prof. Dr. Petra Ahrweiler
EA European Academy of Technology and
  Innovation Assessment GmbH
Bad Neuenahr-Ahrweiler
Germany

ISSN 1860-4803   ISSN 1860-4811 (electronic)
Ethics of Science and Technology Assessment
ISBN 978-3-319-21087-2   ISBN 978-3-319-21088-9   (eBook)
DOI 10.1007/978-3-319-21088-9

Library of Congress Control Number: 2015946585

Springer Cham Heidelberg New York Dordrecht London

Springer International Publishing AG Switzerland is part of Springer Science+Business Media
(www.springer.com)

EA European Academy
of Technology and Innovation Assessment GmbH

The European Academy

The EA European Academy of Technology and Innovation Assessment GmbH deals with the relation of knowledge and society: Science, technology and innovation change our societies rapidly. They open new courses of action and create opportunities but also introduce unknown risks and consequences. As an interdisciplinary research institute, the EA European Academy analyses and reflects these developments. The EA European Academy was established as a non-profit corporation in 1996 by the Federal German state of Rhineland-Palatinate and the German Aerospace Center (DLR).

The Series

The series Ethics of Science and Technology Assessment (Wissenschaftsethik und Technikfolgenbeurteilung) serves to publish the results of the work of the European Academy. It is published by the academy's director. Besides the final results of the project groups, the series includes volumes on general questions of ethics of science and technology assessment as well as other monographic studies and occasionally proceedings.

Preface

This book is a documentation of the summer school "Analyzing the Societal Dimensions of Synthetic Biology" which took place in Berlin in September 2014. It was organized by the EA European Academy of Technology and Innovation Assessment in co-operation with the Center for Literary and Cultural Research Berlin.

Given the wide variety of disciplines and perspectives, it is almost a little surprising that all summer school participants sat together peacefully—and actually with a lot of fun—for one week. This setting made vivid and intensive communication with other disciplines and stances possible and resulted in an atmosphere of mutual learning and appreciation. As one participant has put it retrospectively, "We're being too nice to each other, we're trying hard to downplay our disagreements". Nevertheless, the different views of the participants are certainly reflected in the contributions to this book. We think that this diversity will be of interest to many of those working in synthetic biology, its societal evaluation, and its philosophical scrutiny.

We are grateful to the summer school participants for their insight, enthusiasm and willingness to share their ideas. Special thanks go to the external experts (Michael Bölker, Nediljko Budisa, Ellen-Marie Forsberg, Christian Illies, Sheref Mansy, Arnold Sauter, Röbbe Wünschiers) for their talks, tutoring and—in the case of Nediljko Budisa—for opening the doors of his laboratory. We would also like to mention that we benefited from synergies with the interdisciplinary European Academy project on synthetic biology (funded by the Klaus Tschira Foundation, results forthcoming in this series). The summer school and the publication of its results have been funded by the German Federal Ministry of Education and Research (BMBF).

Bad Neuenahr-Ahrweiler   Kristin Hagen
Bad Neuenahr-Ahrweiler   Margret Engelhard
Berlin   Georg Toepfer

Contents

Contributors

Carlos G. Acevedo-Rocha studied microbiology and biochemistry at the "Escuela Nacional de Ciencias Biólogicas" (ENCB) of the "Instituto Politécnico Nacional" (IPN) and graduated in 2006 with a Diploma in genomics and cancer research under the supervision of Patricio Gariglio at the "Centro de Investigación y Estudios Avanzados del IPN" (CINVESTAV-IPN), Mexico City, Mexico. From 2007 to 2010, he did a Ph.D. in Molecular Biotechnology in genetic code engineering with unnatural amino acids under the guidance of Nediljko Budisa at the Max Planck Institute of Biochemistry in Martinsried and the Technical University Munich, Germany. From 2011 until 2014, he did postdoctoral work in the area of enzyme-directed evolution with Manfred T. Reetz at the Max Planck Institute of Coal Research in Muelheim an der Ruhr and the Chemistry Faculty of Philipps-University Marburg, Germany. He is currently a SYNMIKRO independent postdoctoral fellow focused on microbial genome engineering at the Max Planck Institute of Terrestrial Microbiology and the LOEWE Centre for Synthetic Microbiology (SYNMIKRO) at the Philipps-University Marburg, Germany.

Mirko Ancillotti studied philosophy at the University of Pisa, Italy. His Master thesis regarded John Harris' influence on contemporary bioethical debate on cloning and enhancement. In 2013, he joined the Centre for Research Ethics and Bioethics, Uppsala University, where is currently employed as research assistant working on a project on synthetic biology outreach and the way it is popularized by scientists and the media. Since 2015, he started working also on a project on the ELSI of antibiotic resistance and alternative antibiotic treatments.

Mickael Baqué studied biochemical engineering and microbiology at the Institut National des Sciences Appliquées (INSA) of Toulouse, France. After finishing his studies in 2009, he worked for one year at the Institut des Biomolécules Max Mousseron (IBMM) in Montpellier, France, on an astrobiological project aiming at developing space biochips for interplanetary exploration. He then pursued is interest on astrobiology by a one-year stage at the European Space Research and Technology Centre of the European Space Agency (ESA-ESTEC) in Noordwijk (The Netherlands) as a young graduate trainee. Finally, he recently obtained his

Ph.D. from the University of Rome Tor Vergata under the supervision of Daniela Billi, where he studied the extremotolerant cyanobacterium *Chroococcidiopsis* in the scope of two space exposure experiments Biofilms Organisms Surfing Space (BOSS) and Biology and Mars EXperiment (BIOMEX) onboard the EXPOSE-R2 platform of ESA.

Daniela Billi is assistant professor of botany at the University of Rome Tor Vergata where she is the leader of the Astrobiology and Molecular Biology of Cyanobacteria laboratory. She graduated in biological sciences in 1992 and obtained a Ph.D. in cellular and molecular biology in 1996 and a specialization in biotechnological applications in 1999. She was a postdoctoral research associate at the Virginia Tech Center for Genomics and at the Polar Desert Research Center, FSU. Her researches focus on molecular mechanisms underlying cyanobacterial survival in extreme deserts, under space and Martian simulations and in space (Expose-R2 missions). The aims are to investigate the endurance of life, to identify biosignatures to search for life on Mars, to validate the lithopanspermia hypothesis and to develop life-support systems. Researches are funded by the Italian Space Agency, the National Antarctic Program and the Ministry of Foreign Affairs.

Andreas Christiansen has studied political science and philosophy at the University of Copenhagen and the University of California, Irvine. He is currently doing his Ph.D. on the ethics of synthetic biology at the University of Copenhagen, where he is affiliated with the Center for Synthetic Biology. His research focuses on the use of analogies and cases in ethical reasoning and on how uncertainty, risk and precaution should be managed in social decisions regarding new technologies.

Tobias Eichinger is a senior research assistant at the Institute for Biomedical Ethics and the History of Medicine, University of Zurich, Switzerland. Dr. Eichinger has degrees in philosophy and film studies from the University of Freiburg and the Free University of Berlin. His research interests focus on philosophical and ethical questions of modern biomedicine and life sciences and on issues of the relationship of film and medicine.

Margret Engelhard studied biology at the Philipps-Universität Marburg, the Max Planck Institute for Terrestrial Microbiology, and the University of Edinburgh. She did a Ph.D. at the University of Basel about agriculturally relevant plants that live in symbiosis with nitrogen-fixing bacteria. From 2004 to 2015, she was a member of staff at the EA European Academy of Technology and Innovation Assessment, where she coordinated a number of interdisciplinary projects focusing on the assessment of biotechnologies, and most recently chaired a project on synthetic biology.

Stefan Eriksson is an associate professor of research ethics at Uppsala University and a senior lecturer at the Centre for Research Ethics and Bioethics, Uppsala University, and serves as editor of the Swedish Research Council's website CODEX. His research interests are autonomy and informed consent, dual use research, the regulation of research and publication ethics.

Daniel Falkner studied philosophy, theatre and media science and Ethik der Textkulturen at the Friedrich-Alexander University Erlangen-Nürnberg and University Augsburg. In 2011–2012, he was teaching assistant at Ethik der Textkulturen, and since 2012 he is Ph.D. student and member of the research group Bioethics at the LOEWE Center for Synthetic Microbiology (SYNMIKRO) at the Philipps-University Marburg where he is involved in the development of an stage model for the ethical assessment of synthetic biology. His dissertation project is about the ethical significance of models and metaphors in life sciences and bio-technologies in general, and on "metaphors of life" in the ethical debate on synthetic biology in particular.

Michael Funk studied philosophy and literature at the Technical University of Dresden, Germany. Since 2009, he is working there as research assistant and tea-cher at the Department for Philosophy of Technology. His current research includes intercultural and transdisciplinary philosophy of technologies and sciences, applied ethics, robotics, and life sciences. Latest publications include the books "'Transdisziplinär' 'Interkulturell'. Technikphilosophie nach der akademischen Kleinstaaterei" (M. Funk ed., Königshausen and Neuman 2015) and "Robotics in Germany and Japan. Philosophical and Technical Perspectives" (M. Funk and Bernhard Irrgang eds., Peter Lang 2014). www.funkmichael.com.

Kristin Hagen studied biology, philosophy and agricultural sciences at the University of Tromsø, Norway, and did her Ph.D. on cattle cognition at the University of Cambridge. She did postdoctoral research on cattle behaviour and welfare at the University of Veterinary Sciences, Vienna, and at the Freie Universität Berlin. Since 2006, she has been involved with technology assessment at the EA European Academy of Technology and Innovation Assessment, focusing on ethics and governance of emerging biotechnologies: "pharming", genetic engineering in livestock, concepts of animal welfare, large animals as biomedical models, and synthetic biology.

Inna Kouper studied information systems and applied informatics at the Moscow Institute of Economics and Statistics in Moscow, Russia. In 2001, she received a Ph.D. in sociology from the Institute of Sociology, Russian Academy of Sciences, Moscow, Russia, and then a Ph.D. in information science from the School of Library and Information Science at Indiana University Bloomington, United States of America, in 2011. Her research interests focus broadly on the material, technological and cultural configurations that facilitate knowledge production and dissemination, with a particular emphasis on critical sociotechnical (STS) approaches to the emerging technologies. Since 2014, she is a research scientist at the Data to Insight Center at Indiana University Bloomington, working on the projects to create and promote research data sharing, curation, and reuse.

Leona Litterst studied biology at the University of Hohenheim, Germany. She did her diploma in 2011 with a major in microbiology and molecular biology. Subsequently, she was member of the DFG research training group "Bioethics" at the International Centre for Ethics in the Sciences and Humanities (IZEW),

University of Tübingen, Germany. In 2015, she did her Ph.D. on ethical aspects of synthetic biology at the University of Tübingen.

Ivan G. Paulino-Lima studied biological sciences at State University of Londrina, Brazil, where he also earned his master's degree in genetics and molecular biology. Subsequently, he did his Ph.D. on biophysics at Carlos Chagas Filho Biophysics Institute, Federal University of Rio de Janeiro, Brazil, with one year of internship at the Open University, England. He got a professor position at State University of Londrina where he taught biochemistry for undergraduate courses. He is currently developing his postdoctoral research on radiation-resistant microorganism at Dr. Lynn Rothschild's lab at NASA's Ames Research Center, USA, where he is also involved in a satellite mission.

Martin Müller studied cultural history and theory, comparative religion, education science, sociology, and psychology in Berlin, Paderborn, and Stellenbosch, South Africa. He is researching as scientific staff member in the Cluster of Excellence "Image Knowledge Gestaltung—An interdisciplinary Laboratory" at Humboldt University of Berlin. Recently, he has also been working as scientific staff at the Institute for Cultural History and Theory. Until 2014, he was a member at the graduate school "Automatisms" at the University of Paderborn, Germany. In his Ph.D. project, he investigates epistemological and media theoretical questions on the intersection of current life sciences, biopolitics, and technosciences with special interest for the discourses of synthetic biology. For his research, he has been visiting the Université Paris-1 Panthéon-Sorbonne in 2013, the Columbia University of New York in 2012, and the University of Cape Town in 2009.

Virgil Rerimassie studied constitutional and administrative law (LLM) and science and technology studies (MA) at Maastricht University. Since 2011, Virgil works as a researcher at the Technology Assessment department of the Rathenau Instituut, based in The Netherlands. He is concerned with examining the ethical, legal and societal implications of emerging technologies and stimulating dialogue hereon. He is primarily involved with projects on synthetic biology, nanotechnology and "NBIC-convergence". Before starting at the Rathenau Instituut, Virgil worked as a policy officer at the Dutch Ministry for Housing, Spatial Planning and Environmental Affairs.

Lynn J. Rothschild has been a leader of NASA's efforts in synthetic biology since 2010. She is an evolutionary biologist/astrobiologist at NASA's Ames, and professor (adjunct) at Brown University and the University of California, Santa Cruz. She has degrees from Yale, Indiana University, and a Ph.D. from Brown University in molecular and cell biology. Her research has focused on how life, particularly microbes, has evolved in the context of the physical environment, both here and potentially elsewhere, and how we might tap into "Nature's toolbox" to advance the field of synthetic biology. Field sites range from Australia to Africa to the Andes, from the ocean to 100,000 feet on a balloon. In the last few years, Rothschild has brought her expertise in extremophiles and evolutionary biology to the field of synthetic biology, addressing on how synthetic biology can enhance NASA's missions. Since 2011, she has been the faculty advisor of the Brown-Stanford award-winning iGEM team, which has pioneered the use of synthetic biology to accomplish NASA's mission, particularly focusing on the human settlement of Mars, astrobiology and biodegradable drones. Her laboratory is working on expanding the use of synthetic biology for NASA with projects as diverse as recreating the first proteins *de novo* to biomining to using synthetic biology to precipitate calcite and produce glues in order to make bricks on Mars or the Moon. Rothschild is a fellow of the Linnean Society of London, the California Academy of Sciences and the Explorers Club, and most recently a winner of the Isaac Asimov Award, and the Horace Mann Medal.

Stefanie B. Seitz studied biology at the Friedrich Schiller University of Jena, Germany. Subsequently, she did her Ph.D. on molecular mechanisms of the circadian clock of *C. reinhardtii* with Maria Mittag in Jena as well. Since 2010, she has been working within consultancy projects at the Institute for Technology Assessement and System Analysis (ITAS) at the KIT in Karlsruhe. Here, her research focused on risk assessment and governance of new and emerging science and technologies including nanotechnology, epigenetics and synthetic biology. Moreover, she is interested in public participation research and the role of public engagement in research and innovation processes.

Johannes Steizinger studied philosophy and politic sciences at the University of Vienna. Subsequently, he did his Ph.D. on Walter Benjamin's youth writings with Elisabeth Nemeth at the University of Vienna and Sigrid Weigel at the Center for Literary and Cultural Research (ZfL) in Berlin. In 2011, he was doctoral research fellow at the Franz Rosenzweig Minerva Research Centre, Hebrew University of Jerusalem. From 2012 to 2014, he went on to do postdoctoral research at the ZfL, first on the philosophical writings of Susan Taubes, then on discourses on life around 1900 and their significance for present issues. Since 2014, he is research fellow of the ERC project "The Emergence of Relativism. Historical, Philosophical and Sociological Perspectives" at Department of Philosophy, University of Vienna, working on the history of the philosophy of life with a special focus on the emergence of, and debate over, relativistic themes.

Walburg Steurer studied political science at the University of Vienna, Austria, and the Institut d'Études Politiques in Lyon, France. Since 2013, she holds a position as university assistant (prae doc) at the Department of Political Science, University of Vienna, where she was also member of the research platform "Life-Science-Governance" until 2015. Currently, she is doing her Ph.D. on the meaning of the person in the persistent vegetative state within end-of-life policy controversies in Italy and Germany. Her main research interests are end-of-life issues, interpretive policy analysis, ethics and governance of synthetic biology, and public engagement.

Georg Toepfer is head of the department "Knowledge of Life" at the Centre for Literary and Cultural Research (ZfL) in Berlin. He studied biology and philosophy and received his diploma in biology from the University of Würzburg, his Ph.D. in philosophy from the University of Hamburg and his habilitation in philosophy from the University of Bamberg. Before starting at the ZfL, he was part of the Collaborative Research Centre "Transformations of Antiquity" at the Humboldt University of Berlin. His principal area of research is the history and philosophy of biology, with a special focus on the history and theoretical role of basic biological concepts.

Cyprien N. Verseux is a Ph.D. student co-directed by Daniela Billi, at the University of Rome Tor Vergata (Rome, Italy) and Lynn Rothschild, at NASA's Ames Research Center (Moffett Field, California). He is conducting astrobiology work, more specifically on the detection of life beyond Earth and on the development of biological life-support systems for manned space exploration. Prior to focusing on astrobiology, he obtained master's degrees in systems and synthetic biology from the Institute of Systems and Synthetic Biology (Evry, France) and in biotechnology engineering from Sup'Biotech Paris (Villejuif, France).

Martin G. Weiss is assistant professor of philosophy at the University of Klagenfurt and working on the bioethical and biopolitical implications of biotechnologies. He collaborates with the Interdisciplinary Research Platform Life-Science-Governance of the University of Vienna and is co-director of the trilateral ELSA-GEN Project DNA and Immigration: Exploring the Social, Political and Ethical Implications of Genetic Testing for Family Reunification. Publications include the following: Strange DNA: The rise of DNA analysis for family reunification and its ethical implications. In: Genetics, Society and Politics Journal, Vol. 7 (2011), 1–19; What's wrong with Biotechnology? Gianni Vattimo's Interpretation of Science, Technology and the Media. In: Between Nihilism and Politics. The Hermeneutics of Gianni Vattimo. Ed. by Silvia Benso. New York. Sunny Press 2010, 241–257; Die Auflösung der Menschlichen Natur. In: Bios und Zoe. Die menschliche Natur im Zeitalter ihrer technischen Reproduzierbarkeit. Ed. by M.G. Weiss. Frankfurt/M. Suhrkamp 2009, 34–55; (With T. Heinemann, I. Helén, Th. Lemke, U. Naue) Suspect Families. DNA Analysis, Family Reunification and Immigration Policies. Farnham: Ashgate 2015.

Britt Wray is a science communicator and documentary maker who has studied biology, fine arts and communications in Canada, the Netherlands and Denmark. She is currently a Ph.D. candidate at the University of Copenhagen in the Department of Media, Cognition and Communication, where she studies science communication with a focus on synthetic biology. For the last several years, she has worked as a radio producer at the Canadian Broadcasting Corporation and as a freelance producer with international broadcasters to make radio programmes about science in society. She is writing her first book, to be released in 2016, about the questions that surround species de-extinction.

Röbbe Wünschiers studied biology at the Philipps-University Marburg, Germany. After his Ph.D. on the physiology and biochemistry of green algal hydrogen metabolism at Marburg, he performed postdoctoral research at Uppsala University, Sweden and the University of Cologne, Germany. At that time, his research focused on the molecular biological and bioinformatic analysis of the regulatory network involved in cyanobacterial hydrogen metabolism. Subsequently, he worked for a BASF Plant Science subsidiary. Since 2009, he is professor for biochemistry and molecular biology at the University of Applied Sciences Mittweida, Germany. In his research, he applies experimental and computational methods to analyse and optimize photobiological hydrogen and fermentative methane production.

Editorial: Ambivalences in Societal and Philosophical Dimensions of Synthetic Biology

Kristin Hagen, Margret Engelhard and Georg Toepfer

1 "An Elixir of Eternal Youth"

[Construction of life] can teach us how to program long-lasting synthetic cells, which, in a more human-oriented application of synthetic biology, could provide us with an 'elixir of eternal youth'. In any case, we have just started to explore the exciting scientific and technological prospects of synthetic biology. (de Lorenzo and Danchin 2008, p. 826)

Is quoting grand visions for synthetic biology a good way to begin a book about its societal implications? Metaphors such as "living machines" and "digitizing life" have been ubiquitous in synthetic biology, but with ambivalent effects. For synthetic biology, on the one hand, futuristic—even biblical—visions have helped to establish the field and secure funding. On the other hand, the field needs to deliver; and vivid metaphors as well as "newness" make it an obvious subject for critical voices and regulatory initiatives. Beyond this political dimension, hype and metaphors of synthetic biology—including the label itself—have been inspiring for more nuanced evaluative efforts.

In some cases, the foci for evaluation may have been too strongly influenced by these dynamics. For example, Johannes Steizinger (this book) argues that philosophers evaluating synthetic biology have been overly interested in the concept of "life", with the topics of biosafety and biosecurity being much more urgent.[1]

---

[1]In policy and social sciences, on the other hand, biosecurity has been a rather prominent topic (Jefferson et al. 2014). Like Steizinger, Jefferson et al. suggest that emphases caused by hype and particular research interests may have contributed to unproductive discussions.

K. Hagen (✉) · M. Engelhard
EA European Academy of Technology and Innovation Assessment,
Bad Neuenahr-Ahrweiler, Germany
e-mail: Kristin.Hagen@ea-aw.de

G. Toepfer
Center for Literary and Cultural Research Berlin, Berlin, Germany

© Springer International Publishing Switzerland 2016
K. Hagen et al. (eds.), *Ambivalences of Creating Life*, Ethics of Science
and Technology Assessment 45, DOI 10.1007/978-3-319-21088-9_1

However, despite potential pitfalls, hype and metaphors should be neither ignored nor avoided. Tobias Eichinger (in this volume) acknowledges that while the ethical debate may have been skewed, it is nevertheless importantly symptomatic in pointing to genuine unease with the underlying worldview reflected in the "creating life" metaphor. And Daniel Falkner (this book) argues that we should not forget the functions of metaphors: they have heuristic value, they can open up new perspectives, and they can influence debates in constructive as well as distorting ways (see also de Lorenzo 2011). With regard to synthetic biology, ignoring the metaphors would be to ignore profoundly important aspects of its impact on society. Several contributions to this volume (e.g., by Inna Kouper; Leona Litterst; Martin Müller) take hype and metaphors seriously in this sense.

In calling for balanced approaches to the description and evaluation of synthetic biology, we might end up downplaying its potential achievements and implications. It may well be that synthetic biology does not, at present, "create" living objects and that it has so far only resulted in few products. But nevertheless, there is significant work being carried out within synthetic biology, and new developments may occur sooner than anticipated, as the case of genome editing has recently shown.

Some of the contributions to this book are about current research in synthetic biology: Carlos Acevedo-Rocha gives an overview of its present state and an analysis of the synthetic biology "tribes" and their historical roots. The chapters by Röbbe Wünschiers and Cyprien Verseux et al. give examples at two very different poles of on-going synthetic biology research: a personal down-to-earth account of the work towards replacing today's fuels with bacteria-produced energy carriers, and a consideration of the use of synthetic biology in a human colonization of Mars. For some of us, it requires considerable imagination to think of potential human settlements on Mars, and it may be premature to consider what the ethical consequences of such a scenario would be (engaging, according to Nordmann (2007), in "speculative" ethics). But nevertheless, for an evaluation of potential societal consequences of an emerging technoscientific field, it is necessary to pay attention to its research agendas. And it can be interesting to even engage in science fiction. In a manner reminiscent of Stanislaw Lem's prose, we *could* imagine that all the different synthetic biology research groups had succeeded in developing their methodologies significantly and were cooperating to really create a living cell, or even a multicellular organism. It could be made from non-living material, with genes made of xeno-nucleic acids. The code could be designed to make the organism produce something useful, be perfectly adapted to life on Mars, or maybe to living on an inhospitable Earth...

Futuristic visions—be it in science or in various genres of art—can thus serve to provoke, antagonize, and stir debate. However, they can also serve habituation—and in the long run acceptance. Markus Schmidt, organizer of the bio:fiction synthetic biology film festivals,[2] explicitly welcomed the habituation effect in the context of the first festival and associated bioart exhibition:

---

[2]www.bio-fiction.com. Accessed 24 June 2015.

Die Synthetische Biologie wird uns wahrscheinlich in extreme Bereiche führen, und insofern kann solche Kunst eine gute Vorbereitung darauf sein. (Interview with Schmidt in Karberg 2012, p. 12)[3]

Especially when engagement of art and science fiction—as in this case—is financed in the context of ELSI activities, or when it is financed by natural science research councils (as in the Synthetic Aesthetics project, see Ginsberg et al. 2014) there is therefore some reason to suspect acceptance-creating agendas. In Oron Catts' words:

[...] it is quite striking to see how artists and designers have been opted to engineer public acceptance for a new technology that does not really exist. (Catts personal communication 2012)

If we think that we can—and sometimes should—influence technological developments in the sense of slowing or stopping them, then this is a genuine dilemma that is well-known from the media: criticism can make people aware of problematic aspects of its object of criticism—and at the same time serve habituation to it.

2 The "Synthetic Biology" Label

There is no agreement about the disciplinary boundaries or definition of synthetic biology, but this is not in itself problematic. One approach to this situation has been to specify which particular *area* of synthetic biology we talk about. Different areas can be identified on the basis of their methods and objects as well as their agendas and philosophies, and a number of well-founded classifications with considerable overlap have so far been suggested (e.g., Deplazes 2009; Krohs and Bedau 2013; see also Acevedo-Rocha in this volume). The identification of areas can be seen as reasonably established and allows delineation of the "synthetic biology" label's scope for specific purposes.

From a legal perspective, if protocell biology and xenobiology are excluded, most organisms produced by synthetic biology fall under the legal regime for genetic engineering and are classified as GMOs (Genetically Modified Organisms) in European law (Breitling et al. 2015; SCENIHR et al. 2015; Winter 2015). From the perspectives of discipline building and governance, broader conceptions of synthetic biology have been more common. Carlos Acevedo-Rocha offers a maximally inclusive notion of synthetic biology: to let "synthetic biology" encompass not only xenobiology and protocell biology, but also traditional genetic engineering

---

[3]Translation (KH): Synthetic biology is likely to lead us into extreme areas, and in this sense, art can be a good preparation.

and the novel genome editing techniques. Reasons for this broad stance include the perceived necessity of interdisciplinary collaborations in discipline building and the interpretation of the virtues of synthetic biology as the new biotechnology paradigm (for a philosophical analysis of this shift, see Michael Funk's chapter in this book). Whether or not the label will be retained in the future is from a broad perspective not the point. "Synthetic biology" as an umbrella term will have served to bring together scientists, to draw the attention of funding bodies, to facilitate scientists' involvement with governance—and maybe also to focus governance activities.

Although there can be good reasons for preferring different scopes of the "synthetic biology" label, their ambiguous use (depending on political context) can blur discussions. In promoting the field, very wide definitions of synthetic biology are more often used when it comes to funding policy, but very narrow definitions are common when regulations are discussed.[4] A similar pattern of politically motivated use also holds for the *novelty* of synthetic biology: stakeholders who promote the field sometimes emphasize its specialness in the context of funding and discipline-building, whereas parallels and continuities of synthetic biology with genetic engineering are stressed when the regulatory context is discussed. A similar pattern has previously been observed in nanotechnology (Shelley-Egan 2010). Critics typically focus on novelty and specialness, unless their criticism targets more fundamental issues. The chapter by Andreas Christiansen offers an analysis of the use and validity of the "Argument from Continuity" in this respect.

The use of the label "synthetic biology" is thus sometimes overstretched on the level of political discussions. Differentiating the discussion with regard to the field of research concerned has been one useful approach to preventing misunderstandings, but this is not sufficient. We suggest elsewhere focusing not on the label "synthetic biology", but on the *features* that synthetic biology brings along and that are relevant to the societal evaluation (Engelhard et al. 2016a, b). Many relevant features are well-known, notably from the context of genetic engineering, and some of these are taken to significantly different levels in synthetic biology: for example, the depth of genetic intervention. Among the most novel features of synthetic biology are the engineering paradigm, the digitization, and the "creation" of life.

---

[4]See for example the Convention on Biological Diversity online discussion about synthetic biology, Topic 3: "Operational definition of synthetic biology, comprising inclusion and exclusion criteria", where operational definitions are discussed, bch.cbd.int/synbio/open-ended/pastdiscussions. shtml#topic3, Accessed 19 June 2015. Jim Thomas from the ETC-group (post [#6829]), for example, suggests a very wide definition, whereas Steven Evans from Dow AgroSciences (post [#6877]) writes that "one line in the sand for separating 'traditional' molecular biology and synthetic biology is the point at which the resulting organism, irrespective of how they were inspired or how they were actualized, can no longer exchange information or transcribe/translate information with its originating species strain or any other 'natural' species." Thus, in effect, Evans suggests restricting an operational definition of synthetic biology to xenobiology.

3 Ambivalent Engagement

In the public sphere, too, we can distinguish the term "synthetic biology" from specific topics or features addressed. More may actually be known about (some of) the actual practices of synthetic biology than the expression itself. This hypothesis is suggested by studies (presented in this book) about media coverage in Sweden and Italy (Ancillotti and Eriksson) and public perceptions in Austria (Steurer). It challenges the much-cited phenomenon of extremely low salience of synthetic biology in the public.[5] Irrespectively of this, levels of knowledge about (aspects of) synthetic biology may rise when some individual event or specific development becomes the cue for public attention.[6] In the meanwhile, a number of initiatives aim at engaging people; for overviews, see the contributions by Stefanie Seitz and Walburg Steurer in this book. In Virgil Rerimassie's chapter, we learn how one leading technology assessment institute, the Dutch Rathenau Instituut, has initiated public engagement with synthetic biology. The Rathenau Institute's agenda has been to stimulate political and public discussion "in a timely manner".

The degree to which synthetic biology is known in the public sphere is interesting for science policy and governance as it is part of the picture used to predict future debates. Over the past decade, there has been ample speculation to this effect, including expectations that synthetic biology could lead to a new and less controversial era in the gene technology debate (Kaiser 2012; Schmidt et al. 2013).[7] However, this is by no means sure, and Volker ter Meulen (2014) was described in a Nature Editorial (2014) as seeing that "storm clouds are gathering on the horizon" (p. 133) because of the activities of the Convention on Biological Diversity.

Regarding the scientific research community, it is characteristic for the construction of the synthetic biology field that there were very early (Cho et al. 1999) initiatives to address regulatory and ethical issues together with social scientists and philosophers. This is an on-going activity; for overviews, see, e.g., Torgersen (2009) and Bensaude Vincent (2013). The science chapters in this book offer examples of genuine synthetic biology perspectives on their field's impacts on society. Carlos Acevedo-Rocha exemplifies the self-conception in some parts of synthetic biology as including social sciences, philosophy and art. The shaping of

---

[5]Synthetic biology has a persistently low level of salience in the public sphere as measured in polls and analyses of media coverage: for overviews and interpretations in this volume, see Ancillotti and Eriksson; Seitz; Steurer.

[6]It was Venter's "artificial cell" for a short period of time. Now it could be genetically edited organisms, which is why the relation of genetically edited organisms with the synthetic biology and GMO labels is a political issue.

[7]According to, for example, ter Meulen (2014), this would be adequate because he thinks synthetic biology could reduce many of the perceived risks of genetic modification. In the interest of public acceptance, it would be attractive to repeat the nanotechnology "success story". In initiatives to this effect, the theoretically outdated one-way science communication model still operates with the expectation that "research in social sciences and humanities [...] can [...] find better ways to communicate the issues" (ter Meulen 2014, p. 135).

goals has become part of this approach, because in the technosciences, problems to be fixed are identified by the field itself. The chapter by Martin Müller offers a critical examination of this "promise culture".

From the perspective of social sciences (e.g., Balmer et al. 2012) and ethics (e.g., van der Burg and Swierstra 2013), the intimate engagement with the emerging field of synthetic biology can be very fruitful. Although the integrated critical activities of social scientists and philosophers may also be perceived as uncomfortable (Rabinow and Bennett 2012), it is certainly the trend in current science policy to support "upstream" engagement of the social sciences and humanities as well as the public. However, social scientists and philosophers may "go native" (Zwart et al. 2014), become *too* integrated (Bensaude Vincent 2013; Myskja et al. 2014).[8] According to Jones (2014), "[s]ocial scientists are permanently cautious that they are being co-opted into a project of generating public acceptance for new technologies, and rightly so" (p. 28). As Myskja and Heggem (2006) put it, they can be perceived as "Trojan horses" but also become "useful idiots". Inna Kouper's chapter in this book outlines a critical participatory framework to take these factors into account.

4 Unease

Participants in Walburg Steurer's citizen panels (this volume) expressed discomfort with politics underlying synthetic biology and associated science communication and engagement. There was suspicion that synthetic biology goals might be presented to make people think it is about solving global problems, when the real motive might be to maintain an economical system by creating new markets. Similar problems are touched on by Leona Litterst in her chapter (this volume) about the images of "play" in synthetic biology.

Another source of unease with synthetic biology are its hybrid objects. For Tobias Eichinger (this volume), ethical objections against the concept of creating life are symptomatic for unease with the blurring and transgression of ontological boundaries between technology and nature that are pushed to new extremes in synthetic biology. But this blurring of boundaries does not imply steps towards a biocentric worldview, on the contrary:

> Synthetic biology is thus valued as the continuation of the long-term process of emancipation from nature, equaled with civilization. Far from being a philosophical watershed, it reaffirms that humans are in command of nature. (Bensaude Vincent 2013, p. 373)

The intense efforts of synthesizing life, and the far-reaching aspirations associated with it, are bound to enforce the mind-set from which they result: the attempt

---

[8]This is remeniscent of the classical anthropology dilemma: perspectives from within and from outside of a community cannot be taken at the same time, and experience "from within" will influence later perspectives "from outside". However, in this case, beyond understanding the (research) culture, its critical evaluation is at stake.

to put life under the power of human disposal at all levels. Underlying worldviews fancying synthetic biology may be reductionist or not, and may or may not be a threat to how life is valued (cf. Bensaude Vincent 2013; Eichinger in this volume), but they are definitely views in which humans have dominion over nature and in which the world's "grand challenges" are meant to be solved with the aid of (bio-) technology. One problem is that this agenda fits nicely with the wish of industrialized (including newly industrialized) nations to secure and expand their positions in the global research and development society, develop new markets and ensure economic growth; for Germany, see for example the bioeconomy agenda of the Federal Ministry for Education and Research (BMBF 2010).

According to Martin Weiss (this volume), from a Heideggerian point of view, technology is unavoidable—although there can be "meditative thinking" about it. What can this entail? Britt Wray (this volume) suggests that a polyphony of voices should be included in slowing down our thinking about synthetic biology. In the meanwhile, continued scrutiny can only be encouraged.

References

Balmer A, Bulpin K, Calvert J et al (2012) Towards a manifesto for experimental collaborations between social and natural scientists. http://experimentalcollaborations.wordpress.com. Accessed 24 June 2015

Bensaude Vincent B (2013) Ethical perspectives on synthetic biology. Biol Theory 8:368–375. doi:10.1007/s13752-013-0137-8

Breitling R, Takano E, Gardner TS (2015) Judging synthetic biology risks. Science 347(6218):107

BMBF (2010) Referat Bioökonomie. Nationale ForschungsstrategieBioÖkonomie 2030. Unser Weg zu einer bio-basierten Wirtschaft. BMBF, Bonn, Berlin

Cho MK, Magnus D, Caplan AL et al (1999) Ethical considerations in synthesizing a minimal genome. Science 286(2087):2089–2090

De Lorenzo V (2011) Beware of metaphors: chasses and orthogonality in synthetic biology. Bioeng Bugs 2:3–7. doi:10.4161/bbug.2.1.13388

De Lorenzo V, Danchin A (2008) Synthetic biology: discovering new worlds and new words. EMBO Rep 9:822–827

Deplazes A (2009) Piecing together a puzzle. EMBO Rep 10(5):428–432

Engelhard M, Bölker M, Budisa N (forthcoming 2016a) Everything new or the same old story? In: Engelhard M (ed) Synthetic biology analyzed. Tools for discussion and evaluation. Springer, Berlin

Engelhard M, Bölker M, Budisa N et al (forthcoming 2016b) The new worlds of synthetic biology. In: Engelhard M (ed) Synthetic biology analyzed. Tools for discussion and evaluation. Springer, Berlin

Ginsberg AD, Calvert J, Schyfter P et al (2014) Synthetic aesthetics: investigating synthetic biology's designs on nature. MIT Press, Massachusetts

Jefferson C, Lentzos F, Marris C (2014) Synthetic biology and biosecurity: how scared should we be? King's College London, London

Jones RAL (2014) Reflecting on public engagement and science policy. Public Underst Sci 23:27–31. doi:10.1177/0963662513482614

Kaiser M (2012) Commentary: looking for conflict and finding none? Public Underst Sci 21:188–194

Karberg S (2012) Synthetische Biologie in der Kunst: Spiegel für die Forschung. In: genosphären – Zeitschrift des Österreichischen Genomforschungsprogramms GEN-AU 11/12, pp 12–13. http://www.markusschmidt.eu/wp-content/uploads/2012/03/Genos11_2012.pdf. Accessed 17 June 2015

Krohs U, Bedau M (2013) Interdisciplinary interconnections in synthetic biology. Biol Theory 8:313–317. doi:10.1007/s13752-013-0141-z

Myskja B, Heggem R (2006) The human and social sciences in interdisciplinary biotechnology research: Trojan horses or useful idiots? In: Kaiser M, Lien M (eds) Ethics Polit. Food Prepr. 6th Congr. Int. Soc. Wageningen Academic Pub, pp 138–142

Myskja B, Nydal R, Myhr A (2014) We have never been ELSI researchers—there is no need for a post-ELSI shift. Life Sci Soc Policy 10:9. doi:10.1186/s40504-014-0009-4

Editorial Nature (2014) Tribal gathering. Nature 509(7499):133

Nordmann A (2007) If and then: a critique of speculative nanoethics. Nanoethics 1:31–46. doi:10.1007/s11569-007-0007-6

Rabinow P, Bennett G (2012) Designing human practices: an experiment with synthetic biology. University of Chicago Press, Chicago

Schmidt M, Meyer A, Cserer A (2013) The bio: fiction film festival: sensing how a debate about synthetic biology might evolve. Public Underst Sci. doi:10.1177/0963662513503772 online before print

Shelley-Egan C (2010) The ambivalence of promising technology. Nanoethics 4:183–189. doi:10.1007/s11569-010-0099-2

SCENIHR (Scientific Committee on Emerging and Newly Identified Health Risks), SCHER (Scientific Committee on Health and Environmental Risks), SCCS (Scientific Committee on Consumer Safety) (2015) Synthetic Biology II—risk assessment methodologies and safety aspects, Opinion. European Union

Ter Meulen V (2014) Time to settle the synthetic controversy. Nature 509:135. doi:10.1038/509135a

Torgersen H (2009) Synthetic biology in society: learning from past experience? Syst Synth Biol 3:9–17. doi:10.1007/s11693-009-9030-y

van der Burg S, Swierstra T (2013) Ethics on the laboratory floor. Palgrave-Macmillan, London

Winter G (2015) The regulation of synthetic biology by EU law: current state and prospects. In: Giese B, Pade C, Wigger H, von Gleich A (eds) Synthetic biology. Character and impact. Springer, Berlin

Zwart H, Landeweerd L, van Rooij A (2014) Adapt or perish? Assessing the recent shift in the European research funding arena from "ELSA" to "RRI". Life Sci Soc Policy 10:11. doi:10.1186/s40504-014-0011-x

The Synthetic Nature of Biology

Carlos G. Acevedo-Rocha

1 Introduction

1.1 Synthetic Life Preamble

The basic unit of life is the cell, with the capacities of genetic heredity and evolution as unique hallmarks. For centuries, biology has been the science of life focused on the *analysis* of the microscopic (membranes, cells, tissues, etc.) and macroscopic (insects, animals, plants, etc.) worlds, but it has gradually adopted *synthesis* as a means to understand biological systems since the late 19th century. In 1899, for instance, in a manner that is reminiscent of today's media, the Boston Herald newspaper reported sensationally the work of the US-German biologist Jacques Loeb as the "creation of life" (Ball 2010). Loeb (1899) is known for his invention of artificial parthenogenesis: embryonic development was induced by treating sea urchin eggs with inorganic salts. Loeb conceived of living organisms as chemical machines, and he aimed for a synthetic science of life capable of forming new combinations from the elements of living nature, similar to the way in which an engineer sees his work, as practical, useful and controlled (Fangerau 2009). If we also recall the "Synthetic method to understand life" by the French professor in medicine, Stéphane Leduc (1853–1939), the "Creation of new species by experimental evolution" by the Dutch botanist Hugo de Vries (1848–1935), or the "Synthetic new species by genetic engineering" by the American botanist Albert

C.G. Acevedo-Rocha (✉)
Small Prokaryotic RNA Biology Group, Max-Planck-Institut für terrestrische Mikrobiologie, 35043 Marburg, Germany
e-mail: acevedor@mpi-marburg.mpg.de

C.G. Acevedo-Rocha
Landes-Offensive zur Entwicklung Wissenschafltich-ökonomischer Exzellenz (LOEWE) Centre for Synthetic Microbiology (SYNMIKRO), Philipps-Universität Marburg, 35032 Marburg, Germany

© Springer International Publishing Switzerland 2016
K. Hagen et al. (eds.), *Ambivalences of Creating Life*, Ethics of Science and Technology Assessment 45, DOI 10.1007/978-3-319-21088-9_2

Blakeslee (1874–1954), we can see that ideas of "synthetic life" resounded as early as the 1930s (Campos 2009).

During the second half of the 20th century the field of genetic engineering consolidated, allowing for the emergence of a new era in biotechnology. This was made possible by major breakthroughs including:

1. The elucidation of the structure of the molecule responsible for heredity in all living organisms, the double antiparallel deoxyribonucleic acid (DNA) helix, by James Watson and Francis Crick in the 1950s (Nobel Prize in Physiology or Medicine 1962).
2. The discovery of restriction enzymes by Werner Arber, Daniel Nathans and Hamilton Smith in the 1960s (Nobel Prize in Physiology or Medicine 1978).
3. The application of the restriction enzymes for DNA recombinant technology by Paul Berg and colleagues in the 1970s (Nobel Prize in Chemistry 1980).
4. The development of the technology for oligonucleotide synthesis in the 1980s (fundamental work for modern molecular biology not yet awarded a Nobel Prize).
5. The development of the Polymerase Chain Reaction (PCR) for the specific in vitro amplification of DNA by Kary Mullis in the 1980s (Nobel Prize in Chemistry 1993).

During these decades two journal articles referred to the term "synthetic biology": one to highlight the potential impact of recombinant DNA in biotechnological applications (Szybalski and Skalka 1978) and the other to bring up its relevance in the political debate (Roblin 1979). By the 1980s, synthetic biology was defined as "the synthesis of artificial forms of life" (Hobom 1980), but it also was considered synonymous to "bioengineering" by some scientists (Benner and Sismour 2005). During the 1990s, while "designing synthetic molecules" (Rawls 2000) became a common practice biological research especially in the US, the field of metabolic engineering was emerging (Bailey et al. 1990). From a historical perspective, synthetic biology has been evolving from an old genetic (one gene) era towards a younger metabolic (two or more genes) phase, followed by a current genome (dozens of genes) engineering era that will drive us to a "biosystems engineering" (more than one organism) future (Carr and Church 2009).

In fact, with the increasing sequencing of genomes from different species at the end of the 1990s, the genetic program of complex living systems increasingly came to be regarded as 'digital', a view that had been common for some researchers working in the information technology (IT) sector much earlier (Danchin 2009). Since then biology increasingly depends on computers and mathematics for analysing huge amounts of DNA sequencing data (Shendure and Ji 2008). But it was not until the new millennium that the contemporary field of synthetic biology re-emerged from a community of engineers interested in biology. Their migration into biology (Brent 2004) has enabled the application of "engineering principles" like design, modelling, abstraction and modularity of "circuits" in living systems for useful purposes (Endy 2005). The culmination of synthetic biology as an "engineering" discipline has been considered by some bioengineers to be in 2004

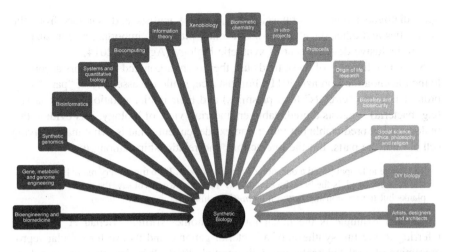

Fig. 1 The different research "tribes" of synthetic biology are, arranged according to the shape of the "penacho" (crest made out of bird feathers) from Moctezuma, the antepenultimate tribe ruler of the Aztec Empire of prehispanic Mexico

during the first international meeting of the BioBricks Foundation at the Massachusetts Institute of Technology (MIT) in Cambridge, USA (Heinemann and Panke 2009).

1.2 Contemporary Synthetic Biology

In 2014, a scientific committee on behalf of the European Union (EU) suggested defining contemporary synthetic biology as *"the application of science, technology and engineering to facilitate and accelerate the design, manufacture and/or modification of genetic materials in living organisms"* (Breitling et al. 2015).[1] In theory, synthetic biology is about engineering and not about science (de Lorenzo and Danchin 2008), but in practice it is composed of different "research tribes" (Nature Editorials 2014) of scientists belonging to large interdisciplinary groups of biologists, chemists, engineers, computer scientists, physicists, mathematicians, scholars from the social sciences, humanities, artists (Reardon 2011) and Do-It-Yourself (DIY) biologists.[2] Figure 1 illustrates the main disciplinary synthetic biology "tribes", but the reader is referred to previous work where the first research networks in synthetic biology were investigated (Oldham et al. 2012). Importantly, an

---

[1]http://ec.europa.eu/health/scientific_committees/emerging/docs/scenihr_o_048.pdf. Accessed 18th May 2015.

[2]A global community of amateur citizens interested in biology, whose garages, closets, and kitchens have been equipped with inexpensive laboratory equipment.

open dialogue among natural scientists, social scientists and scholars from the humanities and other stakeholders has recently played an important role in shaping a more inclusive development of synthetic biology (Agapakis 2014).

Some practitioners of these fields use their own jargon and metaphors according to their agenda (de Lorenzo 2011). For instance, while the assembly of "parts" (e.g. promoters), into "circuits" (e.g. plasmids) and their implementation in a "chassis" (e.g. bacteria) suggests a predictable engineering view of biology, it is evident that biology is not predictable but rather context-dependent. That is, what make a living cell are not the parts, but the interactions and relationships among them:

> The oracle at Delphi posed a question concerning a boat: If, in time, every plank has rotted and been replaced, is the boat the same boat? Yes, the owner will say, the vessel is not its planks but the relationship between them. (Danchin 2003, backcover page).

Another example is the metaphor of "genome writing" (Bedau et al. 2010), referring to the full synthesis of a bacterial genome and its use for cellular reprogramming of a related bacterium (Gibson et al. 2010). It is clear that we are able to copy (DNA sequencing) and print (DNA synthesis) genomes, but we are far away from writing and designing genomes de novo because most of them are not completely understood or the function of many protein-coding genes is still unknown, among other limitations (Porcar and Pereto 2012).

Absolutely, the diversity of the synthetic biology practitioners is vast, and each research "tribe" has an agenda, but instead of promoting particular agendas that could hamper the development of others (i.e. tribalism), it is important to critically assess the various approaches that will evolve into interdependent methodologies in the coming years of research. To this end, synthetic biology can be broadly divided in four main engineering approaches (Fig. 2).

Fig. 2 Proposition of four engineering approaches encompassing all synthetic biology research

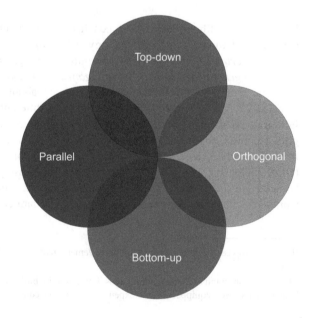

1.2.1 Top-Down Engineering

This approach aims to reduce the complexity of extant cells by comparing universal genes and deleting non-essential ones for constructing a minimal genome (Juhas et al. 2011, 2012). These efforts rest on the idea about the existence of a universal minimal genome that gave rise to all living beings based on the assumption of a unique origin of cellular life, the so-called Last Universal Common Ancestor (LUCA) (Ouzounis and Kyrpides 1996). However, recent research points to the possibility that the tree of life composed of eukaryotic and prokaryotic cells (Eubacteria and Archaea) emerged from a community of primordial cells rather than from a single cell (Kim and Caetano-Anolles 2012). Therefore, while the Holy Grail-like quest for the "minimal genome" has become elusive, eliminating many non-essential functions compromises the fitness of an organism and results in "fragile" genomes (Acevedo-Rocha et al. 2013a).

Another approach in this area involves genome streamlining whereby dispensable DNA elements (mistakenly dubbed "junk DNA" in the past) are deleted to stabilize genomes for optimal performance in many biotechnological applications involving microbes (Leprince et al. 2012; Pal et al. 2014). These efforts coexist with the engineering of genomes (Carr and Church 2009) and epigenomes (Keung et al. 2015) in multiplex or combinatorial manner (Gallagher et al. 2014; Wang et al. 2009; Woodruff and Gill 2011). Importantly, top-down synthetic biology relies on comparative genomics (Abby and Daubin 2007) and proteomics (Nasir and Caetano-Anolles 2013) as well as systems (Lanza et al. 2012) and quantitative (Ouyang et al. 2012) biology. For an example of top-down engineering for the production of hydrogen as clean energy carrier see the chapter by R. Wünschiers in this book.

1.2.2 Bottom-up Engineering

The "bottom-up" synthetic biology approach primarily aims to create "protocells" and find the transition between the non-living and living matter by assembling three components: first, a metabolism for extracting energy from the environment and to construct, salvage and discard aged building blocks; second, an informational program like nucleic acids to control the system; and third, a container bringing these components together for allowing their coordination (Rasmussen et al. 2009). To achieve this, three important principles of self-organization are taken into consideration: reproduction, replication, and assembly. Reproduction (Sole et al. 2007) refers to the ability of a system to reproduce a *similar* copy of itself: cells (composed of container and metabolism) or "hardwares" reproduce; whereas replication (Paul and Joyce 2004) happens when a system replicates an *exact* copy of itself: the genetic program (such as DNA) or "software" is copied. Assembly occurs upon aggregation of vesicles or containers (e.g., Oparin's coacervates) made of small droplets of organic molecules like lipids (Vasas et al. 2012) or liposomes, membrane-like structures containing phospholipids (Oberholzer and Luisi 2002).

Protocell research coexists with other in vitro synthetic biology projects aiming at synthesizing minimal cells (Jewett and Forster 2010), metabolic pathways (Billerbeck et al. 2013) or "never-born proteins" (Chiarabelli et al. 2012), as well as at imitating cellular processes (Forlin et al. 2012) such as cellular division (Schwille 2011) and growth (Blain and Szostak 2014). Although no longer considered synthetic biology research by the current EU definition given above (Breitling et al. 2015), this research, mostly fundamental, deserves proper recognition as synthetic biology research because it has potential impact on other synthetic biology areas such as metabolic engineering by the in vitro optimization of synthetic pathways.

1.2.3 Parallel Engineering or Bioengineering

Parallel engineering research is based on the canonical genetic code that employs standard biomolecules including nucleic acids and the twenty amino acids for engineering biological systems. It includes the standardization of DNA parts (Endy 2005), engineering of switches (Benenson 2012), biosensors (Salis et al. 2009; Zhang and Keasling 2011), genetic circuits (Brophy and Voigt 2014), logic gates (Win and Smolke 2008), and cellular communication operators (Bacchus and Fussenegger 2013) for a wide range of applications in biocomputing (Wang and Buck 2012), bioenergy (Malvankar et al. 2011), biofuels (Kung et al. 2012; Peralta-Yahya et al. 2012), bioremediation (de Lorenzo 2010; Schmidt and de Lorenzo 2012), optogenetics (Bacchus et al. 2013) and medicine (Ruder et al. 2011; Weber and Fussenegger 2012). Most of these applications conventionally rely on the use of one or more vectors (or plasmids) for controlling the expression of two or more genes and/or proteins. Plasmids are small, circular, double-strand DNA molecules that can replicate independently from chromosomal DNA, mostly found in prokaryotic but also sometimes in eukaryotic cells.

A large number of practitioners in this field are engineers aiming to abstract the complexity of biological systems into "parts", "devices" and "systems" whose interactions could be predicted according to the dictum "what I cannot create, I do not understand" by Richard Feynman (Keller 2009). The migration of engineers into biology resulted in the first model-based design of genetic circuits (i.e., "circuit engineering") based on simple mathematical models such as the toggle switch and the "repressilator" (Cameron et al. 2014). In some of these collaborations biologists and engineers work together with computer scientists to develop the next generation computer aided design (CAD) software for engineering-based synthetic biology (MacDonald et al. 2011). In summary, the main goal of bioengineers is to predict the behaviour of living systems for the sake of safety applications, as Drew Endy says:

Engineers hate complexity. I hate emergent properties. I like simplicity. I don't want the plane I take tomorrow to have some emergent property while it's flying.[3]

---

[3]http://www.softmachines.org/wordpress/?p=389. Accessed 20th May 2015.

1.2.4 Orthogonal or Perpendicular Engineering

Also known as "chemical synthetic biology" (Chiarabelli et al. 2012), this approach primarily aims to modify or expand the genetic codes of living systems with unnatural DNA bases (Benner and Sismour 2005; Pinheiro and Holliger 2012) and/or amino acids (Budisa 2004; Liu and Schultz 2010). This subarea also relates to xenobiology, an emergent area at the interface of synthetic biology, exobiology, systems chemistry and origin of life research (Schmidt 2010). In the last decades, scientists have synthesized molecules structurally related to the canonical bases of DNA to test whether those "alien" or xeno (XNA) molecules could be used as carriers of genetic information (Kwok 2012). Similarly, the DNA sugar (desoxyribose) has also been replaced by noncanonical moieties.

The genetic code can also be modified or expanded to express information beyond the 20 canonical amino acids of proteins. One strategy uses orthogonal enzymes and a transfer RNA adaptor from an unrelated organism to incorporate a given unnatural, noncanonical or xeno amino acid (XAA) into one or more proteins at one or more specific sites (Budisa 2014). The orthogonal enzymes are generated by "directed evolution", a method that consists of repeated cycles of gene mutagenesis (genotypic diversity generation), screening or selection (of a particular phenotypic trait), and amplification of an improved variant for the next iterative round (Reetz 2013). Dozens of XAAs have been successfully incorporated into proteins in bacteria, yeast and human cell lines (Liu and Schultz 2010), but also in more complex organisms like worms and flies (Chin 2014). Directed evolution also enables the development of orthogonal ribosomes (based on canonical DNA sequence changes) to facilitate the incorporation of XAAs into proteins (Wang et al. 2007) or of "mirror life", i.e., biological systems endowed with biomolecules composed of enantiomers of opposite chirality (Renders and Pinheiro 2015; Zhao and Lu 2014).

Another method dubbed "experimental evolution" (Kawecki et al. 2012) pushes microorganisms to incorporate XNAs into their genomes (Marlière et al. 2011) or XAAs into their proteomes by serial culturing (Yu et al. 2014). Orthogonal engineering based on experimental evolution also aims for engineering cells that can survive in asteroids, on the moon and even on Mars (Menezes et al. 2015). Although the changes at the DNA level are likely to be canonical, cells suitable for non-terrestrial habitats could be considered as orthogonal life. For more details regarding the colonization of Mars with the aid of synthetic microbes, see chapter by C. Verseux et al. in this book.

2 Synthetic Life Forms

Thanks to the technological advances in molecular biology, organic chemistry, and engineering in the last decades, the ease and speed of genetic modification has enabled the development of emergent Genetically Modified Organisms (GMOs).

Table 1 The various types of emergent GMOs

Abbreviation	Definition	Engineering approach	Reference
GMO	Genetically modified organism	Top-down, parallel, orthogonal	EU committee[a]
GEM	Genetically engineered machine	Parallel	This work
GDO	Genomically designed organism	Bottom-up	This work
GEO	Genomically edited organism	Top-down	This work
GRO	Genomically recoded organism	Top-down, parallel, orthogonal	(Lajoie et al. 2013b)
CMO	Chemically modified organism	Orthogonal	(Marlière et al. 2011)

[a]http://eur-lex.europa.eu/legal-content/EN/TXT/?uri=CELEX:32001L0018. Accessed 19th June 2015

For example, the engineering-inspired design of GMOs has resulted in the crowdsourcing of Genetically Engineered Machines (GEMs), while the decreasing costs of synthetic DNA has enabled the reprogramming of Genomically Designed Organisms (GDOs). More recently, cutting-edge molecular tools have allowed for the arrival of Genomically Edited Organisms (GEOs). Finally, evolutionary approaches have accelerated the development of not only Genomically Recoded Organisms (GROs) harbouring expanded genetic codes, but also Chemically Modified Organisms (CMOs) endowed with unnatural DNA bases or amino acids. In this work, the three first definitions are introduced, whereas the two latter ones have already been proposed by others (Table 1).

In the following sections, the various types of GMOs (mostly microbial[4]) are introduced together with methods for their development and potential applications in biotechnology. Since GMEs, GEOs, GROs are modified using synthetic DNA, all these fall under the GMO definition (see next subsection). CMOs can be considered GMOs or not GMOs depending on whether their genetic changes are only induced by genetic modification or serial cultivation, respectively. The main purpose of showing existing and emergent types of GMOs and non-GMOs (e.g. CMOs) is to illustrate the most recent efforts undertaken by particular research "tribes" and to categorize them according to their engineering approach(es) (for an overview, see Fig. 3). The biosafety and biosecurity issues that these synthetic life forms may represent are subsequently highlighted. In this manner, it is easier to explore the context of a particular GMO/CMO regarding its history, origin, methodology, possible applications and risks in order to enable a better understanding of the organism and possibly an accurate technological assessment.

---

[4]The definition of Genetically Modified Microorganism (GMM) is established: http://www.efsa.europa.eu/en/scdocs/doc/374.pdf.

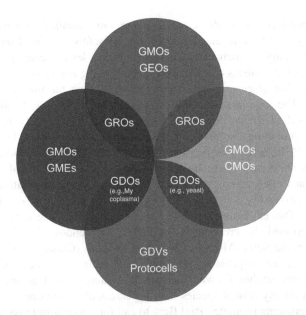

Fig. 3 The synthetic nature of biology. Genetically modified organisms (GMOs) are produced using old and modern genetic engineering tools based on standard nucleic and amino acids as building blocks via top-down (*up*), bottom-up (*down*), parallel (*left*) and orthogonal (*right*) engineering. Engineering-based GMO design and construction has resulted in the crowdsourcing of Genetically Engineered Machines (GEMs) through the famous international GEMs competition (see below). In the bottom-up approach, Genomically Designed Viruses (GDVs) and Organisms (GDOs) can be reprogrammed by assembling commercial synthetic DNA, which can be subsequently transplanted into living cells. The first GDO reported was Mycoplasma, whose genetic code—do not confound with genetic program—was not altered, in contrast to yeast, whose genetic code (and program) is being modified in order to accommodate unnatural amino acids (see below). Genomically Edited Organisms (GEOs) are GMOs whose genetic material has been modified employing cutting-edge molecular tools. Genomically Recoded Organisms (GROs) are GEOs whose genetic code has been modified to accommodate unnatural amino acids (and perhaps unnatural DNA bases in the future). Finally, Chemically Modified Organisms (CMOs) are composed of unnatural nucleic acids and/or amino acids. Note that overlap between top-down (*up*) and bottom-up (*down*) is not shown because no truly synthetic cell, in which all components are synthesized in the lab and assembled into a living organism, thus far exists. There is also no overlap between engineering—(*left*) and evolutionary—(*right*) based approaches because the first aims to predict function from structure, whereas the latter does not predict function because this is extremely challenging if non-additive effects are taken into account (for a discussion on this topic, see Silver et al. 2014)

2.1 Genetically Modified Organisms

The EU law defines Genetically Modified Organisms (GMOs) as living entities whose "*genetic material has been changed in a way that does not occur under natural conditions through cross-breeding or natural recombination*".[5] Internationally, the "Cartagena Protocol on Biosafety to the Convention on Biological Diversity", which includes 170 countries, legally considers GMOs as Living Modified Organisms (LMOs), that is, "*any living organism that possesses a novel combination of genetic material obtained through the use of modern biotechnology*".[6] The forerunners of genetic engineering, Paul Berg, Stanley Norman Cohen and Herbert Boyer, developed molecular tools to introduce foreign genes into bacteria, providing the basis for the development of GMOs. In 1972, Berg combined DNA from the gram-negative model bacterium *Escherichia coli* and the Simian Virus 40 in an exogenous plasmid or vector (Jackson et al. 1972). The following year Cohen demonstrated that DNA from the gram-positive bacterium *Staphylococcus* could be introduced and stably propagated into *E. coli* (Chang and Cohen 1974). This experiment gave rise to the first GMO. The next year, Cohen and Boyer reported the creation of the first *E. coli* endowed with a transgene originally from the South African clawed frog *Xenopus* (Cohen 2013), resulting in the first transgenic organism. By using plasmids as gene vectors, these proof-of-principle studies showed that genetic material could be transferred not only between closely related species but also across unrelated ones.

These experiments prompted Paul Berg to call for a moratorium on recombinant DNA technology to assess its risks (see below), while Cohen and Boyer urged to file a patent for exploiting recombinant DNA technology. At the same time, many biotechnology companies were founded, in 1976, for instance, Boyer co-founded Genentech, one of the first biotech companies that was able to "engineer" GMOs to produce human insulin (Goeddel et al. 1979b) and growth hormone (Goeddel et al. 1979a), among other blockbuster substances including alpha interferon, erythropoietin, and tissue plasminogen activator (Rasmussen 2014). Nowadays, many patients around the globe with diabetes type 1, growth problems and immunological disorders benefit from taking these hormones, sometimes on a daily basis. "*For his fundamental studies of the biochemistry of nucleic acids, with particular regard to recombinant-DNA*", Paul Berg was awarded the Nobel Prize in Chemistry 1980, half of which was shared with Walter Gilbert and Frederick Sanger "*for their contributions concerning the determination of base sequences in nucleic acids*".[7]

This brief introduction to the history of genetic engineering shows not only that scientists have been tinkering with the genomes of microorganisms for almost half a century, but also that this would not have been possible without the development of molecular biology, which was essentially the result of the interdisciplinary collaboration between biologists, physicists, and mathematicians during the second half of the 19th century (Morange 2009).

---

[5]Article 2 of the EU Directive on the Deliberate Release into the Environment of Genetically Modified Organisms (2001/18/EG).

[6]http://bch.cbd.int/protocol/cpb_faq.shtml. Accessed: 4th June 2015.

[7]http://www.nobelprize.org/nobel_prizes/chemistry/laureates/1980/. Accessed 20th May 2015.

During the 1990s, the field of metabolic engineering (two or more genes) emerged as an extension of genetic engineering (one gene) when the chemical engineer James Bailey realized that the microbial production of chemicals and antibiotics could be optimised if the metabolic resources of a cell could be adjusted, concluding that

[...] the emergence of a systematic paradigm for metabolic engineering will transform the present pharmaceutical, food, and chemical industries. (Bailey et al. 1990, p. 15)

The following year, Bailey analysed useful molecules that could be successfully produced in microorganisms (Bailey 1991), while others proposed that the modifications of the carbon metabolic flux should occur only at the principal node of the primary metabolic networks to overproduce desired metabolites (Stephanopoulos and Vallino 1991). These remarkable reviews clearly identified metabolic engineering as an extension of genetic engineering, laying a foundation for the successful field that now allows the production of biofuels (Alper and Stephanopoulos 2009), pharmaceuticals, fine- and bulk-chemicals (Keasling 2010).

Jay Keasling's most recent breakthrough has been the production of a precursor of artemisinin in engineered yeast. Artemisinin is chemically known as a sesquiterpene lactone endoperoxide, the drug of choice to treat malaria, preferentially in combination with other derivatives. Every year, 500 million people become infected with this disease, and 1 million persons die of it in the developing world, mostly in sub-Saharan Africa, Southeast Asia and Latin America. Every 30 s, on average, a child dies from malaria worldwide. Artemisinin is obtained from the leaves of the plant *Artemisia annua*, also known as sweet wormwood. Using its leaves, a simple tea used to be prepared to treat fever, including malaria, in the traditional Chinese medicine but it was not until the 1970s that it was rediscovered for modern medical science. Chinese scientists extracted the active compound artemisinin (known as *arteannuin* in the past or *qinghaosu* in Chinese). However, this drug remained largely unknown to the Western scientific community until 1979 when a review of its application was published in English language (Group 1979). In fact, it took a while for the rest of the world—and especially the World Health Organization (WHO)—to discover the potential of the antimalarial drug (Tu 2011). Unfortunately, it has been argued that there is a global supply shortage of artemisinin owing to an increasing demand and to the environmental factors affecting the harvest of the 14-week-growing-plant, primarily in Chinese and Vietnamese but more recently in Indian and East African farms (Enserink 2005), yet this is a matter of controversy.[8] Artemisinin can also be chemically synthesized, but it is very expensive and non-affordable to most of the patients.

To come up with another source of artemisinin, Keasling and colleagues ingeniously inserted several genes from different species into the baker's yeast *Saccharomyces cerevisiae*, which, fed solely by sugar and basic nutrients, is pushed

---

[8]http://www.etcgroup.org/content/why-synthetic-artemisinin-still-bad-idea-response-rob-carlson. Accessed 20th May 2015.

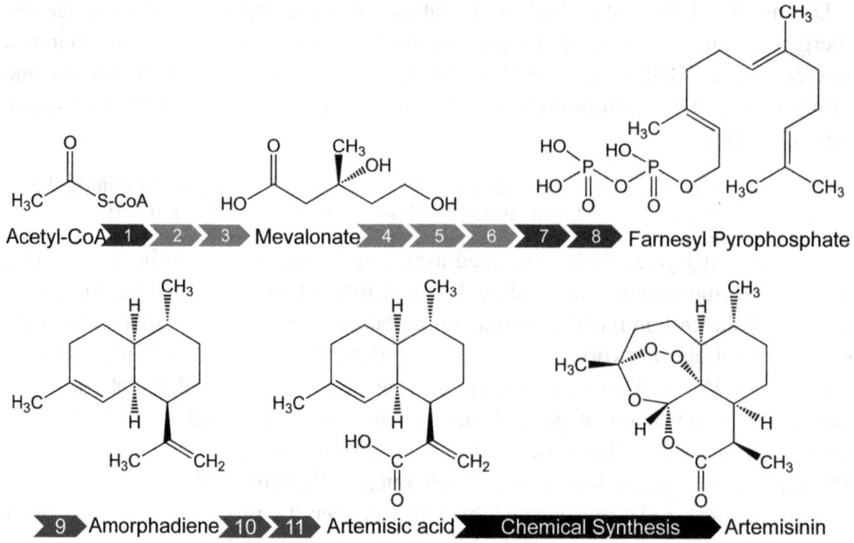

Fig. 4 Keasling's metabolic pathway for semi-synthetic artemisinin. Upon glucose take-up, *E. coli* transforms it into acetyl-CoA via the glycolysis pathway. The introduction of 11 enzymes from *E. coli* (*brown; 1, 7, 8*), *S. cerevisiae* (*orange; 2–6*) and *A. annua* (*green; 9–11*) allows the conversion of acetyl-CoA via the mevalonate pathway into artemisic acid, which can be thereafter chemically converted to artemisinin. Enzymes: *1* AtoB, acetoacetyl-CoA thiolase; *2* HMGS, hydroxymethylglutaryl-CoA (HMG-CoA) synthase; *3* tHMGR, truncated HMG-CoA reductase; *4* MK, mevalonate kinase; *5* PMK, phosphomevalonate kinase; *6* MPD, mevalonate diphosphate decarboxylase; *7* idi, isopentenyl diphosphate isomerase; *8* ispA, Farnesyl pyrophosphate synthase; *9* ADS, amorpha-4,11-diene synthase; *10* CPR, cytochrome p450 redox partner; and *11* P450, monooxygenase (CYP71AV1) (Color figure online)

to start a 3-day-culture production of artemisic acid, a precursor that can be chemically oxidized by standard procedures to artemisinin (Ro et al. 2006). The process of producing artemisinin has been shortened to only 14 days, but this took Keasling and colleagues many years of research with both successes and failures (Keasling 2008). For example, suboptimal expression of exogenous genes (having different genetic code usage) or volatility of molecules were encountered, but these problems were solved by optimizing synthetic genes (Martin et al. 2003) and reaction conditions (Newman et al. 2006). In the end, Keasling's group was able to produce significant levels (>100 mg/L) of artemisic acid by engineering the mevalonate metabolic pathway with a total of 11 enzymes from *E. coli*, the yeast *S. cerevisiae* and the plant *A. annua* in *E. coli* (Fig. 4) (Chang et al. 2007).

The metabolic pathway engineered in *E. coli* was later transferred to a genetically stable yeast strain that can efficiently transport out of the cells up to 115 mg of artemisic acid per litre of culture, allowing therefore a simple and inexpensive purification process (Ro et al. 2006). Recently, German scientists reported the development of an optimized system where the three-step chemical synthesis from artemisic acid to artemisinin can be reduced to a more economical and efficient single-step based solely on oxygen and light (Levesque and Seeberger 2012). This achievement should contribute to increasing the yields (hormones are produced in the gram range) while reducing artemisinin costs. A dose cost $2.40 United States (US) dollars several years ago, but the alliance of the first non-profit pharmaceutical company "Institute for OneWorld Health" and the company co-founded by Keasling "Amyris Biotechnologies" planed to decrease the dose costs ten-fold (ca. $0.25) with aid of $42.6 million from the Bill and Melinda Gates Foundation (Towie 2006), and the cooperation with the French international company Sanofi-Aventis for scaling-up the industrial process. Thus far, more than 1.7 million of semi-synthetic artemisinin doses have been shipped to malaria-endemic countries in Africa including Burkina Faso, Burundi, Democratic Republic of the Congo, Liberia, Niger, and Nigeria.[9]

2.2 Genetically Engineered Machines

Genetically Engineered Machines (GEMs) are herein defined as GMOs that emerge yearly thanks to a crowd-sourcing approach: The international Genetically Engineered Machines (iGEM) competition. Undergraduate students pay a large registration fee for getting standardized DNA parts or "BioBricks" (BioBricks can be thought of as a type of DNA lego; for a discussion about the "playing" component in synthetic biology, see chapter by L. Litterst in this book), mostly originated from the bacterium *E. coli,* at the Registry of Standard Biological Parts (http://parts.igem.org) in MIT to construct at home, in less than a year, biological systems composed of "parts", "devices" and "systems" (Smolke 2009). For example, the bacterium *E. coli* has been converted into "Eau D'e coli", bacteria that smell like wintergreen or bananas depending on the growth state (MIT iGEM team in 2006); *E. chromi,* bacteria that glow in different colours (Cambridge iGEM team in 2009); or *E. cryptor,* a "hard-drive" device to store information (CU-Hong Kong iGEM team in 2010). The main difference between GEMs and GMOs is that the former ones are built with BioBricks and following engineering principles, whereas the latter ones are constructed without these two requirements. The idea behind this distinction is to emphasize the development of potential applications based on GEMs that can be predictable, in contrast to "standard" GMOs where this endeavour is not per se attempted from the onset.

Besides promoting creativity and innovation, iGEM serves as an international platform for fostering in young students self-confidence and awareness of the ethical, legal, and social implications of their synthetic biology project beyond the bench, i.e., "human practices" or more recently "policy and practices". The first real iGEM competition took place in 2004, when 5 teams (Boston University, Caltech, MIT, Princeton University, and The University of Texas at Austin) participated,

---

[9]http://www.path.org/news/press-room/685/. Accessed 3rd June 2015.

Table 2 Evolution of the iGEM competition

Year	Number of teams	Number of new parts	Number of cumulative parts
2004	5	50	50
2005	13	125	175
2006	32	724	899
2007	54	800	1699
2008	84	1387	3086
2009	112	1384	4470
2010	130	1863	6333
2011	165	1355	7688
2012	190	1708	9396
2013	215	1708	11,104
2014	245	n.a.y	n.a.y
2015	281	n.a.y	n.a.y

http://igem.org. Accessed 4th June 2015
n.a.y Not available yet

depositing a total of 50 parts. The University of Texas team designed the first biological photographic film with a bacterial lawn displaying the phrase "Hello World"[10] (Levskaya et al. 2005). Since its first meeting, the iGEM competition has grown steadily in the number of countries, teams and delivered parts (Table 2).

Interestingly, most of the successful iGEM teams tend to avoid using the registry parts and prefer to deposit new parts, perhaps because a major portion of the parts have not been tested or do not work as expected (Vilanova and Porcar 2014). These facts illustrate not only the context-dependency of biology and the importance of molecular relationships beyond synthetic DNA, but also the challenges that the iGEM competition will meet in the coming years: There is a need to better characterize the existing parts and to increase their quality, and there are additional issues regarding industrial applicability, team judgement and research funding transparency (Vilanova and Porcar 2014). From all the winners at each competition, only a few projects have been published, and even less have a potential industrial application (Vilanova and Porcar 2014). This is why the application of engineering principles in biology based on standardized DNA parts is a challenging endeavour that requires understanding the complexity of gene networks and variability of each cell within heterogeneous cellular populations (Kwok 2010).

2.3 Genomically Designed Viruses and Organisms

Genomically Designed Viruses (GDVs) and Genomically Designed Organisms (GDOs) are viruses or living entities, respectively, that have been reprogrammed

---

[10]This phrase has been a tradition for software developers since it was first introduced in the Bell Laboratories in 1974.

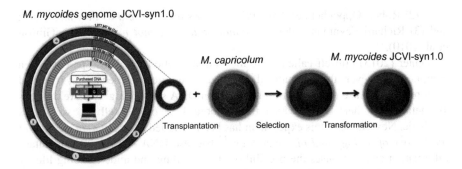

M. mycoides genome JCVI-syn1.0

M. capricolum   M. mycoides JCVI-syn1.0

Transplantation   Selection   Transformation

Fig. 5 Venter's GDO. The genome of *M. mycoides* JCVI-syn1.0 built in vitro by assembling purchased DNA fragments, following by in vivo assembly using yeast. The designed genome was then transplanted into the parent bacterium *M. capricolum*, which upon replication in selective media acquired the phenotype of *M. mycoides*. The genome JCVI-syn1.0 contained four encrypted watermarks sequences indicated by numbers

using a genome that was copied from nature into a computer and later on *designed* to be bottom-up synthetized by chemical means. In this work, these two new definitions are introduced to differentiate from other GMOs in which the genome is borrowed from existing organisms or viruses and is not entirely synthetic (i.e., top-down engineering). The first genome that was chemically synthesized was that of the poliovirus (Cello et al. 2002) that infects humans, followed by that of the bacteriophage Phi X174 (Smith et al. 2003) that normally infects *E. coli*. The genome of the 1918 Spanish' influenza pandemic virus was then synthesized (Tumpey et al. 2005), followed by other human retroviruses (Wimmer et al. 2009). All these GDVs were capable of infecting cells, suggesting that chemically synthesized genomes of higher organisms would be functional if these could be inserted into living cells.

In 2008, the Venter lab reported the synthesis of the complete genome of 582,970 base pairs (bps) of the bacterium *Mycoplasma genitalium* (Gibson et al. 2008), giving rise to the first GDO. In parallel trials, Venter's team transplanted the natural genome of *M. mycoides* into *M. capricolum*, conferring the latter cell the identity of the former upon cell division and genetic selection (Lartigue et al. 2007). Mycoplasmas are parasitic bacteria that cause respiratory and inflammatory diseases in humans. They lack a cell wall but the reason why these bacteria were chosen as model organisms is because they bear the smallest genomes among all bacteria that can support cellular growth in the laboratory. After the two aforementioned breakthroughs, the next logical step was to combine them: In 2010, Gibson et al. reported the genome synthesis of *M. mycoides* genome (1,080,000 bps) and its transplantation into *M. capricolum*, reprogramming again the latter cell into the former, but the difference being that in the genome of the reprogrammed cells there were four encrypted watermark sequences (Fig. 5), indicating the names of 46 persons involved in the project, an email address, a website and three famous quotations: (1) James Joyce: *"To live to err, to fall, to triumph, to recreate life out of life*"; (2) Robert Oppenheimer: "*See things not as they are, but as they might be*"; and (3) Richard Feynman: "*What I cannot build, I cannot understand*" (Gibson et al. 2010).[11]

Venter (2013) himself called these the first synthetic cells whose parents are a computer. However, the computer did not create the genome sequence; it served to store the sequence retrieved from nature upon DNA sequencing. Nor are the cells synthetic: only their DNA, which forms about 1 % of the cell dry weight, is synthetic. Nevertheless, this experiment has been regarded as "*a defining moment in the history of biology and biotechnology*"[12] because DNA controls the hereditary information and this raises the possibility of controlling and understanding life by using synthetic DNA (Bedau et al. 2010).

The synthesis of genomes is possible via "synthetic genomics" (Montague et al. 2012), an established field that emerged with the technological synergies between synthetic organic chemistry and engineering for high-throughput DNA synthesis (Carlson 2009). Synthetic genomics, in turn, has enabled the emergence of not only other GDOs including the baker's yeast (Annaluru et al. 2014) and the bacterium *Vibrio cholera* (Messerschmidt et al. 2015), but also tools in basic research for assembling efficiently synthetic DNA (Gibson 2011). Last but not least, synthetic genomics promises to revolutionize the medical sector by reducing the time needed for the production of synthetic flu vaccines in case of pandemics from months to days (Okie 2011). Another health care application may be the development of synthetic bacteriophages (bacteria-killing viruses) given the recent emergence of antibiotic resistance, a huge global public health concern. Phage therapy is an old treatment that dates back to more than one century, but with the discovery of antibiotics during the first half of the 20th century; its application remained limited in the Western world (Reardon 2014). Thus, there is a potential in synthetic genomics for developing innovative phage therapies, but limited host range and side effects of bacterial lysis remain, among other non-technical issues (Citorik et al. 2014).

2.4 Genomically Edited Organisms

Genomically Edited Organisms (GEOs) are herein defined as those GMOs whose genomes have been modified with advanced molecular engineering tools. Genome editing can be performed in small or large scale as when respectively mutating a single-nucleotide polymorphism to correct a disease genotype (e.g. sickle-cell anemia) like in human cells (Charpentier and Doudna 2013; Doudna and Charpentier 2014; Gaj et al. 2013), or genome regions in multiplex for the combinatorial optimization of

---

[11]Feynman actually wrote: "What I cannot create, I do not understand".

[12]http://www.theguardian.com/science/2010/may/20/craig-venter-synthetic-life-form. Accessed: June 4th 2015.

metabolic pathways in microbes (Gallagher et al. 2014; Wang et al. 2009; Woodruff and Gill 2011). GEOs are also GMOs but the reason behind this differentiation is to indicate the methodological differences: GMOs are usually modified using natural or synthetic DNA sequences encoded in plasmids, but modifying the chromosomal DNA (using or not plasmids) in GEOs usually involves the employment of synthetic DNA that is bought online from biotech companies.[13]

Chromosomal modifications typically include DNA deletions (knock-out), additions (knock-in), or replacements that are crucial in fundamental research for understanding the function of a given gene, protein and/or genetic element in a physiological context. The foundations of genome editing can be traced down to the discovery of the cellular systems in yeast and mammalian cells that repair double-strand DNA breaks (DSBs) that otherwise would be lethal due to cellular death or oncogenic mutations (Doudna and Charpentier 2014). The repair of DSBs is possible by the activation of homologous recombination (HR), which is a "copy and paste" mechanism that requires an undamaged copy of the homologous DNA segment as a template for copying the DNA sequence along the break (Porteus and Carroll 2005).

In the past, UV radiation, chemicals and restriction enzymes were used to induce DSBs, but these were random and could not be directed to predetermined sites (Jasin 1996). Nevertheless, the pioneering work of Mario Capecchi, Martin Evans and Oliver Smithies lead to a basic understanding of the HR mechanism for repairing DSBs. In 2007, they shared The Nobel Prize in Physiology or Medicine *"for their discoveries of principles for introducing specific gene modifications in mice by the use of embryonic stem cells"*.[14] Since the 1990s, new genetic tools have been developed for modifying genomes more precisely, including rare-cutting homing endonucleases (Jasin 1996), Zinc-Finger Nucleases (ZFNs) (Porteus and Carroll 2005) and Transcription Activator-Like Effector Nucleases (TALENs) (Sun and Zhao 2013). By specifically inducing DSBs, these tools enable high HR that can be used in basic biology and to correct various disease-causing mutations associated with haemophilia, sickle-cell disease, and other deficiencies (Gaj et al. 2013).

Although ZFNs and TALENs nucleases provide access to most of the recent health care applications (Gaj et al. 2013), both depend on custom-made proteins for modifying each DNA target, which limits their use due to large costs when multiple genes are mutated simultaneously, as is necessary in more complex diseases involving multiple genes (Cox et al. 2015). Nonetheless, a new tool dubbed CRISPR-Cas9 has recently emerged for editing genes in multiplex with excellent HR efficiencies comparable to ZFNs and TALENs, but lower costs thanks to the programmability at the RNA level (Mali et al. 2013). Composed of clustered regularly interspaced short palindromic repeats (CRISPRs) of DNA and CRISPR-associated

---

[13]Currently there is legal uncertainty, depending on the country, as to whether GEOs can be categorized as GMOs or as organisms originated from breeding. See, for example, Araki and Ishii (2015) as well as the Cibus case in Germany: http://www.nature.com/news/seeds-of-change-1. 17267.

[14]http://www.nobelprize.org/nobel_prizes/medicine/laureates/2007/. Accessed 20th May 2015.

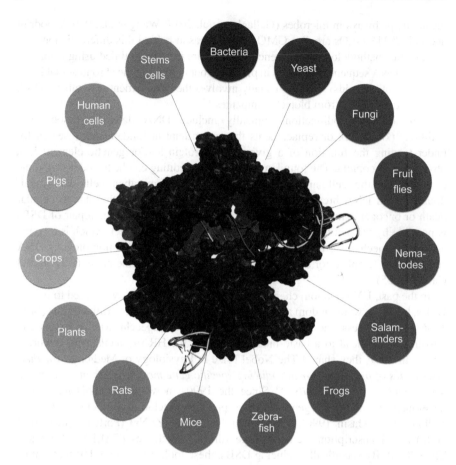

Fig. 6 The CRISPR-Cas9 technology has been used to modify the genomes of bacteria, yeast, fungi, nematodes, salamanders, frogs, fruit flies, zebrafish, mice, rats, plants, crops (rice, wheat, sorghum, tobacco), pigs, animal and human cell lines as well as embryonic stem cells. The *Streptococcus pyogenes* Cas9 nuclease (*dark blue*) in complex with single-guided RNA (*green*) and its target DNA (*red*) was made using the 3D crystal structure PDB (Protein Data Bank) file 4OO8 (Nishimasu et al. 2014) and the PyMOL Molecular Graphics System, version 1.5.0.4 Schrödinger, LLC (Color figure online)

genes (Cas) along the genome, the CRISPR-Cas is an immune system that evolved in bacteria and archaea to combat hostile viruses and foreign plasmids by inducing site-specifically DSBs (Doudna and Charpentier 2014). The type-II CRISPR-Cas9 system has been used for deleting, adding, activating and suppressing target genes with great efficiency in many organisms (Fig. 6) (Charpentier and Doudna 2013; Doudna and Charpentier 2014). The technology has allowed the correction of genetic mutations of diseases such as cataracts and cystic fibrosis as well as the development of cancer models in animal tissues (Doudna and Charpentier 2014), and more recently the generation of human cells resistant to infection by HIV, the Human Immunodeficiency Virus (Liao et al. 2015).

In just two years, (since the beginning of 2013 until the end of 2014) examples of the application of the CRISPR-Cas9 technology in both basic and applied research have been reported and reviewed in more than 1000 publications (Doudna and Charpentier 2014). The revolutionary Cas9 technology has already been compared to the restriction enzymes and the PCR, essential tools in modern molecular biology research, because it promises to accelerate the editing of genomes across the medical, agricultural, environmental, pharmaceutical, chemical, and biotechnological sectors.

2.5 Genomically Recoded Organisms

Genomically Recoded Organisms (GROs) were introduced by the Church lab at Harvard (see below) referring to microbes whose genetic codes have been recoded using advanced genomic tools like those used for GEOs. The genetic code describes a set of rules relating the order of three DNA bases (codon) and a corresponding amino acid to synthesize proteins across all life forms; thus, establishing a universal link between information storage and execution (Fig. 7). For example, a small peptide composed of the amino acids MASTER can be coded at the RNA level by the codons AUG/GCC/AGC/ACC/GAA/AGA, but this order can be modified to synonymous codons (AUG/GCA/UCU/ACG/GAG/CGG) with the same amino acid meaning: MASTER. In addition, non-synonymous ones can also be introduced when another codon (e.g., the amber stop codon UAG) is used to incorporate a non-standard amino acid X as follows: MAXTER (ATG/GCA/UAG/ACG/GAG/CGG).

In 1968, Holley, Khorana and Nirenberg shared the Nobel Prize in Physiology or Medicine "*for their interpretation of the genetic code and its function in protein synthesis*".[15] In the same year, Crick called the genetic code a "frozen accident", implying that "no new amino acid could be introduced without disrupting too many proteins" (Crick 1968, p. 375). Since the 1990s, however, the incorporation of XAAs into proteins has been thawing the "universal" genetic code. There are two basic ways to engineer the genetic code (Bacher et al. 2004): The components involved in the synthesis of proteins are engineered by directed evolution to allow the recognition of specific XAAs (Liu and Schultz 2010). But in these cases the genetic code changes are usually non-heritable, in contrast to the second approach, in which organisms (so far bacteria) are pushed to incorporate XAAs into their proteomes via experimental evolution, resulting in progeny with heritable changes (see CMOs below). Both approaches, nonetheless, could be combined to render offspring dependent on XAAs for survival.

GROs are relatively new organisms that were introduced by George Church and colleagues at Harvard by exploiting the mechanism of HR in microbes. His team built up a device that automates the process of gene delivery, targeting and

---

[15]http://www.nobelprize.org/nobel_prizes/medicine/laureates/1968/. Accessed 20th May 2015.

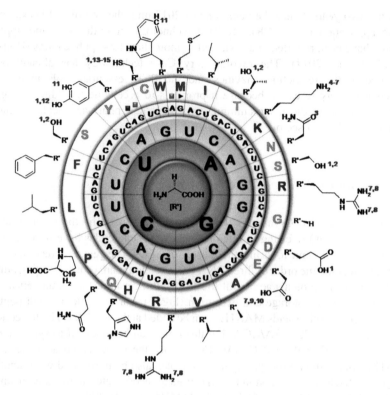

Fig. 7 The universal genetic code in RNA format (bases: AUGC) [RNA uses adenine (A), guanine (G) cytosine (C) and uracil (U) as bases, whereas DNA uses AGC and thymine (T)]. The 20 canonical amino acids are encoded by 61 sense codons. Translation starts (*Black right-pointing pointer*) at AUG codon and terminates (*Black square*) at stop codons UAA (ochre), UGA (opal), or UAG (amber). Amino acids are arranged according to physicochemical properties: polar (*green; T, N, S, G, Q, Y, C*), nonpolar (*red; M, I, A, V, P, L, F, W*), basic (*blue; K, R, H*) and acidic (*pink; D, E*): *M* methionine (AUG); *I* isoleucine (AUA/G/C); *T* threonine (ACG/A/C/U); *K* lysine (AAA/G); *N* asparagine (AAC/U); *S* serine (AGU/C and UCG/A/C/U); *R* arginine (AGA/G and CGG/A/C/U); *G* glycine (GGU/C/A/G); *D* aspartate (GAU/C); *E* glutamate (GAA/G); *A* alanine (GCU/C/A/G); *V* valine (GUU/C/A/G); *H* histidine (CAU/C); *Q* glutamine (CAA/G); *P* proline (CCG/A/C/U); *L* leucine (CUG/A/C/U and UUA/G); *F* phenylalanine (UUU/C); *Y* tyrosine (UAU/C); *C* cysteine (UGU/C); and *W* tryptophan (UGG). Numbers indicate posttranslational modifications, for more details see Acevedo-Rocha (2010) (Color figure online)

replacement as well as microbial recovery and growth called "Multiplex Automated Genomic Engineering" (MAGE) (Wang et al. 2009). MAGE allows performing

> [...] up to 50 different genome alterations at nearly the same time, producing combinatorial genomic diversity. In one instance, Church and Wyss researchers were able to make the bacteria Escherichia coli (E. coli) synthesize five times the normal quantity of lycopene, an antioxidant, in a matter of days and just $1000 in reagents.[16]

---

[16]http://wyss.harvard.edu/viewpage/330/. Accessed 20th May 2015.

Although this machine could accelerate the fields of metabolic and genome engineering for the microbe-based production of biofuels, pharmaceuticals and other chemicals, it is very expensive, and it is still very challenging to know with accuracy what gene(s) and/or protein(s) to target for mutagenesis.

Since 2011, nonetheless, Church and colleagues have reported five "tour-de-force" studies using MAGE with another application in mind: First, they exchanged 314 out of 321 UAG stop codons to synonymous UAA stop ones in several *E. coli* strains, but with some technical hurdles (Isaacs et al. 2011). Second, after overcoming the technical difficulties, they were able to recombine all strains and delete completely the 321 UAG codons as well as the protein that terminates protein synthesis at this signal for yielding the first GRO: *E. coli* C321.ΔA. This bacterium exhibited improved efficiencies for incorporating XAAs into various proteins and increased resistant against infection by the bacteriophage T7 compared to the parental *E. coli* strain MG1655 (Lajoie et al. 2013b). Third, they probed the limits of MAGE by eliminating the most rare codons present in 42 essential genes involved in *E. coli* translation as a means to "emancipate" codons for incorporating further XAAs (Lajoie et al. 2013a). In this work, it was realized that all non-essential genes could be modified with synonymous codons without compromising cellular fitness. Finally, based on the previous work and the *E. coli* C321.ΔA strain, on one hand, Church and colleagues, and, on the other, Isaacs et al. engineered essential genes to be functional by the rational dependence of XAAs, thus imposing the cells to be metabolically dependent on the external supply of XAAs (Rovner et al. 2015; Mandell et al. 2015). Both studies were able to show that *E. coli* can be contained in physical isolation without undetectable growth in liquid media for up to 14 or 20 days, as long as the XAAs was not added to the culture. The authors argue that their strategy *"is a significant improvement over existing biocontainment approaches"* (Rovner et al. 2015) and that it *"provides a foundation for safer GMOs that are isolated from natural ecosystems by a reliance on synthetic metabolites"* (Mandell et al. 2015). Beyond biocontainment, other applications of GROs would be the production of proteins endowed with XAAs for basic research and perhaps biocatalysis (Budisa 2014).

2.6 Chemically Modified Organisms

Chemically Modified Organisms (CMOs) are livings systems endowed with unnatural, noncanonical or xeno building blocks. The CMO term was introduced by Marlière et al. (2011), whose work was highlighted by Acevedo-Rocha and Budisa (2011) referring to *"E. chlori"* (see below). There are two basic approaches for creating CMOs. One strategy tackles proteins, the main executors of information, whereas the other one deals with the information carriers or nucleic acids (NA). As indicated in the previous section, together with some genetic tricks, the cultivation of microbes in the presence of unnatural building blocks allows their introduction in lieu of canonical amino acids or DNA. In 1983, *Bacillus subtilis* strain QB928 was able to grow on 4-fluorotryptophan, a synthetic analogue of tryptophan (Trp), one of the 20 canonical amino acids (Wong 1983). Thirty years later, the same microbe does not require Trp anymore for propagation, but only 4-fluorotryptophan thanks to its adaption that was accompanied by a relatively small degree of genomic changes (Yu et al. 2014). This is the first CMO whose proteome has been chemically modified by experimental evolution. *E. coli* has also been shown to grow on similar fluorinated Trp analogues (Bacher et al. 2004), but it also has the potential for accommodating other XAAs into its proteome (Bohlke and Budisa 2014).

CMOs with unnatural building blocks as genetic polymers have been also reported. DNA is composed of the bases adenine (A), guanine (G) cytosine (C) and thymine (T), each linked to a sugar (deoxyribose), which in turn is connected to two phosphate groups. But scientists have synthesized molecules structurally related to the components of A, T, C and G to test whether these "alien" molecules could be also used as carriers of genetic information, including *iso*-C, *iso*-G, K, X, Q, F, P, Z, NaM and 5SICS (Fig. 8) (Benner and Sismour 2005; Kwok 2012). Likewise, the sugar deoxyribose has been replaced by threose (TNA), arabinose (ANA), glycerol (GNA), hexitol (HNA), cyclohexene (CeNA), fluoro arabinose (FANA), etc. (Pinheiro and Holliger 2012).

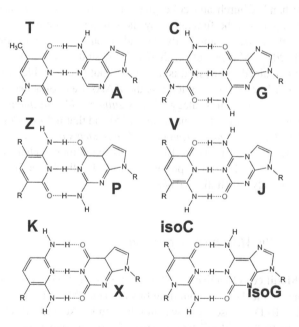

Fig. 8 Alternative genetic systems. On the top, the canonical DNA bases thymine (T) and adenine (A) forming two hydrogen bonds as well as cytosine (C) and guanine (G) forming three hydrogen bonds are shown. In the middle and bottom layers, unnatural DNA bases that are achieved by classical synthetic chemistry are depicted including Z and P, V and J, K and X, as well as *isoC* and *isoG*, all interacting via thee hydrogen bonds

Although most xeno-nucleic acids (XNAs) are incompatible with living systems, very recently Holliger and colleagues described the directed evolution-based engineering of polymerases (enzymes that copy nucleic acids) that were capable of synthesizing XNA from a DNA template and from DNA back to XNA (Pinheiro et al. 2012). This work is very important because it showed that genetic information could be stored in diverse unnatural polymers capable of heredity and evolution. Along the same vein, but using experimental evolution, Marlière and colleagues practically replaced all T bases by 5-chlorouracil in *E. coli* with concomitant dependence on this alien substance for survival (Marlière et al. 2011). These modifications resulted in morphological alterations; yet the *E. coli* cells were able to survive, giving rise to "*E. chlori*", the first CMO bearing a chlorinated genome and potentially endowed with a "genetic firewall" (Acevedo-Rocha and Budisa 2011).

More recently, Romesberg and colleagues reported the addition of the XNA pair NaM and 5SICS to *E. coli* by introducing an algae membrane-protein that allowed the transportation of the XNA pair into the cells, followed by its replication in a plasmid (Malyshev et al. 2014). Although this CMO has been regarded the first living being bearing six bases as genetic code with many potential applications in basic and applied research (Thyer and Ellefson 2014), it should be mentioned that only one base pair per plasmid was present, corresponding to less than 0.0001 % of the total genetic content of that bacterium, which is about 1 % of the total cell dry weight. Thus, the in vivo replication and propagation of a truly six-membered genetic code controlling some biological process remains to be shown.

Several applications of orthogonal engineering have been envisaged. Basic research on "artificial genetic information systems" has already resulted in a more accurate diagnosis of viruses in patients: the inclusion of the bases isoC and isoG (Fig. 8) into "Branched DNA Assays", which detect pathogenic NAs, significantly improves the signal-to-noise detection ratio. These assays are important because the viral load is critical to determine the amount and type of drug necessary in patients with HIV and Hepatitis C, two major health problems with 35 and 180 million cases worldwide, respectively. Using these assays, about 400,000 patients per year can be assigned to a more appropriate treatment and personalized medical care (Benner and Sismour 2005). Similar types of research on other XNAs are being pursued for diagnosing cystic fibrosis, severe acute respiratory syndrome (SARS) triggered by corona-virus, and other pathogens.

Synthetic amino acids have also found applications in human health: clinical trials using a human growth hormone containing a modified XAA recently demonstrated the required safety and efficacy as well as increased therapeutic potency and reduced injection frequency in adults (Cho et al. 2011). In scientific research, while proteins endowed with XAAs have become invaluable tools for unraveling complex cellular processes within living cells (Davis and Chin 2012), unnatural enzymes have potential applications in biocatalysis (Acevedo-Rocha et al. 2013b; Hoesl et al. 2011). That is, the use of enzymes and/or microbes to synthesize useful compounds like food additives, flavors, fragrances, fine chemicals, biofuels, bioplastics, biomaterials (e.g., silk) and many other molecules (Reetz 2013).

3 Biosafety and Biosecurity

3.1 Biosafety: Asilomar Meeting

In February 1975, scientists called for a worldwide moratorium on genetic engineering given the concern of the potential accidental release of GMOs into the environment. Upon the moratorium, some institutions stopped any kind of genetic engineering research, while others took the lead (Danchin 2010), deterring basic science and the development of innovative products in certain regions of the world (ter Meulen 2014). Nevertheless, the moratorium was critical because it allowed the planning of an international meeting in the Asilomar Conference Ground in Pacific Grove, California, USA, where scientists agreed that genetic engineering should continue under stringent guidelines (Berg et al. 1975). This important meeting allowed setting down international guidelines on research involving recombinant DNA (Berg 2008), giving rise to physical containment cautions for biological agents in four "Bio-Safety Levels": 1, 2, 3 and 4.

In the Asilomar meeting, scientists themselves took the initiative to come together. Given the exposure of synthetic biology to the media, and concerns about harmful intentional or accidental misuses of synthetic microorganisms (Ferber 2004), several groups of bioethicists and members of non-governmental organizations have suggested having another Asilomar meeting to discuss the responsibility of synthetic biology research to society. Likewise, debates on synthetic biology under the Convention on Biological Diversity have considered a potential moratorium on the release of synthetic organisms, cells and genomes into the environment (Oldham et al. 2012). Although these debates were triggered after the publication of the first GDO by Venter and colleagues in 2010, biosafety concerns can be justified by the earlier synthesis of pathogenic viruses, such as the deadly 1918 H1N1 "Spanish Flu" (which killed an estimate of up to 50 million people) that was "resurrected" from archaeological samples (Tumpey et al. 2005), or the poliovirus, which was synthesized even without a template (Cello et al. 2002). Even though it has been argued that GDVs offer an opportunity to better understand infectious diseases (Wimmer et al. 2009), these pathogenic viruses also pose potential biosafety risks to the health of the laboratory workers and the public. Two incidents in 2014 at the Centers for Disease Control and Prevention (CDC) in Atlanta, USA (where the Spanish flu virus was synthesized) illustrate this: Samples from a low-virulence flu virus were contaminated with the lethal H5N1 avian flu strain, and bacteria causing deadly anthrax were not properly inactivated before their transportation into a lower biosafety level laboratory (Butler 2014), potentially exposing staff to the dangerous pathogens (Owens 2014).

3.2 Biosecurity: Beyond Asilomar

Whereas biosafety deals with the inherent capability of organisms or viruses to cause disease, biosecurity is mainly concerned with their misuse as biological weapons for bioterrorism. Bioterrorism is nothing new: the intentional release of pathogenic toxins, viruses, bacteria, fungi, and insects is a mode of biological warfare against humans and animals that many countries have used in the past. For example, during World War I, a German secret agent travelled to the USA to infect horses with glanders, a severe infectious disease caused by the bacterium *Burkholderia mallei,* which provokes respiratory ulcers, septicemia and death in horses, donkeys and mules. For this reason, the Geneva Protocol was introduced in 1925 to prohibit the use of chemical and biological weapons in international armed conflicts. Later on, the 1972 Biological Weapons Convention (BWC) and the 1993 Chemical Weapons Convention (CWC) were introduced to regulate the production, storage and transportation of biological and chemical arms.

Concerns over mishandling of harmful biological agents are nothing new to synthetic biology: The CDC in Atlanta, USA, besides harbouring the synthesized "Spanish Flu" virus, also possess dangerous human pathogens such as *Francisella tularensis, Yersinia pestis* and *Bacillus anthracis,* and have had to consider their potential misuse. Indeed, international regulations for gene synthesis were lacking for several years. It was only after 2007, when a journalist ordered online from a biotech company synthetic DNA of the smallpox virus to his home (Grushkin 2010), that the synthetic biology community proposed a set of policies related to the processes involving nucleic acids synthesis (Bugl et al. 2007). Among the new proposed policies is a first set that applies to firms supplying synthetic DNA. Using a database of pathogenic genomes and sequences of toxin genes, the firms should use special software to screen orders for potentially harmful DNA. A second set of proposed rules is aimed at regulating the purchase of DNA synthesizers and the reagent used in synthesis by enforcing the registration of the machines or the distribution of licenses to purchase specific chemicals needed for DNA synthesis.

3.3 New Synthetic Life Forms: Beyond Existing Regulations

Beyond contributing to the development of rules and regulations, the research community has also played a role in shaping biosafety and biosecurity policies, with the usual conjecture that what is natural is potentially much more dangerous and incontrollable than what is artificial (Marlière 2009). Because organisms close to their progenitors are pre-adapted to their native environments, it is logical to assume that GMOs, GEMs, GEOs would pose a mayor threat than GDOs, GROs and CMOs if all of these were accidentally released in the environment. However, since the latter organisms use or potentially can use synthetic building blocks as constituents that could only be synthesized in the lab; they would be quickly outcompeted by other organisms if the chemicals are correctly disposed. Therefore, orthogonal and bottom-up engineering should also follow the guidelines that chemistry labs have established for chemical disposal, together with those for biosafety labs involving GMOs, which is the norm for top-down and parallel engineering approaches. Given that synthetic biology is an extension of genetic engineering, current legislation on GMOs also applies to GEMs, GDOs, GEOs, GROs and those CMOs that have been previously modified to incorporate unnatural DNA bases and/or amino acids, at both the EU[17] and international[18] level. But there have been doubts about whether the current risk assessment procedures will be overburdened by the increasing pace of genetic modification, which is the case for GDOs, GEOs and GROs. In consequence, the latest EU recommendations *"call for research to improve the ability to predict the behavior of complex engineered organisms"*, and for the *"development of additional approaches, including genetic firewalls based on noncanonical genetic material"* (Breitling et al. 2015).

3.4 Genetic Firewall: Data Are Missing

It has been advocated that unnatural DNA would prevent information exchange between natural and synthetic organisms by acting as a genetic firewall (Schmidt 2010). Indeed, Church and colleagues showed that the first *E. coli* GRO was capable of resisting to some extent infections by the T7 bacteriophage because their genetic codes are not compatible (Lajoie et al. 2013b). However, the GRO infection by the virus was only slightly attenuated compared to the parental strain, implying that there are other factors that account for the lack of complete immunity.

So how many changes are needed in the genetic code to render a GRO completely immune to viruses without affecting its fitness? Given the high mutability of viruses, would it be possible that these could hijack the new genetic code if sufficient time is given in a non-controlled environment? If GROs and CMOs are not able to survive outside the lab, what would be the response of a natural organism to their unnatural building blocks? On the other hand, if a GRO or CMO dependent on XAAs manages to survive in the environment by any means, would it pose threats to other natural organisms, and if so, what kind of threats? Would GROs be more robust than CMOs in case of an accidental release in the environment? Are there other means of cellular communication beyond horizontal gene exchange that GROs or CMOs could use to persist or even proliferate? Clearly, many questions remain to be solved to show that a "genetic firewall" could be used as an efficient means to separate the natural from the synthetic world, especially when it is

---

[17]http://ec.europa.eu/health/scientific_committees/emerging/docs/scenihr_o_048.pdf. Accessed 20th May 2015.

[18]http://bch.cbd.int/protocol/cpb_faq.shtml. Accessed: 4th June 2015.

believed that XAA-based biocontainment would be a means to eliminate the fears of the public when it turns to apply GMOs in agriculture, medicine and environmental clean-up (Dolgin 2015).

3.5 Genetically Modified Humans: Napa Meeting

In March 2015, two groups of scientists called for a moratorium (Lanphier et al. 2015) and for a framework for open discourse (Baltimore et al. 2015) on any experiments that involve genome editing in human embryos or cells that could give rise to sperm or eggs. The articles were responses to some rumours by referees who reviewed work by Chinese scientists describing the use of the CRISPR-Cas9 system in human embryos and who fear, it *"could trigger a public backlash that would block legitimate uses of the technology"* (Vogel 2015). The main problem of the technology are the secondary effects (off-target and on-target mutational events with unintended consequences) that are not yet completely understood, for example: *"Monkeys have been born from CRISPR-edited embryos, but at least half of the 10 pregnancies in the monkey experiments ended in miscarriage. In the monkeys that were born, not all cells carried the desired changes, so attempts to eliminate a disease gene might not work"* (Vogel 2015). Despite this, the rumors of the referees became truth after some months: Work concerning the genomic edition of human embryos was recently published, but the authors argue that this was necessary to illustrate that the CRISPR-Cas9 technology is still far from clinical applications (Liang et al. 2015).[19]

During a meeting in Napa in early 2015, Berg, Church and other scientists called urgently for a framework to discuss openly the safe and ethical use of the CRISPR-Cas9 technology to manipulate the human genome (Baltimore et al. 2015). The proposals for regulating "germline engineering" broadly include:

1. The discouragement of any attempts at genome modification for clinical application in humans in those countries where it is allowed (some countries don't allow this kind of research or regulate it tightly).
2. The encouragement of transparent research to evaluate the efficacy and specificity of genome editing in human and non-human models relevant for gene therapy, as well as the implementation of standardized methods to determinate frequency of off-target effects and physiology of cells and issues upon genome editing.
3. The creation of forums for the exchange between scientists, bioethicists, government, interest groups and the general public to shape policy while discussing not only the risks and benefits, but also the ethical, legal, and social implications (ELSI) for curing human genetic disease by genome editing (Baltimore et al. 2015).

---

[19]In fact, the journals *Science* and *Nature* rejected the paper owing to ethical concerns.

The fundamental issue with human-germline engineering is that beyond treating genetic disorders such as Huntington disease to eliminate human suffering, designer or "genetically modified babies" could be likewise engineered, facilitating the arrival of a new eugenics era (Pollack 2015). Although human-germline engineering is banned in several countries (not including the United States), a few labs and a company (primarily based in the United States) are working in this research line (Regalado 2015). One of the plans is to edit the DNA of a man's sperm or woman's egg and use the cells in an in vitro fertilization (IVF) clinic to produce the embryo, followed by its implantation in woman uterus to establish a pregnancy of the foetus.[20] Another strategy that promises to be more efficient aims to edit stem cells, which can divide rapidly in the lab, and then turn them into a sperm or egg. Although the CRISPR-Cas9 technology is still too immature to offer babies "à la carte", 15 % of the adults in a recent survey indicated that it would be appropriate to genetically modify a baby to be more intelligent (Regalado 2015). Besides the engineering of more healthy and intelligent babies, some transhumanists think that the human genome is not perfect and that it could be engineered not only to protect against Alzheimer's disease, but also to create "super-enhanced" individuals to solve complex issues like "climate change" (Regalado 2015).

3.6　Gene-Drive Engineering: Will There Be a Meeting?

The willingness of many scientists to engage in a public discussion with other scholars shows that there is an awareness of the potential negative effects of an emerging technology that has not been shown to be mature for clinical applications, especially when dealing with a delicate topic such as human embryonic stem cells. However, other less-concerned scientists not working with human stem cells have already devised a plan to create an "auto-catalytic" genetic system based on the CRISPR-Cas9 to spread mutated genes across populations of GEOs with high efficiency. For example, mosquitos could be engineered to impair the transmission of genes involved in malaria and dengue fever (Bohannon 2015). Beyond the potential benefits, the problem with this "mutagenic chain reaction" technology is that unintended off-target mutations at essential (or non-essential) genes could be triggered, thus spreading irreparable genetic defects (or traits) across natural populations of organisms and potentially driving populations with limited genetic diversity into extinction.

The potential devastating effects of this technology in the wild have been recently warned by Lunshof (2015). Church commented that this technology "is a step too far" (Bohannon 2015), yet he filled a patent for a more secure gene-drive

---

[20]In 2010, Robert G. Edwards was awarded the Nobel Prize in Physiology or Medicine for the development of IVF, which allows the treatment of infertility. Currently, there are 5 millions of IVF-humans, all below 37 years old: The first "test tube baby" was born in 1978.

technology that, he argues, *"would offer substantial benefits to humanity and the environment"* (Esvelt et al. 2014) by eradicating vectors that spread diseases, insect pests and invasive species not without calling for *"thoughtful, inclusive, and well-informed public discussions to explore the responsible use of this currently theoretical technology"* (Esvelt et al. 2014). In response to an ethical analysis on its regulation (Oye et al. 2014), it has been debated that the dual-use potential of this technology raises strong concerns because gene-drives carrying lethal toxins could be designed to eliminate particular human populations and attack their crops (Gurwitz 2014). In fact, Gurwitz (2014, p. 1010) concluded:

"just as the exact technical instructions for making nuclear weapons remain classified 70 years after the Manhattan Project—as they rightfully should—the gene drive methodological details do not belong in the scientific literature."

However, Oye and Esvelt (2014, p. 1011) disagree with that response arguing that

"classifying information required to build gene drives cannot target potential misuses without also impeding development of defenses, as well as environmental, health, agricultural, and safety applications of CRISPR technology".

These debates show the need of an international meeting involving all stakeholders to regulate the development and deployment of GEOs endowed with gene-drives that cannot be confined physically to a single country.

4 Assessing Synthetic Biology Beyond the Bench

In various surveys, synthetic biology has been perceived by society as an extension of genetic engineering, particularly in the production of GMOs. However, synthetic biology aims to avoid the same criticism that genetically modified plants and animals have triggered in various societies. Beyond the technology itself, in what follows, seven topics are discussed in which the challenges, dilemmas and paradoxes surrounding synthetic biology become obvious.

4.1 Global Social Justice

Keasling's artemisinin technological breakthrough is perhaps the most widely used example to show that synthetic biology (in this case, the metabolic engineering "tribe") can provide solutions to global health issues. However, social justice challenges have emerged because economies and employment in the South can be destabilized by synthetic biology carried out in the North (Engelhard 2009). In other words, although Keasling's breakthrough has been welcomed by almost everyone in the synthetic biology community and other advocates fighting against malaria, there is a rising concern that the farmers who traditionally harvest A. *annua* could be losing their jobs, which would affect the families of more than 100,000 farmers worldwide.[21] Furthermore, the introduction of semi-synthetic artemisinin could further destabilise the already variable prices of botanical artemisinin due to market fluctuations (Peplow 2013). This example illustrates the dilemmas and paradoxes that surround the development of synthetic biology in a globalized world.[22]

4.2 Synthetic Biology Democratization

It has been argued that DIY-biology will help in the worldwide democratization of synthetic biology in the same way that IT was democratized in the garages of computer hobbyists during the last century prior to the emergence of Silicon Valley as innovation hub (Wolinsky 2009). DIY biologists or "biohackers" have equipped their garages, closets, and kitchens with inexpensive laboratory equipment (Ledford 2010). Although DIY biologists performed simple experiments in the past, such as the insertion of a fluorescent protein into bacteria for producing glow-in-the-dark yogurt, it is expected that applications in health, energy and environmental monitoring will emerge (Seyfried et al. 2014). In some instances, however, there has been a rising concern that home-brew drugs could be also produced by DIY-biologists using synthetic microbes. This fear has been recently strengthened when metabolic engineers made yeast strains capable of synthesizing one of the various precursors of the opiate morphine, a precursor of heroin that has a high demand in the illegal drug market (Ehrenberg 2015).[23] Although this research is intended to enable the long-term centralized and legal production of opiates for pain relief, engineered yeasts for opiate biosynthesis could transform into illegal systems for criminal networks in USA and Europe where drug demand is the highest, because yeast is extremely easy to be sent (a few dried cells per post would suffice to start a culture), grown (water and basic nutrients are only required) and processed (basic lab equipment such as centrifuges) to extract any drugs (using basic columns and resins for chromatography). Thus, the democratization of molecular biology might be more difficult to regulate than initially thought.[24]

---

[21]http://www.etcgroup.org/content/why-synthetic-artemisinin-still-bad-idea-response-rob-carlson. Accessed 20th May 2015.

[22]Note that this comment is valid for all technologies beyond synthetic biology replacing work performed in developing countries: see the cases of saffron, vanilla, vetiver, cocoa butter, rubber, squalene, coconut oil, palm kernel oil, babasuu and shikimic acid at: http://www.etcgroup.org/tags/synbio-case-studies. Accessed: 27th July 2015.

[23]All the pieces of the puzzle are in place: The components of a metabolic synthetic pathway for producing morphine have been reported in different stains, so it is a matter of time to combine them in a single strain. Still, the efforts will require strain engineering for increasing the drug titers to be economically competitive.

[24]The researchers behind the work called other scholars for an ethical assessment on their research (Oye et al. 2015), ending with four recommendations: (1) The strains should be not appealing to criminals by focusing on alternative opiates with less "street" demand such as the brain, or they should be difficult to cultivate by relying, for instance, on difficult-to-synthetize XAAs for survival.

4.3 Environmental Concern and Policy Regulation

Medicines, pharmaceuticals, fertilizers, pesticides, additives, cosmetics, plastics, cleaners, clothing, pigments, detergents, electronics parts and many other essential products for human needs are conventionally produced by synthetic chemistry. In fact, chemical companies produce yearly billions of tons of chemicals from about 90,000 different substances. One goal of synthetic biology is to produce enzymes and microbes that could replace the production of many of those compounds because their precursors are usually obtained from fossil fuels and thus unsustainable (Nielsen and Moon 2013). A recent sampling of European rivers, however, illustrates the environmental damage that has been produced with toxic cocktails of hormones, pesticides and hazardous chemicals (Malaj et al. 2014), let alone population declines of bees, birds and other insects that are essential in the global food supply chain (Chagnon et al. 2015). In fact, irreversible damages of many toxic, non-degradable and persistent chemicals on the immune, reproductive, endocrine and nervous systems of many animals (and likely humans) have been found.[25] To avoid similar environmental catastrophes using synthetic biology, proper legislation and regulation on the industrial production and waste of engineered microbes have to be implemented from the onset. A challenge for synthetic biology will be to regulate and assess the long-term effects of GMOs in the environment. To enhance biosafety biocontainment strategies, various genetic safeguard mechanisms could be implemented in each engineered microbe together with a "risk analysis and biosafety data" sheet (Moe-Behrens et al. 2013).

4.4 The GMO Debate

The GMO debate has existed since the Asilomar conference when biotechnology started to be regulated in general, but in the last two decades it shifted towards risks posed by genetically modified food and crops owing to the food crisis in Europe during the late 1990s. Nowadays, GMOs are commonly used in industrial (white) and medical (red) biotechnology for the respective production of chemicals and pharmaceuticals (in physically containments), but the genetic modification of plants and animals in agricultural (green) biotechnology has met more resistance in society

---

(Footnote 24 continued)

(2) Synthetic DNA companies should implement algorithms to avoid users buying gene-specific sequences of opiate biosynthetic pathways in the same way that toxins and other harmful biohazards are currently screened. (3) The strains should be kept under controlled environments such as locks, alarms and monitoring systems to prevent theft of samples. The laboratory personnel should be also subject to security screening. (4) Current laws, at least in the US, do not include opiate-producing yeast, so these should be extended.

[25]http://undesigning.org/cmos.html. Accessed: June 4th 2015.

owing to the potential risks of GMOs for humans, animals and the environment. There are different kinds of risks: real versus perceived, which depending on the stakeholder, can vary significantly (Torgersen 2004). Regardless of the risks and semantics (Holme et al. 2013; Hunter 2014; Nagamangala Kanchiswamy et al. 2015), it is clear that the GMO debate is mostly associated to multicellular organisms such as plants (Boyle et al. 2012) and animals (Markson and Elowitz 2014). Given that synthetic biology is increasingly targeting these multicellular organisms for genomic modifications, its acceptance by the public could be more difficult to gain. The main reason is that plants and animals directly affect human life, in contrast to microbes, which affect human life more indirectly: "*As long as synthetic biology creates only new microbial life and does not directly affect human life, it will in all likelihood be considered acceptable*" (van den Belt 2009, p. 257). Beyond the potential of synthetic biology for solving global issues, society may only accept any kind of GMOs, GEMs, GEOs, GDOs, GROs and CMOs if their products are labelled for the consumer in a transparent way where the freedom of choice is given to the consumers.

4.5 Media Hype, Metaphors and Promises

One of the most common mistakes that scientists often make is to claim that they think they have created something, and that this creation, they believe, will be a panacea to humankind. First of all, scientists do not "create", because this word can be used in different contexts: Creation or "*Creatio ex nihilo*" means to bring someone or something into existence out of nothing, and it is usually reserved for a divine force in religious terms that scholars in social sciences and humanities might use more often. In reality, scientists invent or design something that has not existed before by using something that already existed; they are closer to "*manipulatio*" (manipulating) and "*creatio ex existendo*" (creating something out of existent parts) than to "*creation ex nihilo*" (Boldt and Muller 2008). This is a subtle, yet important difference: Scientists claiming the creation of synthetic life have been accused of "*Playing God in Frankenstein's Footsteps*" (van den Belt 2009), with unfortunate consequences for the reputation of the field and its social acceptance (Schummer 2011). Perhaps the natural scientists do this on purpose to challenge the ethical (Link 2012) and religious (Dabrock 2009) views of other scholars working on synthetic biology. On the other hand, overstating that GMOs, GEMs, GEOs, GDOs, GROs and CMOs will solve all human problems is not beneficial because exaggerating exactly into this direction generates mistrust in the public, as many emergent technologies have done in history (Torgersen 2009). Hence, a lesson to synthetic biologists would be to be more cautious with their metaphors and promises, if they want to gain public acceptance of their technologies. For a more insightful discussion of the impact of metaphors in synthetic biology, see the chapter by D. Falkner in this book.

4.6 Semantics and the Public

Synthetic biology is a term that is not well known to the public (for further information, see the chapters by Ancillotti and Eriksson; Rerimassie; Seitz; Steurer). In fact, a recent survey in Germany confirmed this, but also revealed that people spontaneously perceived this field as an abstract and contradictory expression that they associated with interference against Nature.[26] How can nature be synthetic? Synthesis means to put together two parts to form a new one. This definition suggests that any minor modification of an extant organism would render it 'synthetic', which would be the case for all GMOs, CMEs, GEOs, GDOs, GROs and possibly CMOs. However, a bacterium with a synthetic plasmid is not the same as a bacterium with a synthetic genome, so we should make a distinction. For example, the synthetic genome of GDO *M. mycoides* jcvi-syn1.0 represents about 1 % of the whole dry cell mass. The cell reprogramming was performed using a genome borrowed from a related species, in which about 0.1 % genetic changes (watermarks) were done. So, why should this bacterium be called synthetic when the other 98.9 % components are natural? In reality, this bacterium is more natural than synthetic: Although genomes control gene expression, the truth is that genomes can only be useful if there are proteins that can process their information, as it was the case for *M. mycoides* jcvi-syn1.0. In a contrasting example, when the genome of the cyanobacterium *Synechocystis* was cloned into the bacterium *B. subtilis*, genes from the former could not be expressed in the latter organism because of "incompatibility" or context-dependency issues (Itaya et al. 2005). Thus, synthetic genomes only represent a minor part of cells, in contrast to lipids (which are not encoded in the genome) and proteins, of which there are millions of molecules in a single *E. coli* bacterium, let alone other important components (Fig. 9).

A truly synthetic organism would be that in which all the components shown in Fig. 9 would be synthesized and assembled to generate a living organism. Indeed, synthesizing all the molecules of life was the dream of the famous German chemist Emil Fischer at the beginning of the 20th century, when synthetic chemistry experienced a major revolution almost a century after the breakthrough by Friedrich Wohler in 1828: Urea synthesis, the first synthesis of an organic molecule in the lab (Yeh and Lim 2007). Thus a real challenge for bottom-up synthetic biologists will be to synthetize completely from scratch a living microbe (Porcar et al. 2011).

Given that top-down, parallel and perpendicular synthetic biology researchers mostly borrow microbes for tinkering (i.e. improving something by making small changes), the challenge will be to make the public aware that most of their research is not based on entirely synthetic organisms, but rather that synthetic DNA enables researchers to reprogram certain organisms for useful purposes. This endeavour could be facilitated by clearly distinguishing that DNA is not equal to life

---

[26]http://www.biotechnologie.de/BIO/Navigation/DE/root,did=178894.html. Accessed: June 4th 2015.

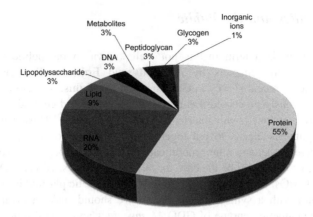

Fig. 9 Pie chart displaying the composition (cell dry mass) of a typical *E. coli* bacterium growing with a doubling time of 40 min. Metabolites include building blocks and vitamins. Note that there are about 2,400,000 protein, 257,500 RNA, 22,000,000 lipid, 1,200,000 lipopolysaccharide, 4400 glycogen, 2 DNA and 1 peptidoglycan molecule(s) per cell (http://book.bionumbers.org/what-is-the-macromolecular-composition-of-the-cell/)

(de Lorenzo 2010), but just another minor component of life, as important as the other ones (Fig. 9). Another manner to engage the public with synthetic biology is by fostering cultural activities around the technology (see the chapter by B. Wray).

4.7 Resurrecting Life

The resurrection of an extinct species has so far been reported for "Celia", the last bucardo (Pyrenenan ibex), which passed away on January 6th 2000. It was resurrected for 7 min on July 30th 2003 by nuclear transfer cloning (the germ cells had been frozen). Nuclear transfer involves the injection of the genetic material to be cloned into an unfertilized DNA-free egg, resulting in cells that have the potential to divide when placed in the uterus of an adult female mammal. This is how Dolly the sheep (5 July 1996—14 February 2003) was cloned from an adult somatic cell (Wilmut et al. 1997), but dozens of other species[27] have also been cloned despite the low efficiency of the technique. Besides nuclear cloning, it has been suggested that synthetic biology could cooperate with the biodiversity conservation community to protect or even resurrect extinct organisms (Redford et al. 2013). However, there are several technical and nontechnical issues that first have to be resolved, and this is clearly reflected in even Venter being cautious:

---

[27]http://en.wikipedia.org/wiki/List_of_animals_that_have_been_cloned. Accessed: 20th May 2015.

I have read too many articles that breezily discuss the reconstruction of a Neanderthal or a woolly mammoth with the help of cloning, even though the DNA sequences that have been obtained for each are highly fragmented, do not cover the entire genome and – as a result of being so degraded – are substantially less accurate than what is routinely obtained from fresh DNA. (Venter 2013, p. 87)

Further, even if fresh intact DNA or "software" of the woolly mammoth or Neanderthals were available, it would be necessary to have the appropriate cells or "hardware" that could interpret the genetic information (Danchin 2009). For example, when the components (nucleus, cytoplasm, and cell membrane) of amoebas of different species were combinatorially reassembled, the only viable organisms were those whose components originated from the same strain (Jeon et al. 1970). This old yet ingenious study shows that reprogramming of life goes beyond pure (synthetic) DNA.

Although it is difficult to define life among scholars from different fields because no consensus has been reached from the up to 123 current definitions (Trifonov 2011), it seems that life requires at least three components: A genetic program, a metabolism and a container (Acevedo-Rocha et al. 2013a). For a discussion about the concept of life, see the chapter by J. Steizinger. Regardless of the definitions, any attempt to reprogram or resurrect life, or construct life-like systems from the bottom-up should consider establishing the connection among these three components (Fig. 10).

Regarding non-technical issues, it is useful to illustrate the pros and cons to better understand the dilemma of resurrecting extinct species with synthetic biology by quoting an expert: *"The 'de-extinction' movement—a prominent group of scientists, futurists and their allies—argues that we no longer have to accept the finality of extinction."* (Minteer 2014). The most persuasive argument is to *"appeal to our sense of justice: de-extinction is our opportunity to right past wrongs and to atone for our moral failings"* (Minteer 2014). De-extinctionists also argue that by resurrecting species, ecological functions could be recovered, increasing the diversity of ecosystems. However, the introduction of revived species could pose disease threats to native species, in a manner reminiscent to the introduction of invasive species into new environments.[28] In addition, some conservationists also worry that "de-extinguished" species would have a limited genetic diversity.

In summary, one should ponder whether it is worthwhile investing huge amounts of human and financial resources in ambitious projects for resurrecting life given not only the technical and nontechnical difficulties, but also the paradoxes of our world: Devastating epidemics, hunting, habitat loss and degradation caused by both human industrial activities and climate change are triggering an unprecedented loss of biodiversity, with estimates of 500 up to 36,000 species of amphibians, birds and mammals disappearing every year (Monastersky 2014). Thus, instead of

---

[28]See for instance: http://www.invasive.org/. Accessed 20th May 2015.

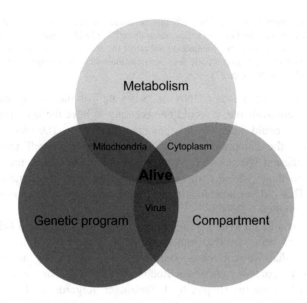

Fig. 10 Life prerequisites. Living systems as we know them have a genetic program (DNA), a metabolism fuelled by enzymatic reactions and proteins that execute this program using RNA as intermediate in protein translation (mRNA, tRNA, rRNA), and a container or membrane formed of lipids. Viruses are proteins (sometimes with lipids) encoding DNA or RNA, but since they have no metabolism, they are not alive, a fact that has puzzled scientists over decades (Villarreal 2004). Organelles such as mitochondria and chloroplasts contain metabolites and usually a minimal genetic program, but they are contained within a larger organism so these alone cannot be alive. Finally, the cytoplasm containing proteins, metabolites and lipids would not be alive without genetic material

resurrecting life, synthetic biologists should help conservation biologists to develop innovative ideas for protecting the already thousands of endangered species: "*Attempting to revive lost species is in many ways a refusal to accept our moral and technological limits in nature.*" (Minteer 2014).

5 Conclusion

I have attempted to outline the most important areas of current research in synthetic biology in a general inclusive framework according to different engineering approaches. Almost all engineering efforts produce protocells, CMOs and GMOs, some of which are developed by well-known genetic engineering methods, and others by using more sophisticated tools. For example, the production of Genetically Engineered Machines (GEMs) is outsourced to undergraduate students every year across the world with the advantage of developing creativity and innovation in younger generations for solving complex problems. Similarly, researchers working in synthetic genomics have produced genomically designed organisms (GDOs) as a result of basic research, spinning-off revolutionary applications for the rapid assembly of synthetic DNA. More recently, the ground-breaking CRISPR-Cas9 tool, product of basic research, has allowed for the emergence of an impressive number of genomically edited organisms (GEOs), and in less than two years to better understand disease models and cellular biology as well as engineering multiple traits in microorganism for the optimal production of drugs, biofuels, and chemicals. However, the use of the CRISPR-Cas9 system should be cautious especially when engineering human embryonic stem cells as well as wild populations of organisms in the open environment (Ledford 2015). Finally, evolutionary approaches have accelerated the construction of microbial GROs and CMOs that promise to shed light on the meaning and evolution of life by endowing living systems with unnatural building blocks.

Importantly, any sort of genetic modification such as a single DNA base mutation, metabolic pathway optimization or whole genome recoding in any of these synthetic life forms will create a novel combination of genes and their products at the level of (pre- and post)-transcription (RNA) and -translation (protein), thus triggering a new network of gene interactions (at the transcriptome and proteome level) in the host organism and across organisms, which are difficult to understand and predict even using the most advanced mathematical algorithms and technological tools. These phenomena, which fall under the scope of epistasis (Phillips 2008), emphasise the evolutionary complexity of gene interactions inherent to biology. For this reason, synthetic biologists should be aware of the limitations that must be overcome to predict the behaviour of living systems. The purpose of illustrating the benefits, but also potential risks of these emergent life forms is to provide scholars from the social sciences and humanities as well as non-scientists with a glimpse of what synthetic biologists are actually doing. The challenges, dilemmas and paradoxes surrounding the field also show that there are already big challenges in a globalized world that cannot be solved exclusively by technological means. Whether synthetic biology will deliver health, food, and energy to societies with very different needs in the South and North without affecting an already polluted environment where biodiversity lost is an everyday phenomenon will be seen in the coming century.

Acknowledgements I am very thankful to Kristin Hagen, Margret Engelhard and Georg Toepfer for feedback on this chapter and for organizing the International Summer School "Analyzing the Societal Dimensions of Synthetic Biology" that was supported by the German Federal Ministry of Education and Research. I also thank Antoine Danchin, Markus Schmidt, Sheref Mansy, and Victor de Lorenzo for the critical reading of this manuscript and insightful discussions. I also thank Paul Berg and Nediljko Budisa for feedback on parts from a previous version of this work. Funding from the SYNMIKRO postdoctoral program is greatly acknowledged.

References

Abby S, Daubin V (2007) Comparative genomics and the evolution of prokaryotes. Trends Microbiol 15(3):135–141

Acevedo-Rocha CG (2010) Genetic code engineering with methionine analogs for synthetic biotechnology. PhD thesis, Technical University Munich

Acevedo-Rocha CG, Budisa N (2011) On the road towards chemically modified organisms endowed with a genetic firewall. Angew Chem Int Ed Engl 50(31):6960–6962

Acevedo-Rocha CG, Fang G, Schmidt M et al (2013a) From essential to persistent genes: a functional approach to constructing synthetic life. Trends Genet 29(5):273–279

Acevedo-Rocha CG, Hoesl MG, Nehring S et al (2013b) Non-canonical amino acids as a useful synthetic biological tool for lipase-catalysed reactions in hostile environments. Catal Sci Technol 3(5):1198

Agapakis CM (2014) Designing synthetic biology. ACS Synt Biol 3(3):121–128

Alper H, Stephanopoulos G (2009) Engineering for biofuels: exploiting innate microbial capacity or importing biosynthetic potential? Nat Rev Microbiol 7(10):715–723

Annaluru N, Muller H, Mitchell LA et al (2014) Total synthesis of a functional designer eukaryotic chromosome. Science 344(6179):55–58

Araki M, Ishii T (2015) Towards social acceptance of plant breeding by genome editing. Trends Plant Sci 20(3):145–149

Bacchus W, Aubel D, Fussenegger M (2013) Biomedically relevant circuit-design strategies in mammalian synthetic biology. Mol Syst Biol 9:691

Bacchus W, Fussenegger M (2013) Engineering of synthetic intercellular communication systems. Metab Eng 16:33–41

Bacher JM, Hughes RA, Tze-Fei Wong J et al (2004) Evolving new genetic codes. Trends Ecol Evol 19(2):69–75

Bailey JE (1991) Toward a science of metabolic engineering. Science 252(5013):1668–1675

Bailey JE, Birnbaum S, Galazzo JL et al (1990) Strategies and challenges in metabolic engineering. Ann NY Acad Sci 589:1–15

Ball P (2010) A synthetic creation story. Nature. doi:10.1038/news.2010.261

Baltimore BD, Berg P, Botchan M et al (2015) A prudent path forward for genomic engineering and germline gene modification. Science 348(6230):36–38

Bedau M, Church G, Rasmussen S et al (2010) Life after the synthetic cell. Nature 465(7297): 422–424

Benenson Y (2012) Synthetic biology with RNA: progress report. Curr Opin Chem Biol 16 (3–4):278–284

Benner SA, Sismour AM (2005) Synthetic biology. Nat Rev Genet 6(7):533–543

Berg P (2008) Meetings that changed the world: Asilomar 1975: DNA modification secured. Nature 455(7211):290–291

Berg P, Baltimore D, Brenner S et al (1975) Asilomar conference on recombinant DNA molecules. Science 188(4192):991–994

Billerbeck S, Harle J, Panke S (2013) The good of two worlds: increasing complexity in cell-free systems. Curr Opin Biotechnol 24(6):1037–1043

Blain JC, Szostak JW (2014) Progress toward synthetic cells. Annu Rev Biochem 83(1):615–640

Bohannon J (2015) Biologists devise invasion plan for mutations. Science 347(6228):833–836

Bohlke N, Budisa N (2014) Sense codon emancipation for proteome-wide incorporation of noncanonical amino acids: rare isoleucine codon AUA as a target for genetic code expansion. FEMS Microbiol Lett 351(2):133–144

Boldt J, Muller O (2008) Newtons of the leaves of grass. Nat Biotechnol 26(4):387–389

Boyle PM, Burrill DR, Inniss MC et al (2012) A BioBrick compatible strategy for genetic modification of plants. J Biol Eng 6(1):8

Breitling R, Takano E, Gardner TS (2015) Judging synthetic biology risks. Science 347(6218):107

Brent R (2004) A partnership between biology and engineering. Nat Biotechnol 22(10):1211–1214

Brophy JA, Voigt CA (2014) Principles of genetic circuit design. Nat Methods 11(5):508–520

Budisa N (2004) Prolegomena to future experimental efforts on genetic code engineering by expanding its amino acid repertoire. Angew Chem Int Ed Engl 43(47):6426–6463

Budisa N (2014) Xenobiology, new-to-nature synthetic cells and genetic firewall. Curr Org Chem 18(8):936–943

Bugl H, Danner JP, Molinari RJ et al (2007) DNA synthesis and biological security. Nat Biotechnol 25(6):627–629

Butler D (2014) Biosafety controls come under fire. Nature 511(7511):515–516

Cameron DE, Bashor CJ, Collins JJ (2014) A brief history of synthetic biology. Nat Rev Microbiol 12(5):381–390

Campos L (2009) That was the synthetic biology that was. In: Schmidt M, Kelle A, Ganguli-Mitra A (eds) Synthetic biology: the technoscience and its societal consequences. Springer, Netherlands, Dordrecht, Heidelberg, London, New York, pp 5–22

Carlson R (2009) The changing economics of DNA synthesis. Nat Biotechnol 27(12):1091–1094

Carr PA, Church GM (2009) Genome engineering. Nat Biotechnol 27(12):1151–1162

Cello J, Paul AV, Wimmer E (2002) Chemical synthesis of poliovirus cDNA: generation of infectious virus in the absence of natural template. Science 297(5583):1016–1018

Chagnon M, Kreutzweiser D, Mitchell EA et al (2015) Risks of large-scale use of systemic insecticides to ecosystem functioning and services. Environ Sci Pollut Res Int 22(1):119–134

Chang AC, Cohen SN (1974) Genome construction between bacterial species in vitro: replication and expression of Staphylococcus plasmid genes in Escherichia coli. Proc Natl Acad Sci U S A 71(4):1030–1034

Chang MC, Eachus RA, Trieu W et al (2007) Engineering Escherichia coli for production of functionalized terpenoids using plant P450s. Nat Chem Biol 3(5):274–277

Charpentier E, Doudna JA (2013) Biotechnology: rewriting a genome. Nature 495(7439):50–51

Chiarabelli C, Stano P, Anella F, Carrara P, Luisi PL (2012) Approaches to chemical synthetic biology. FEBS Lett 586(15):2138–2145

Chin JW (2014) Expanding and reprogramming the genetic code of cells and animals. Annu Rev Biochem 83:379–408

Cho H, Daniel T, Buechler YJ et al (2011) Optimized clinical performance of growth hormone with an expanded genetic code. Proc Natl Acad Sci USA 108(22):9060–9065

Citorik RJ, Mimee M, Lu TK (2014) Bacteriophage-based synthetic biology for the study of infectious diseases. Curr Opi Microbiol 19:59–69

Cohen SN (2013) DNA cloning: a personal view after 40 years. Proc Natl Acad Sci USA 110(39):15521–15529

Cox DB, Platt RJ, Zhang F (2015) Therapeutic genome editing: prospects and challenges. Nat Med 21(2):121–131

Crick FH (1968) The origin of the genetic code. J Mol Biol 38(3):367–379

Dabrock P (2009) Playing god? Synthetic biology as a theological and ethical challenge. Syst Synth Biol 3(1–4):47–54

Danchin A (2003) La Barque de Delphes: what genomes tell us. Harvard University Press, Boston

Danchin A (2009) Bacteria as computers making computers. FEMS Microbiol Rev 33(1):3–26

Danchin A (2010) Perfect time or perfect crime? EMBO Rep 11(2):74

Davis L, Chin JW (2012) Designer proteins: applications of genetic code expansion in cell biology. Nat Rev Mol Cell Biol 13(3):168–182

de Lorenzo V (2010) Environmental biosafety in the age of synthetic biology: do we really need a radical new approach? Environmental fates of microorganisms bearing synthetic genomes could be predicted from previous data on traditionally engineered bacteria for in situ bioremediation. BioEssays 32(11):926–931

de Lorenzo V (2011) Beware of metaphors: chasses and orthogonality in synthetic biology. Bioeng Bugs 2(1):3–7

de Lorenzo V, Danchin A (2008) Synthetic biology: discovering new worlds and new words. EMBO Rep 9(9):822–827

Dolgin E (2015) Synthetic biology: safety boost for GM organisms. Nature 517(7535):423

Doudna JA, Charpentier E (2014) The new frontier of genome engineering with CRISPR-Cas9. Science 346(6213):1258096

Ehrenberg R (2015) Engineered yeast paves way for home-brew heroin. Nature 521(7552):267–268

Endy D (2005) Foundations for engineering biology. Nature 438(7067):449–453

Engelhard M (2009) Synthetic biology gets ethical. Europäische Akademie Bad Neuenahr-Ahrweiler Newsletter 94:1–3. http://www.ea-aw.de/fileadmin/downloads/Newsletter/NL_0094_112009.pdf. Accessed 19 May 2015

Enserink M (2005) Infectious diseases. Source of new hope against malaria is in short supply. Science 307(5706):33

Esvelt KM, Smidler AL, Catteruccia F et al (2014) Concerning RNA-guided gene drives for the alteration of wild populations. Elife 3:e03401

Fangerau H (2009) From Mephistopheles to Isaiah: Jacques Loeb, technical biology and war. Soc Stud Sci 39(2):229–256

Ferber D (2004) Synthetic biology. Time for a synthetic biology Asilomar? Science 303(5655):159

Forlin M, Lentini R, Mansy SS (2012) Cellular imitations. Curr Opin Chem Biol 16(5–6):586–592

Gaj T, Gersbach CA, Barbas CF 3rd (2013) ZFN, TALEN, and CRISPR/Cas-based methods for genome engineering. Trends Biotechnol 31(7):397–405

Gallagher RR, Li Z, Lewis AO, Isaacs FJ (2014) Rapid editing and evolution of bacterial genomes using libraries of synthetic DNA. Nat Protoc 9(10):2301–2316

Gibson DG (2011) Enzymatic assembly of overlapping DNA fragments. Methods Enzymol 498:349–361

Gibson DG, Benders GA, Andrews-Pfannkoch C et al (2008) Complete chemical synthesis, assembly, and cloning of a Mycoplasma genitalium genome. Science 319(5867):1215–1220

Gibson DG, Glass JI, Lartigue C et al (2010) Creation of a bacterial cell controlled by a chemically synthesized genome. Science 329(5987):52–56

Goeddel DV, Heyneker HL, Hozumi T et al (1979a) Direct expression in Escherichia coli of a DNA sequence coding for human growth hormone. Nature 281(5732):544–548

Goeddel DV, Kleid DG, Bolivar F et al (1979b) Expression in Escherichia coli of chemically synthesized genes for human insulin. Proc Natl Acad Sci USA 76(1):106–110

Group CR (1979) Antimalaria studies on Qinghaosu. Chinese Med J 92(12):811–816

Grushkin D (2010) Synthetic Bio, Meet "FBIo". The Scientist 24(5):44

Gurwitz D (2014) Gene drives raise dual-use concerns. Science 345(6200):1010

Heinemann M, Panke S (2009) Synthetic biology: putting engineering into bioengineering. In: FU P, Panke S (eds) Systems biology and synthetic biology. Wiley, New York, pp 387–409

Hobom B (1980) Gene surgery: on the threshold of synthetic biology. Med Klin 75(24):834–841

Hoesl MG, Acevedo-Rocha CG, Nehring S et al (2011) Lipase congeners designed by genetic code engineering. ChemCatChem 3(1):213–221

Holme IB, Wendt T, Holm PB (2013) Intragenesis and cisgenesis as alternatives to transgenic crop development. Plant Biotechnol J 11(4):395–407

Hunter P (2014) "Genetically Modified Lite" placates public but not activists: new technologies to manipulate plant genomes could help to overcome public concerns about GM crops. EMBO Rep 15(2):138–141

Isaacs FJ, Carr PA, Wang HH et al (2011) Precise manipulation of chromosomes in vivo enables genome-wide codon replacement. Science 333(6040):348–353

Itaya M, Tsuge K, Koizumi M et al (2005) Combining two genomes in one cell: stable cloning of the Synechocystis PCC6803 genome in the Bacillus subtilis 168 genome. Proc Natl Acad Sci USA 102(44):15971–15976

Jackson DA, Symons RH, Berg P (1972) Biochemical method for inserting new genetic information into DNA of Simian Virus 40: circular SV40 DNA molecules containing lambda phage genes and the galactose operon of Escherichia coli. Proc Natl Acad Sci USA 69 (10):2904–2909

Jasin M (1996) Genetic manipulation of genomes with rare-cutting endonucleases. Trends Genet 12(6):224–228

Jeon KW, Lorch IJ, Danielli JF (1970) Reassembly of living cells from dissociated components. Science 167(925):1626–1627

Jewett MC, Forster AC (2010) Update on designing and building minimal cells. Curr Opin Biotechnol 21(5):697–703

Juhas M, Eberl L, Church GM (2012) Essential genes as antimicrobial targets and cornerstones of synthetic biology. Trends Biotechnol 30(11):601–607

Juhas M, Eberl L, Glass JI (2011) Essence of life: essential genes of minimal genomes. Trends Cell Biol 21(10):562–568

Kawecki TJ, Lenski RE, Ebert D et al (2012) Experimental evolution. Trends Ecol Evol 27 (10):547–560

Keasling JD (2008) Synthetic biology for synthetic chemistry. ACS Chem Biol 3(1):64–76

Keasling JD (2010) Manufacturing molecules through metabolic engineering. Science 330 (6009):1355–1358

Keller EF (2009) What does synthetic biology have to do with biology? Biosocieties 4(2–3): 291–302

Keung AJ, Joung JK, Khalil AS et al (2015) Chromatin regulation at the frontier of synthetic biology. Nat Rev Genet 16(3):159–171

Kim KM, Caetano-Anolles G (2012) The evolutionary history of protein fold families and proteomes confirms that the archaeal ancestor is more ancient than the ancestors of other superkingdoms. BMC Evol Biol 12:13

Kung Y, Runguphan W, Keasling JD (2012) From fields to fuels: recent advances in the microbial production of biofuels. ACS Synth Biol 1(11):498–513

Kwok R (2010) Five hard truths for synthetic biology. Nature 463(7279):288–290

Kwok R (2012) Chemical biology: DNA's new alphabet. Nature 491(7425):516–518

Lajoie MJ, Kosuri S, Mosberg JA et al (2013a) Probing the limits of genetic recoding in essential genes. Science 342(6156):361–363

Lajoie MJ, Rovner AJ, Goodman DB et al (2013b) Genomically recoded organisms expand biological functions. Science 342(6156):357–360

Lanphier E, Urnov F, Haecker SE et al (2015) Don't edit the human germ line. Nature 519 (7544):410–411

Lanza AM, Crook NC, Alper HS (2012) Innovation at the intersection of synthetic and systems biology. Curr Opin Biotechnol 23(5):712–717

Lartigue C, Glass JI, Alperovich N et al (2007) Genome transplantation in bacteria: changing one species to another. Science 317(5838):632–638

Ledford H (2015) CRISPR, the disruptor. Nature 522(7554):20–24

Ledford H (2010) Garage biotech: life hackers. Nature 467(7316):650–652

Leprince A, van Passel MW, dos Santos VA (2012) Streamlining genomes: toward the generation of simplified and stabilized microbial systems. Curr Opin Biotechnol 23(5):651–658

Levesque F, Seeberger PH (2012) Continuous-flow synthesis of the anti-malaria drug artemisinin. Angew Chem Int Ed Engl 51(7):1706–1709

Levskaya A, Chevalier AA, Tabor JJ et al (2005) Synthetic biology: engineering Escherichia coli to see light. Nature 438(7067):441–442

Liao H-K, Gu Y, Diaz A et al (2015) Use of the CRISPR/Cas9 system as an intracellular defense against HIV-1 infection in human cells. Nat Commun 6:6413

Liang P, Xu Y, Zhang X et al (2015) CRISPR/Cas9-mediated gene editing in human tripronuclear zygotes. Protein Cell 6(5):363–372

Link HJ (2012) Playing god and the intrinsic value of life: moral problems for synthetic biology? Sci Eng Ethics 19(2):435–448

Liu CC, Schultz PG (2010) Adding new chemistries to the genetic code. Annu Rev Biochem 79:413–444

Loeb J (1899) On the nature of the process of fertilization and the artificial production of normal larvae (plutei) from the unfertilized eggs of the sea urchin. Am J Physiol 3(3):135–138

Lunshof J (2015) Regulate gene editing in wild animals. Nature 521(7551):127

MacDonald JT, Barnes C, Kitney RI et al (2011) Computational design approaches and tools for synthetic biology. Integr Biol (Camb) 3(2):97–108

Malaj E, von der Ohe PC, Grote M et al (2014) Organic chemicals jeopardize the health of freshwater ecosystems on the continental scale. Proc Natl Acad Sci USA 111(26):9549–9554

Mali P, Esvelt KM, Church GM (2013) Cas9 as a versatile tool for engineering biology. Nat Methods 10(10):957–963

Malvankar NS, Vargas M, Nevin KP et al (2011) Tunable metallic-like conductivity in microbial nanowire networks. Nat Nanotechnol 6(9):573–579

Malyshev DA, Dhami K, Lavergne T et al (2014) A semi-synthetic organism with an expanded genetic alphabet. Nature 509(7500):385–388

Mandell DJ, Lajoie MJ, Mee MT et al (2015) Biocontainment of genetically modified organisms by synthetic protein design. Nature 518(7537):55–60

Markson JS, Elowitz MB (2014) Synthetic biology of multicellular systems: new platforms and applications for animal cells and organisms. ACS Synth Biol 3(12):875–876

Marliere P (2009) The farther, the safer: a manifesto for securely navigating synthetic species away from the old living world. Syst Synth Biol 3(1–4):77–84

Marliere P, Patrouix J, Doring V et al (2011) Chemical evolution of a bacterium's genome. Angew Chem Int Ed Engl 50(31):7109–7114

Martin VJ, Pitera DJ, Withers ST et al (2003) Engineering a mevalonate pathway in Escherichia coli for production of terpenoids. Nat Biotechnol 21(7):796–802

Menezes AA, Cumbers J, Hogan JA et al (2015) Towards synthetic biological approaches to resource utilization on space missions. J R Soc Interface 12(102):1–20

Messerschmidt SJ, Kemter FS, Schindler D et al (2015) Synthetic secondary chromosomes in Escherichia coli based on the replication origin of chromosome II in Vibrio cholerae. Biotechnol J 10(2):302–314

Minteer B (2014) Is it right to reverse extinction? Nature 509(7500):261

Moe-Behrens GH, Davis R, Haynes KA (2013) Preparing synthetic biology for the world. Front Microbiol 4:5

Monastersky R (2014) Biodiversity: life—a status report. Nature 516(7530):158–161

Montague MG, Lartigue C, Vashee S (2012) Synthetic genomics: potential and limitations. Curr Opin Biotechnol 23(5):659–665

Morange M (2009) A new revolution? The place of systems biology and synthetic biology in the history of biology. EMBO Rep 10(Suppl 1):S50–S53

Nagamangala Kanchiswamy C, Sargent DJ, Velasco R et al (2015) Looking forward to genetically edited fruit crops. Trends Biotechnol 33(2):62–64

Nasir A, Caetano-Anolles G (2013) Comparative analysis of proteomes and functionomes provides insights into origins of cellular diversification. Archaea 2013:648746

Nature Editorials (2014) Tribal gathering. Nature 509(7499):133

Newman JD, Marshall J, Chang M et al (2006) High-level production of amorpha-4, 11-diene in a two-phase partitioning bioreactor of metabolically engineered Escherichia coli. Biotechnol Bioeng 95(4):684–691

Nielsen DR, Moon TS (2013) From promise to practice. The role of synthetic biology in green chemistry. EMBO Rep 14(12):1034–1038

Nishimasu H, Ran FA, Hsu PD et al (2014) Crystal structure of Cas9 in complex with guide RNA and target DNA. Cell 156(5):935–949

Oberholzer T, Luisi PL (2002) The use of liposomes for constructing cell models. J Biol Phys 28 (4):733–744

Okie S (2011) Is Craig Venter going to save the planet? Or is this more hype from one of America's most controversial scientists? The Washington Post

Oldham P, Hall S, Burton G (2012) Synthetic biology: mapping the scientific landscape. PLoS ONE 7(4):e34368

Ouyang Q, Lai L, Tang C (2012) Designing the scientific cradle for quantitative biologists. ACS Synth Biol 1(7):254–255

Ouzounis C, Kyrpides N (1996) The emergence of major cellular processes in evolution. FEBS Lett 390(2):119–123

Owens B (2014) Anthrax and smallpox errors highlight gaps in US biosafety. Lancet 384 (9940):294

Oye KA, Esvelt KM (2014) Gene drives raise dual-use concerns–response. Science 345 (6200):1010–1011

Oye KA, Esvelt K, Appleton E et al (2014) Biotechnology. Regulating gene drives. Science 345 (6197):626–628

Oye KA, Lawson JC, Bubela T (2015) Drugs: regulate 'home-brew' opiates. Nature 521 (7552):281–283

Pal C, Papp B, Posfai G (2014) The dawn of evolutionary genome engineering. Nat Rev Genet 15 (7):504–512

Paul N, Joyce GF (2004) Minimal self-replicating systems. Curr Opin Chem Biol 8(6):634–639

Peplow M (2013) Malaria drug made in yeast causes market ferment. Nature 494(7436):160–161

Peralta-Yahya PP, Zhang FZ, del Cardayre SB et al (2012) Microbial engineering for the production of advanced biofuels. Nature 488(7411):320–328

Phillips PC (2008) Epistasis–the essential role of gene interactions in the structure and evolution of genetic systems. Nat Rev Genet 9(11):855–867

Pinheiro VB, Holliger P (2012) The XNA world: progress towards replication and evolution of synthetic genetic polymers. Curr Opin Chem Biol 16(3–4):245–252

Pinheiro VB, Taylor AI, Cozens C et al (2012) Synthetic genetic polymers capable of heredity and evolution. Science 336(6079):341–344

Pollack R (2015) Eugenics lurk in the shadow of CRISPR. Science 348(6237):871

Porcar M, Danchin A, de Lorenzo V et al (2011) The ten grand challenges of synthetic life. Syst Synth Biol 5(1–2):1–9

Porcar M, Pereto J (2012) Are we doing synthetic biology? Syst Synth Biol 6(3–4):79–83

Porteus MH, Carroll D (2005) Gene targeting using zinc finger nucleases. Nat Biotech 23(8): 967–973

Rasmussen N (2014) Gene jockeys: life science and the rise of biotech enterprise. Johns Hopkins University Press, Baltimore

Rasmussen S, Bedau MA, Chen L et al (2009) Protocells: bridging nonliving and living matter. MIT Press, Cambridge

Rawls R (2000) Synthetic biology makes its debut. Chem Eng News 78(17): 49–53

Reardon S (2011) Visions of synthetic biology. Science 333(6047):1242–1243

Reardon S (2014) Phage therapy gets revitalized. Nature 510(7503):15–16

Redford KH, Adams W, Mace GM (2013) Synthetic biology and conservation of nature: wicked problems and wicked solutions. PLoS Biol 11(4):e1001530

Reetz MT (2013) Biocatalysis in organic chemistry and biotechnology: past, present, and future. J Am Chem Soc 135(34):12480–12496

Regalado A (2015) Engineering the perfect baby. MITS Technol Rev 118(3):27–33

Renders M, Pinheiro VB (2015) Catalysing mirror life. ChemBioChem 16(6):899–901

Ro DK, Paradise EM, Ouellet M et al (2006) Production of the antimalarial drug precursor artemisinic acid in engineered yeast. Nature 440(7086):940–943

Roblin R (1979) Synthetic biology. Nature 282:171–172

Rovner AJ, Haimovich AD, Katz SR et al (2015) Recoded organisms engineered to depend on synthetic amino acids. Nature 518(7537):89–93

Ruder WC, Lu T, Collins JJ (2011) Synthetic biology moving into the clinic. Science 333 (6047):1248–1252

Salis H, Tamsir A, Voigt C (2009) Engineering bacterial signals and sensors. Contrib Microb 16:194–225

Schmidt M (2010) Xenobiology: a new form of life as the ultimate biosafety tool. BioEssays 32 (4):322–331

Schmidt M, de Lorenzo V (2012) Synthetic constructs in/for the environment: managing the interplay between natural and engineered biology. FEBS Lett 586(15):2199–2206

Schummer J (2011) Das Gotteshandwerk: Die künstliche Herstellung von Leben im Labor. Suhrkamp, Berlin

Schwille P (2011) Bottom-up synthetic biology: engineering in a tinkerer's world. Science 333 (6047):1252–1254

Seyfried G, Pei L, Schmidt M (2014) European do-it-yourself (DIY) biology: beyond the hope, hype and horror. BioEssays 36(6):548–551

Shendure J, Ji H (2008) Next-generation DNA sequencing. Nat Biotechnol 26(10):1135–1145

Silver PA, Way JC, Arnold FH et al (2014) Synthetic biology: engineering explored. Nature 509 (7499):166–167

Smith HO, Hutchison CA 3rd, Pfannkoch C et al (2003) Generating a synthetic genome by whole genome assembly: phiX174 bacteriophage from synthetic oligonucleotides. Proc Natl Acad Sci USA 100(26):15440–15445

Smolke CD (2009) Building outside of the box: iGEM and the BioBricks Foundation. Nat Biotechnol 27(12):1099–1102

Sole RV, Munteanu A, Rodriguez-Caso C et al (2007) Synthetic protocell biology: from reproduction to computation. Philos Trans R Soc Lond B Biol Sci 362(1486):1727–1739

Stephanopoulos G, Vallino JJ (1991) Network rigidity and metabolic engineering in metabolite overproduction. Science 252(5013):1675–1681

Sun N, Zhao H (2013) Transcription activator-like effector nucleases (TALENs): a highly efficient and versatile tool for genome editing. Biotechnol Bioeng 110(7):1811–1821

Szybalski W, Skalka A (1978) Nobel prizes and restriction enzymes. Gene 4(3):181–182

ter Meulen V (2014) Time to settle the synthetic controversy. Nature 509(7499):135

Thyer R, Ellefson J (2014) Synthetic biology: new letters for life's alphabet. Nature 509 (7500):291–292

Torgersen H (2009) Synthetic biology in society: learning from past experience? Syst Synth Biol 3 (1–4):9–17

Torgersen H (2004) The real and perceived risks of genetically modified organisms. EMBO Rep 5 (Spec No):S17–S21

Towie N (2006) Malaria breakthrough raises spectre of drug resistance. Nature 440(7086): 852–853

Trifonov EN (2011) Vocabulary of definitions of life suggests a definition. J Biomol Struct Dyn 29 (2):259–266

Tu Y (2011) The discovery of artemisinin (qinghaosu) and gifts from Chinese medicine. Nat Med 17(10):1217–1220

Tumpey TM, Basler CF, Aguilar PV et al (2005) Characterization of the reconstructed 1918 Spanish influenza pandemic virus. Science 310(5745):77–80

van den Belt H (2009) Playing God in Frankenstein's footsteps: synthetic biology and the meaning of life. Nanoethics 3(3):257–268

Vasas V, Fernando C, Santos M et al (2012) Evolution before genes. Biol Direct 7:1

Venter JC (2013) Life at the speed of light. Little Brown, London

Vilanova C, Porcar M (2014) iGEM 2.0–refoundations for engineering biology. Nat Biotechnol 32 (5):420–424

Villarreal LP (2004) Are viruses alive? Sci Am 291(6):100–105

Vogel G (2015) Bioethics. Embryo engineering alarm. Science 347(6228):1301

Wang BJ, Buck M (2012) Customizing cell signaling using engineered genetic logic circuits. Trends Microbiol 20(8):376–384

Wang HH, Isaacs FJ, Carr PA et al (2009) Programming cells by multiplex genome engineering and accelerated evolution. Nature 460(7257):894–898

Wang K, Neumann H, Peak-Chew SY et al (2007) Evolved orthogonal ribosomes enhance the efficiency of synthetic genetic code expansion. Nat Biotechnol 25(7):770–777

Weber W, Fussenegger M (2012) Emerging biomedical applications of synthetic biology. Nat Rev Genet 13(1):21–35

Wilmut I, Schnieke AE, McWhir J et al (1997) Viable offspring derived from fetal and adult mammalian cells. Nature 385(6619):810–813

Wimmer E, Mueller S, Tumpey TM et al (2009) Synthetic viruses: a new opportunity to understand and prevent viral disease. Nat Biotechnol 27(12):1163–1172

Win MN, Smolke CD (2008) Higher-order cellular information processing with synthetic RNA devices. Science 322(5900):456–460

Wolinsky H (2009) Kitchen biology. The rise of do-it-yourself biology democratizes science, but is it dangerous to public health and the environment? EMBO Rep 10(7):683–685

Wong JT (1983) Membership mutation of the genetic code: loss of fitness by tryptophan. Proc Natl Acad Sci USA 80(20):6303–6306

Woodruff LB, Gill RT (2011) Engineering genomes in multiplex. Curr Opin Biotechnol 22 (4):576–583

Yeh BJ, Lim WA (2007) Synthetic biology: lessons from the history of synthetic organic chemistry. Nat Chem Biol 3(9):521–525

Yu AC, Yim AK, Mat WK et al (2014) Mutations enabling displacement of tryptophan by 4-fluorotryptophan as a canonical amino acid of the genetic code. Genome Biol Evol 6 (3):629–641

Zhang F, Keasling J (2011) Biosensors and their applications in microbial metabolic engineering. Trends Microbiol 19(7):323–329

Zhao L, Lu W (2014) Mirror image proteins. Curr Opin Chem Biol 22:56–61

Making-of Synthetic Biology: The European CyanoFactory Research Consortium

Röbbe Wünschiers

1 Introduction

A Personal Account Science is meant to be objective. Here, I try to give an as objective as possible view about my subjective experiences in the field of synthetic biology research. This certainly is a non-conventional way, though, to my opinion a valid one when it comes to the questions why and how a scientist is doing research and how a new research field emerges. Thus, in contrast to the results of other studies presented within this book, this one is meant to be a personal account. Currently, I am part of the European Commission funded research consortium named CyanoFactory. Thus, I shall exemplify my viewpoint of the interdisciplinary nature of synthetic biology by describing the research objective of the CyanoFactory consortium and how it relates to synthetic biology.

The Science of Man Agenda and Synthetic Biology A lot has already been said about what synthetic biology is. For me as a trained biologist with a favour for computing and with research experience in both academia and industry, synthetic biology is not only a current buzz word to praise my research ideas to funding agencies, but also a logical continuation of what started off as molecular biology in the Science of Man Agenda, initially funded by the Rockefeller Foundation from 1932 to 1938 (Kay 1992; Weaver 1970). From 1932 to 1959 90,000,000 US$ were spent for the joint effort to analyze physical, chemical and mathematical aspects of life. Its director at that time, Warren Weaver (1894–1978), stated about the Science of Man Agenda:

In memory of my scientific teacher, Horst Senger (14. Aug 1931–7. Feb 2015), professor of plant physiology, who guided me into the world of photosynthesis and photobiological hydrogen production.

R. Wünschiers (✉)
University of Applied Sciences Mittweida, Mittweida, Germany
e-mail: wuenschi@hs-mittweida.de

© Springer International Publishing Switzerland 2016   55
K. Hagen et al. (eds.), *Ambivalences of Creating Life*, Ethics of Science and Technology Assessment 45, DOI 10.1007/978-3-319-21088-9_3

[…] new branch of science […] which may prove as revolutionary […] as the discovery of the living cell […] A new biology – molecular biology – has begun as a small salient biological research […] (cited in Olby 1990, p. 505)

The goal of the Science of Man Agenda was to bring together researches from different fields of mainly natural sciences for the analysis of the molecular basis of life. Milestones in the chain of subsequent discoveries were the separation of molecular structures by ultracentrifugation and the discovery of the molecular basis of heredity and the genetic code. From 1953 to 1965, 17 out of 18 Nobel Prizes that were awarded and touched the field of molecular biology went to scientists funded by the Rockefeller foundation. This highlights the impact of its underlying funding policy.

Why do I see the Science of Man Agenda as a cradle of synthetic biology? First of all, the transition of biology to molecular biology laid the basis for an "internal" view at biological processes, e.g. metabolism and genetic regulation. These processes in turn are the basis for engineering microorganisms instead of engineering their environment. Another important aspect is true interdisciplinary research: Warren Weaver forced natural scientists and mathematicians to conceive cross-border research projects and write joint applications in order to receive joint funding.

Systems Biology Another important basis for synthetic biology is systems biology (Trewavas 2006). Modern systems biology has had its recognition in life sciences since the late 1990s. It comprises all methods to analyze and model living systems as a whole. Since then, all kinds of -omic research projects (genomics, transcriptomics, metabolomics, etc.) have dominated the scene and led to the development of medium- and high-throughput experimental designs as well as to the emergence of the big-data challenges. High-throughput experiments allow the measurement of many parameters at once, e.g. the DNA-microarray or RNASeq (RNA sequencing) based analysis of the transcription of all genes, or the mass-spectroscopic analysis all cellular metabolites. The directed engineering of organisms is an important basis of synthetic biology, which in turn is based on predictions from models. These models need to be trained with real data. Methods from systems biology provide these data.

2 CyanoFactory

Aims The vision of the CyanoFactory research consortium is to carry out integrated, fundamental research aimed at applying synthetic biology principles towards a cell factory notion in microbial biotechnology. Building on recent progress in synthetic biology we try to engineer photosynthetic cyanobacteria as chassis to be used as self-sustained cell factories in generating a solar fuel like biodiesel or hydrogen gas. The photo-production of hydrogen gas is the focus of the on-going research. However, the measure of success is not only the production of hydrogen gas but also the development and application of molecular and computational tools in order to demonstrate the applicability of synthetic biology for the directed engineering of cyanobacteria.

Funding The CyanoFactory research project (cyanofactory.eu) is funded by the European Commission within the Framework Programme 7 with roughly three Mio. Euro over a period of three years from December 2012 to November 2015. It is a collaborate research project for the design, construction and demonstration of solar biofuel production using photosynthetic cell factories. Six universities, two research institutes and two companies from seven European countries have joined forces for the twelve work packages (Figs. 1 and 2; Table 1).

Fig. 1 Geographic distribution of the European CyanoFactory research consortium. *A* Coordinator Peter Lindblad; Uppsala University, Sweden; work packages (WP, see Table 1) 1, 4, 11, 12 *B* Röbbe Wünschiers; University of Applied Sciences Mittweida, Germany; WP 2 *C* Paula Tamagnini; Instituto de Biologia Molecular e Celular, Portugal; WP 3 *D* Matthias Rögner; Ruhr-University Bochum, Germany; WP 5 *E* Marko Dolinar; University of Ljubljana, Slovenia; WP 6 *F* Phillip C. Wright; University of Sheffield, United Kingdom; WP 7 *G* Javier F. Urchueguía; Universidad Politécnica de Valencia, Spain; WP 8 *H* Hans-Jürgen Schmitz; KSD Innovation GmbH, Germany; WP 9 *I* Giuseppe Torzillo; CNR-ISE, Italy; WP 10 *J* Marcello M. Diano; M2M Engineering S.A.S., Italy; WP 10

Fig. 2 Principal investigators of CyanoFactory. The letter code is the same as in Fig. 1

Table 1 List of work packages. The letter code is the same as in Fig. 1

(A)	WP1	ToolBox for cyanobacterial synthetic biology
(B)	WP2	DataWarehouse/Bioinformatics
(C)	WP3	Improvement of chassis growth, functionality and robustness
(A)	WP4	Introduction of custom designed hydrogen producing units
(D)	WP5	Improvement of photosynthetic efficiency towards H2-production
(E)	WP6	Biosafety
(F)	WP7	Analyses of the purpose designed cyanobacterial cells to identify bottlenecks and suggest further improvements
(G)	WP8	Metabolic modelling of the engineered cells
(H)	WP9	Development of an efficient photobioreactor unit
(I)(J)	WP10	Assembly and performance assessment of a larger prototype photobioreactor system
(A)	WP11	Management
(A)	WP12	Dissemination

The project is placed under the umbrella of the Future and Emerging Technologies (FET) programme, which "invests in transformative frontier research and innovation with a high potential impact on technology, to benefit our economy and society" (European Commission 2015). This translates into the classification as being research that potentially changes the world—but might also completely fail to do so. The reason is that synthetic biology is a rather new research field and that it has to be elucidated whether it meets its expectations.

How the Consortium came Together Each member of the consortium has a background in working either with cyanobacteria, biohydrogen or photobioreactors. Some of the members have already collaborated in previous research consortia, while others met for the first time within the CyanoFactory framework. Personally, I was asked by the project coordinator Peter Lindblad to join the newly formed consortium in December 2010. We knew each other from my PostDoc time, which I spent in his laboratory in Uppsala/Sweden from 1999 to 2001. We had been in contact since and he recognized my expertise in bridging the gap between experimental and computational biology.

Figure 3 outlines the formation of the consortium. Five members had already collaborated in a jointly funded research project. The topic was biohydrogen production but not synthetic biology. These five partners decided to reassemble for a new application directed towards synthetic biology and biohydrogen and invited five new research groups to join for a common application. A first application was put together for January 2011. All discussions about work packages and research goals were negotiated via Skype video calls. As a matter of fact, the online submission of this application at the web-portal provided by the European Commission failed due to a crash of the uploader's computer. Nevertheless, the consortium stood together and successfully submitted an application draft for a two-stage evaluation process in October 2011. After getting green light to hand in a full application, all partners met in Uppsala/Sweden in January 2012. As a result of this physical meeting and subsequent regular video calls, the full application was submitted in April 2012 and accepted in June 2012. After a negotiation phase the research project began in December 2012.

3 Cyanobacteria and Hydrogen: Scientific Background

Cyanobacteria Cyanobacteria, also known by the outdated term blue-green algae, are prokaryotic microorganisms, unicellular or filamentous, with the same type of photosynthesis as higher plants. They are present in highly diverse and even extreme environments with significant tolerance towards temperature, salinity or pH fluctuations, and water availability. Since 1996, when the first cyanobacterial genome was published, a large number of cyanobacterial genomes have been sequenced.

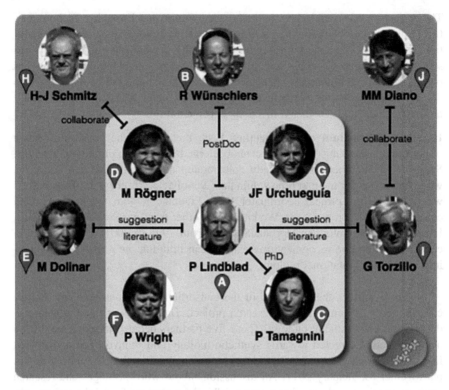

Fig. 3 Formation of the CyanoFactory consortium. Five partners around Peter Lindblad collaborated in a previous jointly funded research project. The new members were known from either previous collaboration, PostDoc times or suggestions from colleagues and the literature. The letter code is the same as in Fig. 1

The ability of cyanobacteria to use solar energy and atmospheric carbon dioxide as energy and carbon sources, respectively, together with their fast growth rates (compared to plants) and the relative ease with which they can be genetically engineered (compared to the difficulties faced using algae), make cyanobacteria stand out from all other organisms used in biotechnological applications (Wijffels et al. 2013). Cyanobacteria have thus become model organisms, and genetic tools for molecular technologies in cyanobacteria are readily available and constantly being developed further. In recent years, there has been a very strong trend, both academic and commercial, to use a standardized genetic engineering methodology, i.e. synthetic biology, to develop efficient photosynthetic microbial cell factories for generation of a portfolio of biofuels directly from solar energy (Angermayr et al. 2009; Ducat et al. 2011; Wijffels et al. 2013). One promising biofuel is hydrogen gas.

Hydrogen as an Energy Carrier Hydrogen is believed to be a potential future energy carrier (Züttel et al. 2008). Already in 1874 Jules Verne proposed in his novel "The Mysterious Island":

> Yes, my friends, I believe that water will one day be employed as fuel, that hydrogen and oxygen which constitute it, used singly or together, will furnish an inexhaustible source of heat and light, of an intensity of which coal is not capable. Some day the coalrooms of steamers and the tenders of locomotives will, instead of coal, be stored with these two condensed gases, which will burn in the furnaces with enormous calorific power. There is, therefore, nothing to fear. As long as the earth is inhabited it will supply the wants of its inhabitants, and there will be no want of either light or heat as long as the productions of the vegetable, mineral or animal kingdoms do not fail us. I believe, then, that when the deposits of coal are exhausted we shall heat and warm ourselves with water. Water will be the coal of the future. (Verne 1874, Chap. 11)

Today, more than 140 years later, not only the depletion of fossil energy sources forces us to investigate new energy concepts. Public acceptance and environmental aspects further push international research for alternative energy sources. Increasingly, renewable and sustainable sources like sunlight, wind power, biomass, geothermal resources, and hydroelectric power, are contributing to our energy supply. Most of these energy sources are restricted to certain localities like sun- or wind-rich areas. The energy needs to be transferred from the place of production to where it is needed. Furthermore, the energy must be stored, since most renewable energy sources are not available continuously. Since many renewable energy sources produce electricity, hydrogen gas can be produced electrolytically from water. Technologies for stationary storage and transport of hydrogen, generally as compressed gas or cryogenic liquid, are commercially available and already in use. Due to its high energy density when liquefied and its versatile use in either combustion (generating mechanical energy) or fuel cells (generating electrical energy), hydrogen is being used in space travel since its beginnings. Developments like metal hydrides, carbon nanotubes or glass microspheres are promising to further increase the energy content per weight and volume of storage system. Fuel cells, internal combustion engines and hydrogen burners are sophisticated devices, which convert hydrogen's energy content into electricity, mechanical work or heat, respectively. The principle by-product is water that can be safely returned to the environment. In fact, projects in several countries already demonstrate the safe and efficient use of hydrogen gas to drive public busses or generate heat and electricity in pilot power plants.

Hydrogen Production The major current use of hydrogen is for ammonia production in fertilizer industries or fat refinery in food industries. Still more than 96 % of worldwide hydrogen production depends on fossil resources (gas, oil, and coal) and consumes as much as 2 % of the world's energy demand. Alternatively, hydrogen can be produced from water using renewable electricity (via electrolysis, as described above) or microorganisms (via hydrogenase or nitrogenase enzymes coupled either to fermentation or to photosynthesis). CyanoFactory is one of many projects seeking for alternative ways to produce hydrogen gas biologically.

Hydrogen from Cyanobacteria Hydrogen gas is of major importance in the biosphere. Several archaebacteria, bacteria, cyanobacteria and green algae contain enzymes known as hydrogenases that either oxidize hydrogen to protons and electrons or reduce protons and thus release molecular hydrogen. The natural physiological functions and biochemical characteristics of these hydrogenases are diverse (Vignais and Billoud 2007; Wünschiers and Lindblad 2003; Wünschiers 2003). Most biologically produced hydrogen in the biosphere is evolved in microbial fermentation processes, e.g. by rumen bacteria in ruminants. Ultimately, these organisms decompose organic matter to carbon dioxide and hydrogen as was shown already over 100 years ago by the biochemist Hoppe-Seyler (1887).

The reduction of protons to hydrogen serves to dissipate excess electrons within the cell and generally permits additional energy generating steps in metabolism. The produced hydrogen gas is usually taken up directly by hydrogen consumers within the same ecosystem. These organisms use the reducing power of hydrogen to drive metabolic processes. Hydrogen bacteria (German: "Knallgas" bacteria) can even grow autotrophically with hydrogen gas as sole reducing power and energy substrate. In these bacteria oxygen serves as terminal electron acceptor, thus, water is formed in a biological oxy-hydrogen ("Knallgas") reaction. Although it is estimated that more than 200 million tons of hydrogen gas are cycled within the biosphere every year, the atmosphere only harbors some 0.000078 % hydrogen (Bélaich et al. 1990).

The only other known enzyme that metabolizes hydrogen gas is the nitrogenase (Peters et al. 1995; Raymond et al. 2003). It occurs only in few bacteria and cyanobacteria. Nitrogenases catalyze the conversion of atmospheric nitrogen gas to ammonia and are thus responsible for nitrogen fixation, i.e. they convert the inert to bioactive nitrogen. The biochemical reduction of nitrogen to ammonia is a highly endergonic reaction requiring metabolic energy in the form of adenosine triphosphate. The catalytic activity of the nitrogenase is accompanied by an obligatory reduction of protons to hydrogen gas, which is released from the enzyme into the cell. However, there has not been a compelling demonstration of an obligatory mechanism responsible for coupling hydrogen evolution with nitrogen reduction, although hydrogen is always observed as a product with ammonia. At least two hydrogen molecules are released per synthesized ammonium molecule, which adds to the huge energy amount required by the nitrogenase enzyme complex. Recycling the released hydrogen gas by uptake-hydrogenases in vivo usually counteracts the low efficiency and energy loss by the process. However, it makes its application less attractive for biotechnological hydrogen production.

There are many ways to utilize hydrogenase or nitrogenase containing microorganisms for hydrogen gas production (Pandey et al. 2013). The aim of the CyanoFactory research project is to couple a hydrogen-producing hydrogenase enzyme to water-splitting photosynthesis (Fig. 4). The result would be a photobiological equivalent of electrical water-electrolysis. This, of course, involves a number of challenges.

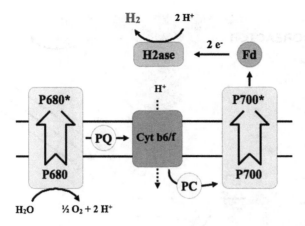

Fig. 4 The electron transport chain from water to hydrogen gas. *P680* Photosystem II; *P700* photosystem I; *PQ* plastoquinone; *Cyt b6/f* cytochrome b6/f complex; *PC* plastocyanin; *Fd* ferredoxin; *H2ase* hydrogenase

4 Challenges for the CyanoFactory Work

The principle behind research collaboration is usually that a consortium should be better than the simple sum of its partners. In the case of CyanoFactory the interdisciplinary consortium tries jointly to solve the problem of bioengineering cyanobacteria. Below, I will describe some important biological and technical challenges that are taken on in this connection (see also Fig. 5).

Everyday life shows that collaboration also has drawbacks. At certain points the individual has to step back for the better of the group. Thus, potentially successful approaches to solve a problem may be turned down by the majority. Interdisciplinary research also requires the formulation of standards, e.g. for measurements or data formats. I will give some examples of these management challenges below, too.

4.1 Biological Challenges

DNA-parts One of the huge promises of synthetic biology is to make molecular bioengineering easier. This is the whole idea behind the BioBrick concept (Shetty et al. 2008), i.e. the creation of biological DNA-parts that can be combined for new predicted functions in host organisms. While a considerable amount of research has been performed in this field with the bacterium *Escherichia coli*, almost nothing had been tried with cyanobacteria when we started in 2012 (Berla et al. 2013; Camsund and Lindblad 2014; Heidorn et al. 2011). When our partner in Uppsala tried to use BioBricks developed for *Escherichia coli* in cyanobacteria like

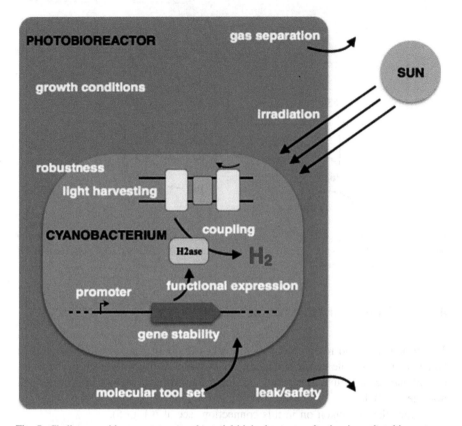

Fig. 5 Challenges with respect to cyanobacterial biohydrogen production in a photobioreactor

Synechocystis PCC 6803—the agreed-on chassis organism in CyanoFactory—they did not work well (Huang et al. 2010). Thus, it is important to develop new molecular tools in order to transfer the concept of DNA-parts to cyanobacteria. Another important requirement in genetic engineering is the need to fine-tune the expression, i.e. the activity, of genes. This requires engineering of tunable promoters, about which little has been known for cyanobacteria. As the work by one partner has shown, cyanobacteria-specific or even strain-specific adoptions of generic expression systems have to be performed (Camsund et al. 2014).

Genomic Integration and Biosafety Any genetic construct that is introduced into the host organism has to be stably integrated into its genome. The genomic integration must not disturb the integrity of the host and needs to be transferred from one generation to the next. Typically, foreign genetic add-ons are lost for several reasons, not least because they are a metabolic burden to the host organism. This requires the adoption of methods known from genetic engineering to stabilize the genetic construct. On the other hand, both the host and the construct need to disintegrate and become bio-inactive in case they should escape into the environment. Thus, measures for biosafety need to be taken and according techniques are being developed within CyanoFactory.

Metabolic Modelling As described above, although huge amounts of hydrogen gas are synthesized by the biosphere, almost no hydrogen is released into the atmosphere. The obvious reason is that hydrogen constitutes a valuable energy source for microorganisms, too. To make cyanobacteria release hydrogen gas into the environment, in our case a photobioreactor, they have to be engineered. As already mentioned above, the aim of our research consortium is to couple a hydrogen-producing hydrogenase to photosynthesis. It could be shown earlier that hydrogenases with the highest activity are not encoded by cyanobacteria. Thus, an appropriate enzyme has to be chosen and adopted in order to be functionally expressed in the cyanobacterial host. Concurrently, the foreign hydrogenase needs to be connected to photosynthesis such that the maximum amount of electrons derived from photosynthetic water splitting are redirected to hydrogen production. This requires a whole range of metabolic optimizations that are not tested by trial-and-error but computed in advance from specific mathematical models developed within CyanoFactory (Fig. 6). These models allow the creation of virtual mutants and computational hypothesis testing (Karr et al. 2012). Of course, these models do only approximate the complex metabolism of a living cell, but they help to restrict the number of experiments required and illustrate the usefulness of interdisciplinary research.

Metabolic Design A by-product from photosynthesis is oxygen gas. Certainly, oxygen is a very important by-product that many organisms in the biosphere thrive on. However, for the oxygen-producing organism it is a burden because it easily

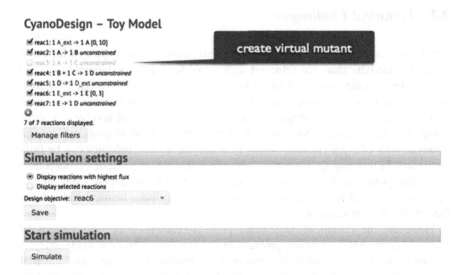

Fig. 6 The web interface of CyanoDesign with a loaded example (toy model). Single reactions can be deleted, which corresponds to genetic knockouts

transforms to destructive, highly reactive oxygen species and—important to our research—it inactivates the hydrogenase. This is a major challenge to be resolved: while photosynthesis is supplying the hydrogenase with electrons and protons for hydrogen gas synthesis, it also produces oxygen gas that irreversibly inactivates the hydrogenase. Nature presents us different solutions to circumvent this problem, mainly by temporal or spatial separation of both reactions. These solutions, however, are not feasible for a biotechnological approach. Thus, one engineered solution is designing an oxygen-insensitive hydrogenase (Bingham et al. 2012; Schäfer et al. 2013) or to add oxygen-consuming reactions.

Chassis Development Additionally, the consortium wishes to engineer a salt-tolerant chassis. The cyanobacterium *Synechocystis* PCC 6803 has been chosen by the consortium because it is a well characterized model organism. However, it is a fresh-water bacterium with moderate salt-tolerance. The development of biotechnological applications based on mass cultivation of salt-tolerant microorganisms is very desirable, since water shortage is already a major problem affecting worldwide agricultural productivity. Therefore, the possibility to engineer *Synechocystis* PCC 6803 in order to function in salt water would allow its use as a robust chassis.

The actual industrial style production with the engineered cyanobacteria will take place in photobioreactors (see below). Here, an important issue is light absorption and shading. Hence, the consortium seeks for ways to engineer cyanobacteria with smaller light harvesting complexes.

4.2 Technical Challenges

For the production of biofuels with cyanobacteria, cultivation systems with much higher productivity than that achieved with actual facilities are mandatory for two reasons. First, competing fossil-based fuels dictate the cost of biofuels. Therefore, the production cost must drop at least by one order of magnitude below the present cost. Second, to get any ecological benefit, new biofuels should involve a positive energy balance, other than for food or high value products. Both goals can only be achieved with efficient culture systems in which the culture behavior can be fully controlled. In this respect, photobioreactors facilitate a high yield of cyanobacteria on limited ground area and avoid competition with food production. Compared to open ponds, closed reactor cultivation minimizes fresh water demand and reduces the dependency on seasonal variations.

However, the design of large-scale photobioreactors is not straightforward (Torzillo et al. 2003). A major problem in current design and life cycle analysis studies is the lack of comparable experimental results. Therefore, two industrial partners of the consortium develop an one liter indoor and an 1000 liter outdoor photobioreactor (Fig. 2). Aspects to consider to reach a high photosynthetic efficiency are: (a) uniform illumination of the culture; (b) an efficient mixing system without stressing the cells; (c) a low mixing time of the photobioreactor in order to guarantee a fast hydrogen degassing of the cultures; (d) complete automation of the process.

4.3 Management Challenges

Some challenges that affect our research consortium as a whole are difficult to categorize as either biological or technical. The examples given below probably have parallels in many research consortia:

Common Experimental Basis One important decision that had to be taken at the very beginning of our joint research and that had strong impact on some partners was the selection of the bacterial strain to work with. The model organism *Synechocystis* PCC 6803 is available in different strains. The "Kazusa" strain was the first photosynthetic prokaryote of which the genome sequence was determined (Kaneko et al. 1996). Like the sister strains PCC ("Pasteur Culture Collection"), ATCC ("American Type Culture Collection") and GT ("Glucose Tolerant"), the "Kazusa" strain was derived from the "Berkeley" strain, which was isolated from California freshwater in 1971 (Stanier et al. 1971). Recently, it has been shown that all sister strains can be distinguished by single nucleotide polymorphisms and insertion/deletion mutations (Ikeuchi and Tabata 2001; Kanesaki et al. 2012; Trautmann et al. 2012). Furthermore, many sub- or laboratory strains have been derived from all four strains. This leads to experimental and computational results based on different genetic backgrounds in different laboratories. Ultimately, this may lead to non-comparable results. Thus, it was an important decision to agree on working with one substrain from one laboratory. This is specifically important for a synthetic biology project, where experimental results are used to train a genome-wide metabolic flux model to predict metabolism.

Of equal importance is the issue of comparable measurements, i.e. hydrogen evolution rates. Browsing the scientific literature reveals many different units and relations, such as evolution rates based on dry weight biomass, fresh weight biomass, chlorophyll *a* content, total chlorophyll content, and others. Often, measurement techniques have a long tradition in laboratory practice, and who likes to change a winning horse? However, joint research of course requires easily comparable results.

Data Exchange and Knowledge Management This brings me to yet another not so obvious issue: data exchange and knowledge management. Life science research is dominated by two conditions: interdisciplinarity and high-throughput. Interdisciplinarity leads to datasets with highly diverse types of content, while high-throughput yields massive amounts of data. Both aspects are reflected by the byte-growth of public bio-databases and the diversity of specialised databases.

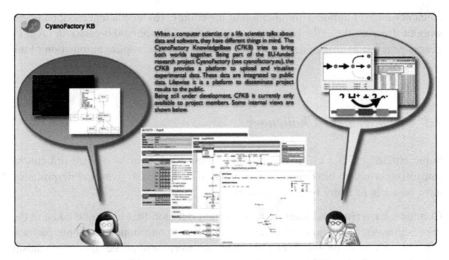

Fig. 7 The CyanoFactory knowledgebase brings computational and experimental scientists together. The web-based user interface provides tools for data visualization

With the rising amount of biological data and the increasing capabilities of computer hardware, many attempts have been undertaken to automatically harvest, store, cross-link and provide biological data in databases and databases of databases, i.e. data warehouses. Their task is to integrate and recombine new and old, internal and external data and to provide a data exchange platform. The CyanoFactory research consortium is unique in trying to bring all experimental data produced by the consortium under one hood, in the web-based "CyanoFactory Knowledgebase" (Kind et al. 2015). It is meant to be the place to collect internal data, integrate them with public data and allow data visualization (Fig. 7). It is also where CyanoDesign can be accessed, and it will be the platform for data dissemination when the project finishes. The goal, which is also explicitly required by the European Commission, is to get away from research consortia where each lab has its own collection of data files and Excel sheets that are neither accessible to partners, nor to the public. The development of the CyanoFactory Knowledgebase reflects the demands of high-throughput measurement techniques with their huge data sets, as in next generation sequencing and modern proteomics.

5 Conventional Versus Synthetic Biology

Conventional Biology Classical biotechnological approaches to improve microorganisms to fulfill novel tasks include (a) screening methods for either strains or enzymes with improved activities; (b) tuning the growing conditions by optimizing exogenic, abiotic parameters such as temperature, light-conditions, ionic concentrations etc.; (c) the creation of knockout mutants to, e.g., turn down competing reactions; (d) the introduction of genetic modifications in order to engineer proteins or act on gene regulation; or (e) the introduction of trans- or cis-genes, like heterologous or homologous hydrogenase operons, respectively, to modify metabolism (Pacheco et al. 2014).

Synthetic Biology The CyanoFactory project claims to fall into the field of synthetic biology by applying and—importantly—extending classical biotechnological approaches. We do so by the development of novel tools to rationally engineer cyanobacteria. In line with synthetic biology standards, the consortium has developed an approach to design, engineer, construct and analyze cyanobacteria for biofuel production (Lindblad et al. 2012). The first step in this experimental flow is the formulation of hypotheses based on bioinformatics investigations. Computer modelling of biological processes, e.g., helps to perform targeted instead of trial-and-error genetic engineering (Kind et al. 2015). This is followed by the synthesis of DNA parts such as promoters, ribosomal binding sites, genes, and terminators to fit the cyanobacterial host. The consortium sticks to the BioBrick standards in order to facilitate the assembly of the novel genetic parts into a genetic vector. This genetic construct is then introduced into the cyanobacterial host, which is in our case a genetically modified cyanobacterium (chassis) that we design to withstand varying salt concentrations and other relevant changing conditions (Lopo et al. 2012; Pinto et al. 2012).

The concept of a chassis that carries basic desired functionality for growth and propagation and that can be expanded with "metabolic plugins" is another typical synthetic biology approach. The engineered cyanobacterium is then grown in indoor and/or outdoor photobioreactors under controlled conditions and quantitatively analyzed by high throughput data acquisition at multiple scales. Based on the complete genome sequence (chromosomal DNA plus plasmids) of the engineered strain under investigation, the complete sets of all transcripts (the transcriptome consisting of messenger-RNAs, ribosomal-RNAs, transfer-RNAs, and non-coding-RNAs) and all proteins and enzymes (proteome) are elucidated by DNA- and RNA-sequencing and mass-spectroscopy, respectively. Likewise, the quantity (metabolomics) and dynamics (fluxomics) of a wide range of organic metabolites is studied. The data obtained from these experiments is used for whole-genome genetic and metabolic modelling and form the basis for hypothesis validation. Overall, this rational design based process shall lead to an iteratively improved cyanobacterial biosystem (Pacheco et al. 2014).

6 Conclusion

When focusing on individual research groups or work packages, it is sometimes difficult to judge whether the methodology applied or scientific question tackled belongs to the field of synthetic biology or is "merely" gene technology, or not even that—as in the case of bioinformatics. Still, these groups and their applied methodology may well contribute to and be part of synthetic biology. Looking at the research of a whole consortium with a large-scale research goal brings about another picture. In this chapter I have shown that synthetic biological research can only be seen as an interdisciplinary quest—as envisioned by Warren Weaver's Science of Man Agenda in the 1930s (Olby 1990; Weaver 1970).

The fascinating nature of life itself attracts scientists from other disciplines, as Francis Crick noted almost 50 years ago in a paper about the future of molecular biology:

> Not only are biologists themselves increasing in number, but fairly large numbers of people are moving into biology from other scientific disciplines. [...] In spite of this it is rare for biologists to leave biology and take up problems in chemistry and physics proper. (Crick 1970, p. 613)

In the same paper, Crick states

> [...] that molecular biology can be defined as anything that interests molecular biologists. (Crick 1970, p. 613)

Reversed and transferred to todays science funding policy one might say that synthetic biologists are scientists, from any discipline, that are funded under the umbrella of synthetic biology. A similar type of definition has been given for nanotechnology (Schummer 2009). Anyway, I hope that I could convince the reader that there is, despite any policy or definition, a new way, a new synthesis of methods, to approach the engineering of living organisms or, more generally, biological systems. I wish to conclude with a citation from Sir Fransis Bacon's book "The New Atlantis", which to my mind describes well what synthetic biologists are dreaming of: rational engineering in life sciences.

> We have also means to make divers plants rise by mixtures of earths without seeds; and likewise to make divers new plants, differing from the vulgar; and to make one tree or plant turn into another. [...] Neither do we this by chance, but we know beforehand, of what matter and commixture what kind of those creatures will arise. (Bacon 1627)

Acknowledgements The research described in this chapter has received funding from the European Union Seventh Framework Programme (FP7/2007-2013) under grant agreement number 308518 (CyanoFactory). The author wishes to thank Kristin Hagen for her helpful comment.

References

Angermayr SA, Hellingwerf KJ, Lindblad P, Teixeira de Mattos MJ (2009) Energy biotechnology with cyanobacteria. Curr Opin Biotechnol 20:257–263

Bacon F (1627) The New Atlantis. Available via the Gutenberg project. http://www.gutenberg.org/ebooks/2434. Accessed 17 Mar 2015

Berla BM, Saha R, Immethun CM et al (2013) Synthetic biology of cyanobacteria: unique challenges and opportunities. Front Microbiol 4:1–14

Bélaich J-P, Bruschi M, Garcia J-L (eds) (1990) Microbiology and biochemistry of strict anaerobes involved in interspecies hydrogen transfer. Plenum Press, New York

Bingham AS, Smith PR, Swartz JR (2012) Evolution of an [FeFe] hydrogenase with decreased oxygen sensitivity. Int J Hydrogen Energy 37:2965–2976

Camsund D, Lindblad P (2014) Engineered transcriptional systems for cyanobacterial biotechnology. Front Bioeng Biotechnol 2:40

Camsund D, Heidorn T, Lindblad P (2014) Design and analysis of LacI-repressed promoters and DNA-looping in a cyanobacterium. J Biol Eng 8:4

Crick F (1970) Molecular biology in the year 2000. Nature 228:613–615

Ducat DC, Way JC, Silver PA (2011) Engineering cyanobacteria to generate high-value products. Trends Biotechnol 29:95–103

European Commission (2015) Future and emerging technologies (FET). http://ec.europa.eu/digital-agenda/en/future-emerging-technologies-fet. Accessed 18 Mar 2015

Heidorn T, Camsund D, Huang H-H et al (2011) Synthetic biology in cyanobacteria engineering and analyzing novel functions. Syn Biol 497:539–579

Huang HH, Camsund D, Lindblad P, Heidorn T (2010) Design and characterization of molecular tools for a synthetic biology approach towards developing cyanobacterial biotechnology. Nucleic Acids Res 38:2577–2593

Hoppe-Seyler F (1887) Die Methangärung der Essigsäure. Z Phys Chem 2:561–568

Ikeuchi M, Tabata S (2001) Synechocystis sp. PCC 6803—a useful tool in the study of the genetics of cyanobacteria. Photosyn Res 70:73–83

Kaneko T, Sato S, Kotani H et al (1996) Sequence analysis of the genome of the unicellular cyanobacterium Synechocystis sp. strain PCC6803. II. Sequence determination of the entire genome and assignment of potential protein-coding regions. DNA Res 3:109–136

Kanesaki Y, Shiwa Y, Tajima N et al (2012) Identification of substrain-specific mutations by massively parallel whole-genome resequencing of Synechocystis sp. PCC 6803. DNA Res 19:67–79

Karr JR, Sanghvi JC, Macklin DN et al (2012) A whole-cell computational model predicts phenotype from genotype. Trends Genet 150:389–401

Kay LE (1992) The molecular vision of life. Oxford University Press, Oxford

Kind G, Zuchantke E, Wünschiers R (2015) CyanoFactory knowledge base and synthetic biology: a plea for human curated bio-databases. In: Pastor O, Sinoquet C, Fred A et al (eds) 6th international conference on bioinformatics: models, methods and algorithms. Scitepress, Lisboa/Portugal, pp 237–242

Lindblad P, Lindberg P, Oliveira P et al (2012) Design, engineering, and construction of photosynthetic microbial cell factories for renewable solar fuel production. Ambio 41:163–168

Lopo M, Montagud A, Navarro E et al (2012) Experimental and modeling analysis of Synechocystis sp. PCC 6803 growth. J Mol Microbiol Biotechnol 22:71–82

Olby R (1990) The molecular revolution in biology. In: Olby RC, Cantor GN, Christie JRR, Hodge MJS (eds) Companion to the history of modern science. Routledge, London, pp 503–520

Pacheco CC, Oliveira P, Tamagnini P (2014) H_2 production using cyanobacteria/cyanobacterial hydrogenases: from classical to synthetic biology approaches. In: Zannoni D, De Philippis R (eds) Microbial BioEnergy: hydrogen production. Springer, Dordrecht, pp 79–99

Pandey A, Chang J-S, Hallenbeck PC, Larroche C (2013) Biohydrogen. Elsevier Science Limited, Burlington/USA

Peters JW, Fisher K, Dean DR (1995) Nitrogenase structure and function: a biochemical-genetic perspective. Annu Rev Microbiol 49:335–3663

Pinto F, van Elburg KA, Pacheco CC et al (2012) Construction of a chassis for hydrogen production: physiological and molecular characterization of a Synechocystis sp. PCC 6803 mutant lacking a functional bidirectional hydrogenase. Microbiology 158:448–464

Raymond J, Siefert JL, Staples CR, Blankenship RE (2003) The natural history of nitrogen fixation. Mol Biol Evol 21:541–554

Schäfer C, Friedrich B, Lenz O (2013) Novel, oxygen-insensitive group 5 [NiFe]-hydrogenase in *Ralstonia eutropha*. Appl Environ Microbiol 79:5137–5145

Schummer J (2009) Nanotechnologie. Suhrkamp, Frankfurt a.M

Shetty RP, Endy D, Knight TF (2008) Engineering BioBrick vectors from BioBrick parts. J Biol Eng 2:5

Stanier RY, Kunisawa R, Mandel M, Cohen-Bazire G (1971) Purification and properties of unicellular blue-green algae (order Chroococcales). Bacteriol Rev 35:171–205

Torzillo G, Pushparaj B, Masojidek J, Vonshak A (2003) Biological constraints in algal biotechnology. Biotechnol Bioprocess Eng 8:338–348

Trautmann D, Voss B, Wilde A et al (2012) Microevolution in cyanobacteria: re-sequencing a motile substrain of *Synechocystis* sp. PCC 6803. DNA Res 19:435–448

Trewavas A (2006) A brief history of systems biology. Plant Cell 18:2420–2430

Verne J (1874) The mysterious island. Part II, chapter 11. Available via the Gutenberg project. http://www.gutenberg.org/ebooks/1268. Accessed 17 Mar 2015

Vignais PM, Billoud B (2007) Occurrence, classification, and biological function of hydrogenases: an overview. Eur J Biochem 107:4206–4272

Weaver W (1970) Molecular biology: origin of the term. Science 170:581–582

Wijffels RH, Kruse O, Hellingwerf KJ (2013) Potential of industrial biotechnology with cyanobacteria and eukaryotic microalgae. Curr Opin Biotechnol 24:405–413

Wünschiers R (2003) Photobiological hydrogen metabolism and hydrogenases from green algae. In: Nalwa HS (ed) Handbook of photochemistry and photobiology. American Scientific Publishers, Valencia/USA, pp 353–382

Wünschiers R, Lindblad P (2003) Light-dependent hydrogen uptake and generation by cyanobacteria. In: Nalwa HS (ed) Handbook of photochemistry and photobiology. American Scientific Publishers, Valencia/USA, pp 295–328

Züttel A, Borgschulte A, Schlapbach L (2008) Hydrogen as a future energy carrier. Wiley-VCH, Weinheim

Synthetic Biology for Space Exploration: Promises and Societal Implications

Cyprien N. Verseux, Ivan G. Paulino-Lima, Mickael Baqué, Daniela Billi and Lynn J. Rothschild

1 Introduction

Can humanity develop sustainable and autonomous colonies beyond Earth? We landed humans on the Moon during the Apollo program and now, thanks to recent technological advances, sending humans to Mars is a realistic medium-term goal (e.g., Horneck et al. 2006). Several projects designed for this purpose are currently under development. In 1990, Mars Direct, a project initiated by Robert Zubrin (then engineer at Martin Marietta), was designed to bring humans to Mars in a decade (Baker and Zubrin 1990; Zubrin and Wagner 1996). NASA aims at having technologies ready to land humans on Mars in the mid-2030s[1] and, even though many unknowns remain in the agency's plans, its deep space crew capsule successfully made its first in-space test on December 5 2014.[2] Space Exploration Technologies Corporation (SpaceX) targets 2026, and its CEO Elon Musk will unveil his Mars

[1]http://www.nasa.gov/content/nasas-human-path-to-mars. Accessed 15 Mar 2015.
[2]http://www.nasa.gov/press/2014/december/nasa-s-new-orion-spacecraft-completes-first-spaceflight-test. Accessed 15 Mar 2015.

C.N. Verseux (✉) · M. Baqué · D. Billi
Department of Biology, University Tor Vergata, Rome, Italy
e-mail: cyprien.verseux@gmail.com

C.N. Verseux
NASA EAP Associate, NASA Ames Research Center, Moffett Field, CA, USA

I.G. Paulino-Lima
NASA Ames Research Center, Moffett Field, CA, USA

L.J. Rothschild
NASA Ames Research Center, Earth Sciences Division, Moffett Field, CA, USA
e-mail: lynn.j.rothschild@nasa.gov

© Springer International Publishing Switzerland 2016
K. Hagen et al. (eds.), *Ambivalences of Creating Life*, Ethics of Science and Technology Assessment 45, DOI 10.1007/978-3-319-21088-9_4

colonization plans by the end of 2015.[3] Mars One, a private company aiming at sending humans to Mars in a one-way mission, targets 2027 and is in the process of selecting crewmembers.[4] These might fail to meet their announced deadlines, or at all; postponing and cancelling is common in space missions. But other Mars colonization projects and private spaceflight companies are emerging, who can benefit from the advances of their predecessors. Your direct descendants may walk on red dust and contemplate a blue sunset. And going beyond Mars is likely to be possible in the more distant future. But there is a big gap between short-term missions and permanent colonies: the longer you stay and the farther you go, the less you can depend on Earth.

Indeed, launch costs do not allow realistic plans for a continuous resupply of colonies beyond the Moon. Let's take food as an example. If all food is to come from Earth, and assuming the easiest option of providing shelf-stable, pre-packaged food similar to provisions of the International Space Station, about 1.8 kg per day and per crewmember should be sent (Allen et al. 2003). Adding the needed vehicle and fuel weight to carry the food, and assuming a 10:1 vehicle-to-payload ratio (Hoffman and Kaplan 1997), a 1000-day food supply for a crew of six would add more than 108 metric tons to the initial mass of the transit vehicle. Worse, these figures are largely under the food minimum needed for a healthy diet: even though shelf-stable items are convenient due to their reduced need for storage facilities and contamination risks, a diet composed exclusively of this type of food would be nutritionally incomplete and thus not adequate in the long term. Given the technical challenges and costs associated with leaving Earth and landing on planetary bodies (e.g., in the order of $300,000 per kg sent to Mars, according to Massa et al. 2007), sending all consumables needed to sustain crews is unrealistic in the long term.

Thus, the time we can stay in remote settlements will depend on our ability to be independent of Earth. This can be achieved by relying on resources found on site, through an approach referred to as in situ resource utilization (ISRU). ISRU can partially rely on physicochemical processing but some necessary products such as high-protein food can currently not be produced or recycled without biological processing (Drysdale et al. 2003; Montague et al. 2012). Besides, many components of physicochemical life support systems are heavy, bulky, consume large amounts of energy and require high temperatures for processes to occur. Even in the case where physicochemical processes are the backbone of life support systems, these could be complemented by biological ones. Besides, overlapping functions would provide a safe redundancy.

Microorganisms, in particular, could be extremely useful. Humans have been consuming and otherwise using microorganism-produced resources on Earth throughout their history: oxygen produced by cyanobacteria and eukaryotic microalgae, food and drinks as edible microorganisms and fermented products (e.g.,

---

[3]http://www.huffingtonpost.com/2015/01/06/elon-musk-mars-colony_n_6423026.html. Accessed 15 Mar 2015.

[4]http://www.mars-one.com. Accessed 15 Mar 2015.

wine and yoghurt), drugs, various chemicals, biomaterials, biofuels, mined metals and so on. We also rely on them for many critical processes such as, for instance, waste recycling. So, why not use microorganisms to cover our daily needs in space and on foreign planetary bodies as we do on Earth? Here is the beginning of an answer: because all organisms we currently know have evolved on Earth and are not adapted to most environments found beyond it. First, most substrates they usually rely on are absent. If we need to bring from Earth all starting compounds needed for microbial processes to occur there, we slightly move the mass problem but certainly not fix it. As an example, the mass of metabolic consumables needed to sustain a crew of six during a 1000-day Mars mission using the European life support system MELiSSA (Godia et al. 2002) has been estimated to be about 30 metric tons, hygiene water not included (Langhoff et al. 2011). Then, the conditions found outside Earth are generally extremely harsh to all known microorganisms, and reproducing Earth-like conditions within a large volume and surface would be extremely costly.

A solution could be to use synthetic biology to increase the fitness of, and to confer new functions to, the organisms in extraterrestrial outposts. Given its potential for enabling space exploration, synthetic biology has aroused NASA's interest (Cumbers and Rothschild 2010; Langhoff et al. 2011; Menezes et al. 2014). In this paper, a brief overview of the possible applications of synthetic biology within extraterrestrial outposts is given (efforts were done to keep it general and easy to understand by interdisciplinary readers) and the resulting impacts on our society are briefly discussed. Focus is here given to Mars, as it is very likely to be the first planet beyond Earth where autonomous outposts are established, but the general ideas apply to other destinations. The Moon could also have been taken as an example, but Mars colonization was preferred to illustrate the concepts outlined below as (i) travel time, costs and difficulty, as well as scientific work that could potentially be conducted (e.g., search for life) justify the establishment of permanent human outposts there, (ii) since it is much farther from Earth than the Moon and has a higher gravity, sending supplies would be more expensive and challenging, increasing the need for exploiting on-site resources instead, and (iii) key resources for biological systems (e.g., water, carbon dioxide and dinitrogen) are widely available on-site. A consideration about the use of synthetic biology for ascribing value to lunar resources can be found elsewhere (Montague et al. 2012).

2 Providing "Off the Land" Substrates for Microbial Growth

Does Mars contain the substrates we need for feeding microorganisms without sending materials from Earth? Obviously you won't find, waiting under a rock, bottled culture media as those used in laboratories to grow microbes. Yet, most elements needed to support life have been detected in the Martian soils and rocks, including all the basic building blocks (C, H, O, N, P, S) and other elements needed in smaller amounts (Mg, Fe, Ca, Na, K, Mn, Cr, Ni, Mo, Cu, Zn...). There is gaseous carbon (in carbon dioxide but also, as recently evidenced, in methane—see Webster et al. 2015) and nitrogen in the atmosphere, and additional carbon atoms can be found in the CO_2 ice caps, in the surface and subsurface regolith (the loose soil that can be seen on photographs of Martian landscapes) due to exchange with the atmosphere, possibly in reservoirs formed when the atmosphere was thicker (Kurahashi-Nakamura and Tajika 2006). Fixed nitrogen compounds have also been detected (Ming et al. 2014), even though what exactly they are and whether or not they could be used by living organisms is not defined yet.

Thus, Martian rocks (Cockell 2014) and atmosphere seem to contain all the basic elements needed to support life. Water is also there: it has been detected in large amounts (Tokano 2005) as ice at the north polar ice cap, under the south carbon ice cap and in the subsurface at more temperate latitudes, as mineral hydration, and as vapor in the atmosphere, even though at low concentrations. It will also be a by-product of human metabolism and industrial activity. Solar energy is of course present. As Mars is approximately 1.5 AU from the Sun, the average radiation flux is 43 % that of Earth's.

So, while all needed elements are naturally present and some additional sources will come from human activity (Table 1), they are in a form that most organisms cannot use. In particular, many organisms—qualified as heterotrophic and including animals such as us humans, as well as most microorganisms—need organic compounds as carbon and energy sources, and their state and availability on Mars remain poorly known (Ming et al. 2014) but is likely low. Fixed nitrogen, such as nitrate (NO_3^-), ammonia (NH_3) and amino-acid chains (but not atmospheric nitrogen which is in the form of dinitrogen, N_2), is also needed for most organisms. The main limitation is consequently not the lack of life-supporting elements, but the

Table 1 Main sources of nutrients for cyanobacterium-based biological processes on mars

Source	Elements
Atmosphere[a]	CO_2, N_2
Soil, rocks[a]	P, S, Mg, Fe, Ca, Na, K, and metal micronutrients
Ice caps, subsurface ice, atmosphere, hydrated minerals[a]	H_2O
Solar radiation[a]	Energy for photosynthesis, heat
Human waste	Fixed N, organic material, CO_2, H_2O
Side effects of other artificial processes (fuel combustion, manufacturing...)	CO_2, H_2O
Cyanobacteria (fed with the above)	O_2, fixed N, organic material, metal nutrients

[a]Naturally present, independently of human activity

abilities of microorganisms to use them under the form they are encountered on Mars's surface.

That being said, not all microorganisms need organic compounds to grow; autotrophs such as cyanobacteria don't. Just like plants, cyanobacteria can photosynthesize—they use CO_2 and solar radiation as carbon and energy sources to produce their own organic material. In a nutrient desert such as Mars, this would give them a strong advantage over heterotrophic organisms. In addition, some can fix N_2, which like CO_2 is present in the Martian atmosphere. On top of this, some have the ability to extract and use nutrients from analogues of Martian rocks and have consequently been suggested as a basis for systems producing life-sustaining compounds from local resources (Brown et al. 2008; Brown 2008a, b). Most—if not all—nutrients needed to cover their needs could be directly provided from Mars's resources. Some cyanobacteria (e.g., *Anabaena cylindrica*) are capable of growing in distilled water containing only powdered Mars basalt analogues, under terrestrial atmosphere (Olsson-Francis and Cockell 2010a). Other studies showed that the growth of several species of cyanobacteria isolated from iron-depositing hot springs in Yellowstone National Park was stimulated by the presence of Martian soil analogues in culture media (Brown and Sarkisova 2008) and that a strain called *Nostoc* sp. HK-01 could grow on a Mars regolith stimulant for at least 140 days, without any other nutrient source besides atmospheric gas (Arai et al. 2008).

As cyanobacteria produce organic compounds, why not use them for feeding heterotrophic organisms? Cultures could be used after simply destroying the cyanobacterial cells; researchers have successfully used lysed cyanobacterial biomass as a substrate for ethanol-producing yeasts (Aikawa et al. 2013; Möllers et al. 2014). However, if we could harvest nutrients without killing cells, processes could be much more efficient. This could be achieved by having cyanobacteria release substrates in the extracellular medium, and this solution has been investigated in the Rothschild laboratory since the 2011 Brown-Stanford iGEM team engineered *Anabaena* PCC7120 to secrete sucrose.[5] Heterotrophic bacteria from a common soil species, *Bacillus subtilis*, were able to grow in filtered medium in which the engineered *Anabaena* had grown but no additional organic compounds were added (unpublished data). Previously, the cyanobacterial strain *Synechococcus elongatus* PCC7942 had been engineered to produce and secrete either glucose and fructose, or lactate, which then served as a substrate for growing the model bacterium *Escherichia coli* (Niederholtmeyer et al. 2010). Ammonium (NH_4^+; a fixed nitrogen compounds that can be used by most microorganisms) is naturally released by some cyanobacteria. The extracellular concentration of ammonia can reach more than 10 mM in cultures of *Anabaena* species relying on atmospheric nitrogen as a sole nitrogen source, without killing the cyanobacteria (Subramanian and Shanmugasundaram 1986). Ammonia becomes limiting only when at extremely low concentrations; for *Escherichia coli*, for instance, it is below a few µM (Kim et al. 2012), several orders of magnitudes below the above mentioned concentrations in

---

[5]http://2011.igem.org/Team:Brown-Stanford/PowerCell/Introduction. Accessed 6 Mar 2015.

cyanobacterial cultures. Then, cyanobacteria grown using Martian rocks as a substrate would release inorganic elements (Ca, Fe, K, Mg, Mn, etc.) into water, as shown in a study performed with terrestrial analogues of Martian basalt (Olsson-Francis and Cockell 2010a), making them available to species which cannot extract them from rocks. Taken as a whole, these studies suggest that using cyanobacteria to produce substrates for microorganisms from Martian resources (see Fig. 1) may be a viable option.

In addition to that of other microorganisms, cyanobacterial cultures could be used to support the growth of plants. Even though basalt is the dominant rock type in Martian regolith and weathered basalt can yield extremely productive soils on Earth (Dahlgren et al. 1993), regolith would probably need a physicochemical and/or biological treatment before it can be used as growth substrate for plants. Reasons for this include its poor water-holding properties (due to its low organic carbon contents), and that regolith nutrients are hardly available to plants (Cockell 2011; Maggi and Pallud 2010). Besides carbon, the soil will need to be enriched in other elements, including nitrogen, as most plants cannot fix atmospheric nitrogen (even though symbiotic nitrogen fixation occurs in some plants, mainly legumes, due to harboring specific bacteria in their tissues). It has already been proposed to

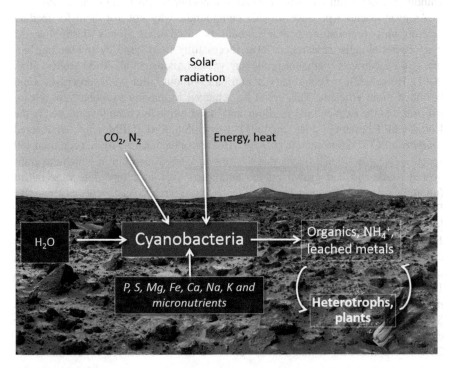

Fig. 1 Using cyanobacteria to process Martian resources into substrates for other organisms. Reproduced from Verseux et al. (2015) with permission from the editor of the International Journal of Astrobiology

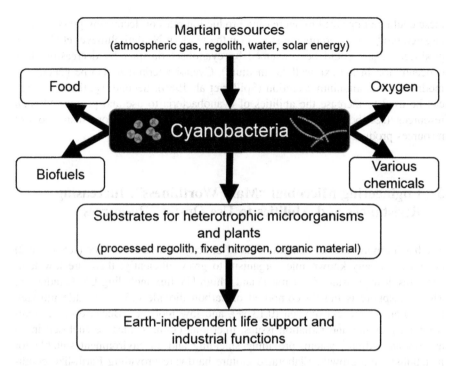

Fig. 2 Cyanobacteria as a link between Martian resources and on-site resource production systems—a simplified overview

use cyanobacteria to release chemical elements from extraterrestrial rocks to an aqueous phase, aimed at being incorporated in a substrate for hydroponic cultivation (Brown and Sarkisova 2008). In addition to these elements, fixed nitrogen and biomass resulting from cyanobacterial cultures could be used as substrates for plant cultivation, as a hydroponic substrate and/or to make a fertile soil.

More information about the potential of Mars-specific cyanobacterium-based biological life support systems (CyBLiSS) is given elsewhere (Verseux et al. 2015) and an overview is given in Fig. 2. The key point is that, thanks to their photosynthetic, rock leaching and nitrogen fixing abilities, cyanobacteria could be used for processing inorganic compounds found on Mars into a form that is available to other microorganisms and to plants. Additional nutrients could come from the recycling of human waste. Finally, if some micronutrients (e.g., some metal ions) could not be mined or biologically synthesized on site, bringing them from Earth would only add negligible mass to the initial payload, as they are needed in trace amounts only.

Why would synthetic biology be useful here? First, as illustrated above, cyanobacteria can be engineered to secrete organic substrates; proofs-of-concept have been done with the secretion of sucrose, lactate, glucose and fructose. Even though growth rates of the heterotrophic organisms were quite low due to low sugar yields, these could be improved by increasing production rates or decreasing processing of targeted products by producing strains (in the work of Niederholtmeyer et al. 2010, produced sugars could be consumed by cyanobacteria, thereby decreasing their concentration in the extracellular medium). Cyanobacteria can also be genetically modified for ammonium secretion (Spiller et al. 1986). Second, synthetic biology can be used to increase the abilities of cyanobacteria to use and process Martian resources (see below), as well as the ability of other microorganisms to use resources produced by cyanobacteria.

3 Engineering Microbial "Mars Worthiness": Increasing Resistances and Abilities to Use On-Site Resources

The harsh environmental conditions faced directly on Mars's surface (see Table 2) do not allow any known microorganism to grow efficiently: there are low temperatures, low pressure (5–11 mbar) and a high UV flux including UV-C radiation. The atmosphere is mostly composed of carbon dioxide (95.3 %), little nitrogen (2.7 %) and even less oxygen (0.13 %), and low moisture. Because of the temperature, pressure and radiation issues, microbial cultures must be enclosed in an appropriate culture system, providing shielding and an environment suitable for metabolism and growth. Elaborated culture hardware providing Earth-like conditions could be suggested, but they would be highly energy consuming, very massive and consequently extremely costly to send to Mars (Lehto et al. 2006) even for small-scale cultures, and would have many possible causes of failure due to relying

Table 2 Environmental parameters on Mars and Earth surfaces

Parameter		Mars	Earth
Surface gravity		$0.38g$	$1.00g$
Mean surface temperature		$-60\ ^{\circ}\text{C}$	$+15\ ^{\circ}\text{C}$
Surface temperature range		-145 to $+20\ ^{\circ}\text{C}$	-90 to $+60\ ^{\circ}\text{C}$
Mean PAR photon flux		8.6×10^{19} photons m^{-2} s^{-1}	2.0×10^{20} photons m^{-2} s^{-1}
UV radiation spectral range		>190 nm	>300 nm
Atmospheric pressure		5–11 hPa	1013 hPa (mean at sea level)
Atmospheric composition (average)	N_2	0.189 hPa, 2.7 %	780 hPa, 78 %
	O_2	0.009 hPa, 0.13 %	210 hPa, 21 %
	CO_2	6.67 hPa, 95.3 %	0.38 hPa, 0.038 %
	Ar	0.112 hPa, 1.6 %	10.13 hPa, 1 %

Adapted from Kanervo et al. (2005) and Graham (2004), and reproduced from Verseux et al. (2015) with permission from the editor of the International Journal of Astrobiology on complex technologies (that being said, the potential of ultimately creating many of these facilities with in situ resources is being explored in Rothschild's lab).

However, culture conditions do not have to be exactly as on Earth: microorganisms have specific ranges of tolerance for environmental factors. In particular, some thrive in extreme environments, including some deserts considered as Mars analogues due to radiation, rock composition, drought and extreme temperatures. Their resistance to conditions found on Mars's surface have been extensively studied, using low Earth orbit- and Earth-based simulations (e.g., Baqué et al. 2013; de Vera et al. 2013, 2014; Olsson-Francis and Cockell 2010b; Rothschild 1990).

The more the organisms can withstand conditions found on site, the simpler the culture system can be. Besides, having resistant organisms would allow loss risk to be minimized (Cockell 2010; Olsson-Francis and Cockell 2010a), both during the journey (the organisms' tolerance to long periods of dehydration, possibly in a differentiated state such as spores or akinetes, would allow a safe and freezing-independent storage) and on site: high resistance to Martian surface conditions would provide safety in case of system malfunction during which cultures could be exposed to less attenuated conditions (e.g., desiccation, low pressure, high radiation levels, altered pH and sudden temperature shift), both when stored and grown.

Synthetic biology could be used to increase the resistance of microorganisms to Martian conditions, probably not enough to make them thrive at the surface but enough to reduce both hardware needs and risks of culture loss. A strategy could be to express genes from other organisms known to confer an advantage in coping with targeted stresses to increase microorganism's fitness under conditions found beyond Earth (Cumbers and Rothschild 2010). This approach has been successful in other contexts, mainly with *E. coli* (see, e.g., Billi et al. 2000; Ferrer et al. 2003; Gao et al. 2003, and the 2012 Stanford-Brown iGEM team[6]). Once specific genes have been shown to confer an advantage to the targeted stress when expressed in the microorganism, they can be improved using various computational and molecular biology tools and methods. This approach is becoming more and more efficient, with notably a sharp decrease in DNA synthesis cost, the improvement of automated gene assembly methods, knowledge gained from systems biology, and the development of biological computer aided design (BioCAD) and other computational tools.

While expressing genes from other organisms (or overexpressing genes from the organism itself) might confer a significant advantage in coping with some environmental stressors, this approach may not be suitable for resistance features that are highly multifactorial, each individual factor having a relatively weak impact. For instance, there is not a single factor that confers *Deinococcus radiodurans* (one of the most radioresistant known microorganism) its extreme radiation resistance, but a very wide combination of features including efficient DNA repair mechanisms, anti-oxidation defenses and specific morphological characteristics (e.g.,

---

[6]http://2012.igem.org/Team:Stanford-Brown/HellCell/Introduction. Accessed 6 Mar 2015.

Slade and Radman 2011). For such multifactorial features, dramatic changes in phenotype through computational genetic engineering are more challenging.

Instead, directed evolution provides an alternative approach that can allow complex modifications at an organismal level without an a priori knowledge of mechanisms. To accomplish this, a parental population is subjected to iterations of mutagenesis and artificial selection. Genetic diversity is created presumably at random, and the mutated population is subjected to selection for the best adapted progeny. To witness the power of selection to shape microbial populations, look no farther than the battle between microbes and antibiotics. Microbial evolution is the reason why doctors insist the patient comply fully with antibiotic regimens. If the antibiotic dosage fails to kill even a few infectious bacteria that happened to be more resistant than the others, they may proliferate and generate a new population of bacteria which will be more resistant than the previous one. Repeat the cycle and you'll need a different treatment.

In the laboratory, similar processes can be exploited to confer on organisms new or improved functions. The dynamics of directed evolution have been widely studied in the last decades and have been used successfully to increase organisms' specific properties (see, e.g., Conrad et al. 2011; Elena and Lenski 2003), including radiation resistance in bacteria (Ewing 1995; Goldman and Travisano 2011; Harris et al. 2009; Wassmann et al. 2010). One of the main issues when designing an optimization process based on directed evolution is the need for linking the optimized function (e.g., production of a compound of industrial interest) to the organism's fitness: strategies must be designed to make the cells of interest thrive while eliminating the others. When increasing resistance, the process is more straightforward: selection can be accomplished by applying increasing levels of the targeted stress and selecting survivors. Directed evolution can be improved by automation (de Crecy et al. 2009; Dykhuizen 1993; Grace et al. 2013; Marlière et al. 2011; Toprak et al. 2013) and recent methods such as, for instance, the so-called genome shuffling (Patnaik et al. 2002) and multiplex genome engineering (Wang et al. 2009).

Once a microbial production system is well established and automated, directed evolution to increase adaption to the Mars environment could be performed on Mars. It can be much faster to evolve organisms on-site than it is on Earth, provided screening methods and adequate organisms are available there (Way et al. 2011). Indeed, Earth-based simulations of some of the factors encountered on Mars (and their combination) are difficult, expensive and cannot faithfully reproduce all their effects.

Computational genetic engineering and directed evolution are not mutually exclusive, in fact, they can be used in combination (Rothschild 2010). Engineered organisms can be submitted to directed evolution for optimization and, conversely, data obtained from genome sequencing of evolved organisms (so as to understand what mutations are responsible for improved properties) can give gene targets for design. An example of strategy combining both is illustrated in Fig. 3. Briefly, natural and evolved gene libraries are generated using directed evolution followed by sequencing and/or comparative gene expression assays in the presence or

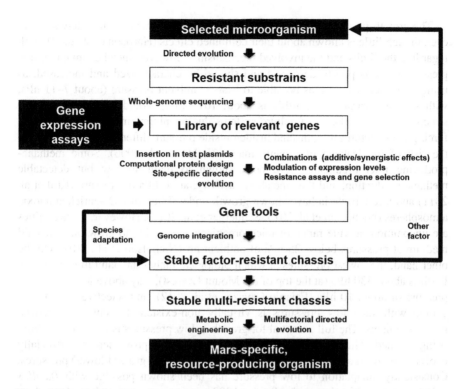

Fig. 3 Theoretical example of a synthetic biology-based strategy combining genetic engineering and directed evolution to improve the suitability of selected microorganisms for resource production in Martian outposts

absence of the environmental stressor. Genes selected for the properties they confer (e.g., metabolic pathway based on local resources or resistance to a target stress) are then engineered using synthetic biology to increase their impact. They are finally adapted to the target organism, the "chassis", which will be used for on-site resource production. Again, this scheme is an example and different workflows can be considered.

Of particular interest would be to increase resistance to long-term dehydration (for storage, as cells enter a state in which they do not need to be fed, and resting stages are often more resistant to environmental extremes), to radiation (ionizing radiation and UV radiation, especially for photosynthetic microorganisms due to the need to access radiation from other parts of the solar spectrum; that being said, ionizing radiation is not extremely high when compared to microorganisms' resistance and UV can easily be blocked or filtered), to a wide range of temperatures, to low pressures, to large and brutal shifts in these parameters (in case of system failure), and to combinations of them. Then, tolerance to a wide range of physicochemical parameters (e.g., high and low pH, presence of oxidative species) would allow constraints on culture conditions to be relaxed.

The growth-permissive limits of most of these factors have been widely studied, even though little is known about their combined effects (Harrison et al. 2013), both regarding the limits and the involved mechanisms. However, metabolism under low pressure remains poorly described and should be characterized and increased, as using a pressure as close as possible to Mars's ambient pressure (about 7–11 hPa, with seasonal variations; Earth's is about 1013 hPa) would greatly lower construction weight and cost of culture systems due to minimizing the need for reinforcing the structures to withstand inside/outside pressure differences, and minimize the risk of the leakage of organic matter (Lehto et al. 2006). Some methane-producing microorganisms have been shown to maintain low but detectable methane production, and thus metabolic activity, at 50 hPa of pressure (Kral et al. 2011) and a few bacteria have shown growth under 7 hPa of CO_2-enriched anoxic atmospheres (Nicholson et al. 2013; Schuerger et al. 2013). However, these abilities are uncommon: a wide range of microorganisms are unable to grow on semisolid medium at pressures below 25 hPa of ambient air (Nicholson et al. 2010). On the other hand, the lowest pressures at which biological niches are naturally present on Earth is about 330 hPa (at the top of the Mount Everest), way above Mars's surface pressure of about 10 hPa (Fajardo-Cavazos et al. 2012), and selective pressure on coping with such low pressures is virtually non-existent for current terrestrial microorganisms. The full potential for growth at low pressures is probably far from being reached. Thus, there might be much room for improvement by artificially evolving microorganisms to grow faster under low (and to grow at lower) pressures. Consistently, adaptation to low pressure has been shown possible with *Bacillus subtilis*: after cultivation at 5 kPa for 1000 generations, one isolate showed an increased fitness at this pressure (Nicholson et al. 2010). It should however be noted that a physical limitation to the pressure range that can be used at growth-permissive temperatures comes from the need to maintain a liquid phase.

Besides resistance, increasing the abilities of microorganisms—especially cyanobacteria if they are used for processing raw resources—to use resources found on Mars's surface would allow yields to be increased, while relaxing culture constraints and the need for materials imported from Earth. In particular, increasing their abilities to leach rocks and to get most of their nutrients from these within a wide range of pH, and to fix molecular nitrogen at low partial pressure, would be highly beneficial. In that case, genetic engineering might be more efficient than it is for increasing their resistance. Some clues have been given regarding the engineering of microorganisms with increased bioleaching abilities (Cockell 2011). However, even though metabolomics has made great advances in the last decade and synthetic biology strongly benefits from it (see for instance Ellis and Goodacre 2012 and Lee 2012), the complex interactions occurring in cells are still hard to predict and model, and whole-cell scaled directed evolution will here again be very useful. In that case, selection can be done by cyclically growing microorganisms and diluting them, in presence of the target nutrient source, and letting the fastest-growing mutants become dominant.

Culture conditions in human outposts on Mars should thus result from a compromise between conditions that provide enough support for microbial metabolism while minimizing costs, initial mass, energy consumption and reliance on materials sent from the Earth. Synthetic biology can be used to push this compromise towards the most sustainable solution, while decreasing risks of losing cultures—what could have terrible consequences if humans rely on them—and increasing yields of the processes by increasing microorganisms' abilities to use on-site resources.

4 Roles of Engineered Organisms in Martian Outposts

Once microorganisms can be grown on Mars, what can they be used for? The most obvious applications will be basic life support functions. Various bioregenerative life support systems (BLSS) based on microorganisms and/or plants have been proposed (and some are under development) for recycling gas, liquid and solid wastes, thereby extending their usage, during beyond-Earth missions (e.g., Drysdale et al. 2004; Giacomelli et al. 2012; Gitelson et al. 2003; Godia et al. 2002; Lobascio et al. 2007; Nelson et al. 2010), including within lunar and Martian outposts (e.g., Blüm et al. 1994; Gitelson 1992; Nelson et al. 2010; Tikhomirov et al. 2007). The most advanced BLSS projects under design depend heavily on materials imported from Earth (even though a recently patented theoretical physicochemical/biological resource production system relies on Martian resources; see Cao et al. 2014). In spite of the potential of some of them to lead to extremely helpful systems in various space mission scenarios, they are consequently not suitable for autonomous, long-term human bases on Mars. Their limitations come, first, from the fact that their running time without resupply is limited by the decreasing amount of materials cycling in the system: recycling cannot reach 100 % efficiency and some losses are unavoidable (for quantitative information regarding the theoretical recycling efficiencies of MELiSSA, see for instance Poughon et al. 2009). Second, they cannot be expanded since the mass of cycling components is at most equal to their initial mass. Currently developed BLSS technologies consequently need regular re-supply of all materials from the ground, which is unrealistic when targeting Mars because of the high costs and high risks of delivery failure. Third, their power consumption is generally high and they represent a large volume and initial mass.

As outlined in the above sections, BLSS could be linked to resources found on site using selected—and possibly genetically enhanced—cyanobacteria. This would allow them to become sustainable, expandable (as new material would enter the loop, processes could be scaled up) and to dramatically decrease the mass of the payload to be sent from Earth. The first point (sustainability) is particularly critical: a permanent human presence on Mars requires colonies that are autonomous.

BLSS could be further improved by the introduction of engineered microorganisms, provided they are genetically stable over the long term. They could perform the functions they are selected for faster and using fewer resources. The system could also be simplified by reducing the number of organisms performing a given set of functions, by increasing organisms' versatility. A smaller set of compartments could thus be used, thereby reducing the variety of resource requirements and the set of possible failure causes.

Another dramatic simplification could be the minimization of the role of plants which, among other drawbacks, are much less efficient than cyanobacteria (which could also perform photosynthesis-related tasks) regarding surface, CO_2 and mineral use (Langhoff et al. 2011; Way et al. 2011), are much more sensitive to environmental conditions, require more manpower, are harder to genetically engineer, take more time to regrow in case of accidental loss (Kanervo et al. 2005; Lehto et al. 2007), are less manageable (Horneck et al. 2003) and contain inedible and hard-to-recycle parts (Hendrickx and Mergeay 2007). The most critical role of plants in BLSS is oxygen and food production, which can also be performed by cyanobacteria (e.g., Hendrickx et al. 2006; Lehto et al. 2006).

However, even though some edible cyanobacterial species such as *Arthrospira* spp. have excellent nutraceutical properties (e.g., Henrikson 2009), they can currently not be used as a staple food due to their unpleasant and unvaried taste, lack of vitamin C and possibly essential oils, and low carbohydrate/protein ratios. These limitations could be addressed using synthetic biology (Way et al. 2011). First, taste, smell and color molecules have already been, or could be, expressed in bacteria. Then, modifying the sugar, protein and lipid ratios, as well as introducing essential molecules (e.g., vitamin C) could be achieved using metabolic engineering and, more generally, nutraceutical properties could be improved by genetic engineering. Preliminary work has been done in this direction; for instance, mutant strains of *A. platensis* have been selected that contained higher contents than the wild-type in essential amino acids, phycobiliproteins and carotenoids, among other nutrients (Brown 2008b). Cyanobacteria could also be used for food complementation without being directly eaten: they can be engineered to secrete nutritional compounds, so used culture media could be harvested without lysing cells and added to food (Way et al. 2011). Besides, as mentioned above, the possibility of engineering cyanobacteria to produce and secrete sugars has already been demonstrated. The use of plants could thus be restrained to applications where no large amounts are needed and where they could be grown within habitats (thus relieving the need for large-scale areas under highly controlled parameters): ornament and horticulture—which have beneficial psychological impact on crewmembers (Allen 1991)—and occasional provision of comfort food.

Another vital resource of human bases will be energy. Solar, wind and nuclear energy are potential sources of on-site (or durable) energy, and these could be complemented by biofuels produced on Mars for, for instance, powering vehicles. Microorganisms are well studied as biofuel producers. Some species studied for their abilities to generate biofuel precursors are heterotrophic and could have organic substrates generated by phototrophic microorganisms: for instance, yeasts are efficient ethanol producers and, even though usually relying on plant agricultural products as a carbon feedstock, cyanobacterial biomass (Aikawa et al. 2013; Möllers et al. 2014) or sugars secreted by cyanobacteria could be used as a substrate generated on-site. However, cyanobacteria could also be used to directly convert solar energy into biofuels, without relying on organic precursors: they can produce various energetic compounds such as alkanes and lipids (see for instance Quintana et al. 2011) and dihydrogen (Raksajit et al. 2012), which can in turn be used for reducing locally available CO_2 to hydrocarbons to produce fuel (Hepp et al. 1993). For both heterotrophic and autotrophic microorganisms, microbial engineering using synthetic biology tools and methods for producing energetic biofuel precursors which are not naturally produced by the hosts, as well as for increasing yields, is a very active field boosting an increasing number of achievements (e.g., Ducat et al. 2011; Hallenbeck 2012; Peralta-Yahya et al. 2012; Radakovits et al. 2010; Zhang et al. 2011).

Engineered microorganisms could also be used for producing a pipeline of drugs on-site. In addition to the antibiotics, therapeutic peptides, antioxidants and other nutraceutical compounds that they naturally produce, microorganisms can be engineered to contain new metabolic pathways leading to the production of various drugs. The most famous example is the production in yeast and bacteria of a direct precursor of artemisinin, an antimalarial drug (Martin et al. 2003; Ro et al. 2006), but much more is to come (see, e.g., Folcher and Fussenegger 2012; Ruder et al. 2011; Weber and Fussenegger 2011).

Then, once all basic physiological requirements are covered (through BLSS processes, and drug and energy production), engineered microorganisms can be used for sustaining industrial processes. Metals could be extracted from Martian rocks by bioleaching: microorganisms are used on Earth to extract metals of industrial interest (e.g., copper and gold) from rocks, and their use on Mars to mine basalt and potential ores has been suggested, as well as ways of engineering microorganisms to increase their abilities to extract (and possibly sort) elements of interest (Cockell 2010, 2011). Extracted elements could be used within many chemical and manufacturing processes such as carbon dioxide cracking, electroplating, production of alloys and manufacturing of solar cells (see Cockell 2011; Dalton and Roberto 2008). Engineered microorganisms could also be used for producing or improving building materials such as bioplastics (Hempel et al. 2011; Osanai et al. 2014) and concrete-like materials (de Muynck et al. 2010; Jonkers et al. 2010).

This paper focuses on applications considered as realistic (even though ambitious) in the short- to medium-term (based on currently available techniques and past successes) and avoids reliance on too tentative speculations on the development of our biological engineering abilities. However, with the advances of synthetic biology, new methods for the improvement of biological strains will likely appear. One can for instance imagine changing the chemical composition of DNA (something that begins to be possible; see, e.g., Marlière et al. 2011) to make it less prone to radiation-induced lesion, or even synthesizing an artificial organism gathering the most relevant features from several microbes, leading to optimized metabolic abilities and outstanding resistance to the Martian environment. A byproduct of the development of such technologies will probably be the wish to use them for ecopoiesis by, for instance, increasing the abilities of selected microorganisms to spread on Mars's surface, to dissolve carbonates for thickening the atmosphere and to produce large amounts of oxygen (Thomas et al. 2006). The implications of such a choice, as well as whether it is desirable or not, are beyond the scope of this paper. Modifications of multicellular organisms can also be thought of: plants, but also animals—including humans (and/or their microbiomes)—could be engineered to increase their fitness under extraterrestrial environments (Langhoff et al. 2011).

Even on a shorter term, many other applications of synthetic biology can be considered in human extraterrestrial outposts. Some are similar to what is expected on Earth (see, e.g., Khalil and Collins 2010 and Church et al. 2014 for examples of potential future applications), but many biotechnologies will probably be specific to space exploration. Indeed, some processes will benefit from synthetic biology there whereas they can be more economically performed by other means on Earth, for example by chemical methods, by using the natural host or by simply harvesting rather than producing the targeted compounds. Besides, metabolic pathways might differ: some substrates that are cheap and abundant on Earth will be extremely hard to provide on Mars. The key point here is that once organic substrates can be obtained from on-site resources (e.g., using cyanobacteria) and that microorganisms can be grown on Mars at low costs (e.g., due to genetically increased resistance), metabolic engineering opens the way to a wide range of potential applications.

There is still a long way to go: the brief overview given above doesn't reflect the work needed to overcome the obstacles lying between us and functional systems. Even though techniques are developed that allow efficient engineering of resistance and metabolic features, the increase in resistance needed to make a significant difference in extraterrestrial environments (e.g., the Martian surface) are huge, and engineering efficient processes for a wide range of products starting from Martian resources is currently at the edge of our abilities. Other tasks unrelated to synthetic biology must be conducted, such as selecting the most promising organisms according to their relevance for biological processes in Martian bases, thoroughly characterizing them to identify a compromise between, on one side, minimal requirements (radiation shielding, atmospheric composition and pressure, gravity, nutrient supply etc.) required for efficient growth and metabolism in planetary bases and, on the other side, conditions that can be provided on site (e.g., low atmospheric pressure composed of CO_2 and N_2) at minimal cost, designing a culture hardware that can provide these conditions directly on Mars's surface, linking the mentioned systems to full BLSS, and extensively testing all the components of the system to demonstrate its operational capability. However, the pace at which our abilities in synthetic biology increase make the design of suitable organisms reasonably achievable in the short- to medium-term; the other tasks are engineering or strategy issues, which are not obviously insurmountable given enough effort is dedicated to them. First manned missions to Mars will likely be short-duration (a few years) missions, and for these astronauts can rely on resources imported from Earth. These missions will be an opportunity to test BLSS technologies on-site while back-up resources are available. As our confidence in BLSS—and in life support technologies in general—increases, longer-duration missions can be planned where our dependency on local resources supplants that on imported materials.

In summary, naturally evolved cyanobacteria or those engineered for increased resistance and abilities to use in situ resources could be used for processing on-site materials and turning them into forms available to other microorganisms. From this basis, synthetic biology could open the way to a wide range of applications including the production of vital resources, energy, drugs, building materials, industrial reagents and comfort goods—all starting from resources found on site. Other potential applications of synthetic biology for space exploration (including modifications to non-microorganisms) are discussed elsewhere (Cumbers and Rothschild 2010; Langhoff et al. 2011; Menezes et al. 2014; Montague et al. 2012). Combined with physicochemical technologies (both completing each other and providing a safe redundancy for vital processes), synthetic biology can thus lead to the development of complex, Earth-independent human bases on Mars and beyond.

5 Societal Implications of Extraterrestrial Human Colonies Relying on Synthetic Biology

The societal ramifications of synthetic biology are being discussed intensely and at the highest levels worldwide. For example, in the US, the first report released by the Presidential Commission for the Study of Bio-ethical Issues was focused on synthetic biology (Presidential Commission for the Study of Bioethical Issues 2010). This report identified five guiding principles: (1) public beneficence, (2) responsible stewardship, (3) intellectual freedom and responsibility, (4) democratic deliberation, and (5) justice and fairness. Neither these principles nor the abundant discussion of the topic in other fora will be detailed here; some aspects are covered in other contributions to this book. Here, we rather focus on issues arising from the use of synthetic biology in support of human space colonization. Some are simply related to the fact that the former is fostered, thereby raising issues related to space exploration itself. Conversely, others come from the stimulation of synthetic biology by its application to the space sector. Finally, very specific issues are raised by the potential development of human colonies beyond Earth that rely on modified organisms.

As humans settle in destinations beyond Earth, a number of social, ethical and psychological concerns arise. There are those who doubt that investment should be made in space exploration: why should resources be spent on such a futuristic plan when there are problems to be solved on Earth?[7] But investment in space exploration is… an investment. Even a very rough assessment of the long-term economic benefits of colonizing new worlds would deserve a dedicated paper, but an interesting analogy can be made with colonization of America in the 17th century, or

---

[7]See for instance R. Hanbury-Tenison's opinion in E&T Magazine (Issue 10, October 2011, p. 28), also available at http://eandt.theiet.org/magazine/2011/10/debate.cfm. Accessed 12 Mar 2015.

Australia in the 19th century.[8] Thinking of money spent in space as a net loss for other concerns on Earth can be compared to misevaluations of the value of these settlements by European governments, which now appear as absurd (Zubrin 1995). On the shorter term, economic benefits can be assessed based on past experience. Measuring economic returns from investment in the space sector is a complex task given that, to be accurate, one should take into account technologies indirectly derived from space innovations, and even technologies created by inventors inspired to pursue a career in science or engineering by space exploration. However, economic returns can be estimated from results of technology transfers; the "Space Economy" was assessed at $180 billion in 2005 by the U.S. Space Foundation.[9]

That being said, benefits brought by space exploration are not best described in economic terms. One of the most commonly mentioned arguments in favor of becoming a multi-planet species is the risk of large-scale disasters such as an asteroid impact, which justifies the development of our ability to reach and settle on other planetary bodies (see for instance Baum 2010; Matheny 2007). As pointed out in a talk at NASA's Goddard Space Flight Center by Michael Griffin, then NASA Administrator, "[t]he history of life on Earth is the history of extinction events, and human expansion into the Solar System is, in the end, fundamentally about the survival of the species" (Griffin 2006, p. 24). On a less extreme side, our everyday life has been greatly enhanced by technological advances brought by the Space Age.[10] It has led the development of a wide range of technologies such as telecommunications, GPS, weather prediction, large-scale environmental analyses, and disaster prediction and monitoring. To these can be added those that come indirectly from the space industry, such as artificial hearts and other medical improvements, innovations in the automotive and home industries, or land mine removal devices (see, e.g., Dick and Launius 2009; ESA and IAA 2005). Imagine our modern society bereft of the contributions derived from space exploration! And many more benefits are expected in the future, including our reliance on clean energies, breakthroughs in transport technologies, low gravity manufacturing, mining in space, increased knowledge of the universe, space tourism and stimulating challenge. In addition, humans have always been explorers, venturing forth out of Africa and beyond. They have a primal wish to discover new lands, which drives inspiration, creativity and discoveries. This sense of curiosity of what is beyond results in a sense of wonder and belonging to the greater spirit and purpose of humanity as a whole.

---

[8]This analogy here refers to economic aspects only and do not represent the authors' opinions on other elements of past colonization such as, for example, the way indigenous people were treated.

[9]http://www.nasa.gov/pdf/189537main_mg_space_economy_20070917.pdf. Accessed 13 Mar 2015.

[10]An analogy can be made with the military sector, which also leads to technological innovations due to the urgent need for technological advances (see for instance Perani 1997). Whether space exploration objectives are preferable to military ones as driving force for innovation is left to the reader's judgement.

The philosophical impacts of space exploration have been less tangible but no less momentous: space exploration humbles us by increasing our knowledge of the universe. It also fosters collaborations and forges agreements among countries, and attracts youth into careers in science and technology. More subtle consequences can also be expected. Space exploration may for instance shift our frame of reference from Earth to the solar system or even to the universe, which is likely to change radically our vision of our planet of origin and of the importance of preserving it. Different scenarios can be expected, which are not necessarily exclusive and may vary among individuals. One possible outcome is a decrease in Earth's perceived value—it would, after all, be just one of the places where we can live—but the opposite is also possible: by contrast to Mars and other planetary bodies, Earth may appear as incredibly hospitable and rich in diversity. Space exploration may also increase our awareness of how vulnerable Earth is, switching the status of global disasters from an abstract concept to a concrete risk, thereby fostering environmental protection policies and individual sustainable behaviors. It is also likely to affect the perception people have of "home"; a good model to predict such reactions can be drawn from emigrants' experience on Earth. People living outside their country generally develop a sense of belonging at a broader scale, defining themselves according to their nationality rather than, for instance, their city. Few people think of themselves as earthlings; what would be the consequences of such an extended sense of belonging? What effects would it have on global decisions? Would people consider interests at a larger scale? Then, adapting organisms to conditions previously considered as inhabitable extends the limits of what we consider as being the envelope for life. Such an artificial adaptation shows that this envelope's limits are not defined by life's absolute potential but rather by the limited evolutionary constraints in presence of which terrestrial life was shaped. The vision we have of the uniqueness of Earth as a life nursery and harbor may be affected, which may encounter resistance from people whose religious beliefs or worldviews are incompatible with it. Similar but more extreme reactions can be expected if space exploration leads us to the discovery of life beyond Earth (Connell et al. 1999; Race et al. 2012).

Outposts on Mars and other planetary bodies may also create a diversity of new cultures and sociopolitical forms. Their development will be facilitated by the freedom and self-reliance of pioneers, by the current lack of legislative framework on site and by the lack of means of remotely enforcing laws. New forms of society may emerge as an adaptive response to local environments and living conditions. They may take an original form, as created in unusual contexts and driven by isolated people sharing very specific traits such as, for instance, high intellectual abilities and a passion for science and engineering. Analogies can be found in human history. One of the most obvious examples comes from the first steps of America's colonization by Europeans: a relatively small number of people, gathered in a remote environment which was original to them, and sharing strong features such as a taste for novelty, a high adaptability and a project-rather than people-oriented mind gave rise to a society with a set of values differing from that of their countries of origin. With the colonization of another planetary body, remoteness, novelty of the environment and strength of the shared features will be pushed to an extreme.

Synthetic biology could, as discussed in this paper, become a powerful tool for human space colonization; it could thereby hasten the benefits mentioned above. If this scenario occurs, it will also lead to the reverse: space exploration will become a driving force for synthetic biology. Little in human history has been such a powerful drive for science and technological breakthroughs as space exploration. In the past 60 years, technologies were developed that we didn't imagine before and would certainly not have expected in such a close future. If the next part of the story relies on synthetic biology, a wealth of game-changing innovations is to be expected in this field. The establishment of human colonies relying to a large extent on modified organisms will require dramatic advances in synthetic biology, as permanent access to many compounds taken for granted on Earth will depend on our ability to generate them on site. With the additional constraints to generate them from a limited set of starting compounds (on-site resources) and possibly under unusual environmental conditions, the limits of synthetic biology (noteworthy metabolic engineering) will have to be pushed to a new level. Its miniaturization and automation will likely be extensively improved as well, due to the need for minimizing mass and manpower requirements in space missions. Space synthetic biology will likely inspire a new generation of biologists, as the space race inspired many physicists and engineers, providing the brainpower needed to support this innovation wave.

By enhancing the pace of achievements in synthetic biology, space exploration can greatly speed the consequences—promises and perils—expected to come from the former. These range from extrapolations by media of talks from synthetic biology public figures, such as Craig Venter and George Church, where all plagues are cured by drugs produced by chemically synthesized microbes and all cars are powered by eco-friendly fuels, to apocalyptical "green goo" scenarios where the world ends after bioterrorists lose control of their creations. There is however a continuum of more plausible scenarios in-between these fantasized extremes, with society-impacting breakthroughs and catastrophes that can realistically be foreseen in the relatively short term. These are largely discussed elsewhere (see for instance Aldrich et al. 2008 and other chapters of this book) and won't be detailed here.

Unexpected applications can also arise on Earth from the fact that synthetic biology is specifically driven by space exploration. Extensive effort is made by space-related institutions to ensure that the developed technologies are transferred to the civil world; in the case of space synthetic biology, what does this imply? Our abilities to design organisms able to grow under modified environmental conditions will increase the efficiency of microorganism-dependent industrial processes which are often limited by the absence of microorganisms that can efficiently catalyze the appropriate chemistry under conditions (e.g., high temperature, high concentration of a reagent or product and specific pH) needed to optimize the process, and relax the need for providing energy-consuming culture conditions. In other words, pushing the bounds of terrestrial life can result in economic benefits, as the more hardy synthetic biology-enhanced life forms may be able to improve, or even revolutionize, the bio-based manufacturing sector. Then, tools and methods developed to engineer the production of specific compounds starting from a very limited and constrained set of substrates could both decrease costs (starting from cheap substrates) and valorize specific products (starting from waste or troublesome compounds, such as glycerol which is largely generated as a by-product of the biofuel industry—see Pagliaro and Rossi 2010). Industrial processes could be more easily combined so that the end product of one becomes the substrate of another, creating a balanced "ecosystem" of biotechnological processes. In addition, transferring to Earth our abilities to design biological systems to live "off the land" in harsh environments and relying on a minimal set of resources could bring much more inspirational applications: generating resources in agriculturally poor countries. Rocky deserts could start to be considered as fertile lands, gathering all elements needed to sustain prospering human civilizations and to generate valuable resources.

The public's opinion of synthetic biology might also be fundamentally altered by its use as a tool in space pioneering. On Earth, practical applications of synthetic biology are often challenged by the public's opinion. Whether the benefits of a given use justify the perceived risks is not always clear and varies according to each individual's values. On other planetary bodies, however, alternatives are often not available and the use of synthetic biology may turn from being a luxury to being a necessity. For instance, a healthy diet can be obtained by traditional agriculture methods on Earth but is unlikely to be entirely produced on Mars without engineering organisms to produce some needed nutrients while relying on local resources. Besides, much of the opposition to synthetic biology is driven by fear of potential negative consequences, be they rational or not. Opposition may thus be reduced if applications are carried on far from Earth, making feared consequences much less likely to affect the layman. The public image of synthetic biologists might also change. Currently, it is mostly based on a very limited number of publicized researchers, and synthetic biologists are sometimes depicted as narcissistic scientists working not for the benefit of the society or for knowledge, but to feel empowered or for "playing God". In the context of space exploration, this image may switch to that of pragmatic scientists doing the necessary to support the expansion and long-lastingness of our species.[11]

There is, however, a major concern raised by the use of synthetic biology in space: contaminating extraterrestrial bodies with terrestrial life. This could jeopardize the search for potential extraterrestrial life and life precursors, extinct or extant. Measures should be taken to evaluate the planetary protection-associated risks of BLSS such as those described above (in addition to those of human-associated microbiomes and microbes present on all imported materials; see for instance McKay and Davis 1989, DeVincenzi 1992, and Debus and Arnould 2008) and strategies should be developed to mitigate them. Current international

---

[11]That being said, the possibility of this being seen as "playing God" cannot be ruled out.

treaties and policies related to planetary protection[12] provide a basis, even though requirements need to be extended to be relevant for manned missions (Horneck 2008). This risk is present even when synthetic biology is out of the equation, but increasing the abilities of microorganisms to withstand the explored environments increases the risk of contamination in case of microorganisms' accidental release. Issues associated with a targeted release of microorganisms, as could be considered for geoengineering, won't be discussed here; first, because the concepts mentioned in this paper assume that used organisms are contained and will not be purposely released and, second, because the ethics of implanting terrestrial life on other planetary bodies, Mars in particular, have been extensively discussed elsewhere (see for instance McKay and Marinova 2001).

6 Conclusion

Advances in applied physics allow us to go farther and farther from the Earth, with NASA's Voyager 1 spacecraft entering interstellar space on 25 August 2012; advances in applied biology can allow us to settle elsewhere in our solar system. Synthetic biology can increase our abilities to live "off the land" on other planetary bodies, thereby increasing our abilities to colonize them, to learn about them and to develop along the way technologies finding applications in our everyday lives. Ultimately, mastering the design of organisms capable of producing life-support resources from local substrates can be a key step in the path leading humans to become a multi-planetary species.

Conversely, using synthetic biology as a tool for space exploration can increase dramatically the development of the former, due to the resulting need for innovation and more generally to the technology drive of the space sector. The development of human colonies depending heavily on synthetic biology for survival and comfort would create an urging need for new discoveries and engineering work in this field. This can bring to a closer future the foreseen consequences—be they greatly beneficial, catastrophic or barely noticeable to the layman—of synthetic biology on Earth.

Besides, in addition to issues raised by the development of both fields independently, the possibly synergistic impacts of both becoming a multi-planetary species and of relying on extensively modified life forms are unprecedented and hard to predict. These can be economical, technological but also philosophical and psychological: how will our vision of our importance in the universe, of the preciousness of Earth and of the value of life be affected? And how will that translate into our everyday lives?

The path we have started to follow can dramatically affect the evolution of our society. Decisions should be taken well ahead of time to ensure that where we are going is not being defined only by what we *can* do, but by what we *decide* to be a desirable future.

---

[12]See for instance http://planetaryprotection.nasa.gov/documents/. Accessed 6 Mar 2105.

Acknowledgements The authors are grateful to those who made possible the TASynBio Summer School where this book chapter was presented: the organizers Kristin Hagen, Margret Engelhard and Georg Toepfer, the attendees for their insightful comments and friendly conversations, and the funding body: the German Federal Ministry of Education and Research. They are thankful to the editors, whose comments and suggestions lead to significant improvements to the manuscript. They also thank the Italian Space Agency for supporting the BIOMEX_Cyano and BOSS_Cyano experiments. This work was supported by IGPL's appointment to the NASA Postdoctoral Program at NASA Ames Research Center, administered by Oak Ridge Associated Universities through a contract with NASA, and by CV's appointment to the NASA Education Associates Program managed by the Universities Space Research Association. Therefore the authors are also greatly thankful to the then Center Director S. Pete Worden.

References

Aikawa S, Joseph A, Yamada R et al (2013) Direct conversion of Spirulina to ethanol without pretreatment or enzymatic hydrolysis processes. Energy Environ Sci 6:1844

Aldrich S, Newcomb J, Carlson R (2008) Scenarios for the future of synthetic biology. Ind Biotechnol 4:39–49

Allen CS, Burnett R, Charles J et al (2003) Guidelines and capabilities for designing human missions. NASA/TM–2003–210785

Allen JL (1991) Biosphere 2: the human experiment. Penguin Books, New York

Arai M, Tomita-Yokotani K, Sato S et al (2008) Growth of terrestrial cyanobacterium, *Nostoc* sp., on Martian Regolith Simulant and its vacuum tolerance. Biol Sci Sp 22:8–17

Baqué M, de Vera J-P, Rettberg P, Billi D (2013) The BOSS and BIOMEX space experiments on the EXPOSE-R2 mission: endurance of the desert cyanobacterium *Chroococcidiopsis* under simulated space vacuum, Martian atmosphere, UVC radiation and temperature extremes. Acta Astronaut 91:180–186

Baum SD (2010) Is humanity doomed? Insights from astrobiology. Sustainability 2:591–603

Baker D, Zubrin R (1990) Mars direct: combining near-term technologies to achieve a two-launch manned mars mission. J Br Interplanet Soc 43:519–526

Billi D, Wright DJ, Helm RF et al (2000) Engineering desiccation tolerance in *Escherichia coli*. Appl Environ Microbiol 66:1680–1684

Blüm V, Gitelson J, Horneck G, Kreuzberg K (1994) Opportunities and constraints of closed man-made ecological systems on the moon. Adv Sp Res 14:271–280

Brown II (2008a) Cyanobacteria to link closed ecological systems and in-situ resources utilization processes. 37th COSPAR Sci. Assem., Montréal, Canada, 13–20 July 2008

Brown II (2008b) Mutant strains of *Spirulina (Arthrospira) platensis* to increase the efficiency of micro-ecological life support systems. 37th COSPAR Sci. Assem., Montréal, Canada, 13–20 July 2008

Brown II, Sarkisova S (2008) Bio-weathering of lunar and Martian rocks by cyanobacteria: A resource for moon and mars exploration. Lunar Planet. Sci. XXXIX, League City, Texas, 10–14 Mar 2008

Brown II, Garrison DH, Jones JA et al (2008) The development and perspectives of bio-ISRU. Jt. Annu. Meet. LEAG-ICEUM-SRR, Cape Canaveral, Florida, 28–31 Oct 2008

Cao G, Concas A, Corrias G et al (2014) Process for the production of useful materials for sustaining manned space missions on Mars through in-situ resources utilization. US Pat. App. US20140165461 A1, 24 July 2012

Church GM, Elowitz MB, Smolke CD et al (2014) Realizing the potential of synthetic biology. Nat Rev Mol Cell Biol 15:289–294

Cockell CS (2010) Geomicrobiology beyond Earth: microbe-mineral interactions in space exploration and settlement. Trends Microbiol 18:308–314

Cockell CS (2011) Synthetic geomicrobiology: engineering microbe–mineral interactions for space exploration and settlement. Int J Astrobiol 10:315–324

Cockell CS (2014) Trajectories of Martian habitability. Astrobiology 14:182–203

Connell K, Dick SJ, Rose K et al (1999) Workshop on the societal implications of astrobiology. NASA Technical Memorandum. NASA Ames Research Center, Moffet Field, California, pp 16–17

Conrad TM, Lewis NE, Palsson BØ (2011) Microbial laboratory evolution in the era of genome-scale science. Mol Syst Biol 7:509

Cumbers J, Rothschild LJ (2010) BISRU: Synthetic microbes for moon, Mars and beyond. In: LPI Contrib. League City, Texas, 26–20 Apr 2010

Dahlgren R, Shoji S, Nanzyo M (1993) Mineralogical characteristics of volcanic ash soils. In: Shoji S, Nanzyo M (eds) Volcanic ash soils—genesis, properties and utilization. Elsevier Science Ltd, Amsterdam, pp 101–143

Dalton B, Roberto F (2008) Lunar regolith biomining: workshop report. NASA/CP-2008-214564. NASA Ames Research Center, Moffet Field, California, 5–6 May 2007

Debus A, Arnould J (2008) Planetary protection issues related to human missions to mars. Adv Sp Res 42:1120–1127

de Crecy E, Jaronski S, Lyons B et al (2009) Directed evolution of a filamentous fungus for thermotolerance. BMC Biotechnol 9:74

de Muynck W, Verbeken K, De Belie N, Verstraete W (2010) Influence of urea and calcium dosage on the effectiveness of bacterially induced carbonate precipitation on limestone. Ecol Eng 36:99–111

de Vera J-P, Dulai S, Kereszturi A et al (2013) Results on the survival of cryptobiotic cyanobacteria samples after exposure to mars-like environmental conditions. Int J Astrobiol 13:35–44

de Vera J-P, Schulze-Makuch D, Khan A et al (2014) Adaptation of an antarctic lichen to Martian niche conditions can occur within 34 days. Planet Space Sci 98:182–190

DeVincenzi DL (1992) Planetary protection issues and the future exploration of Mars. Adv Space Res 12:121–128

Dick SJ, Launius RD (eds) (2009) Societal impact of spaceflight. NASA SP-2007-4801, Washington

Drysdale A, Ewert M, Hanford A (2003) Life support approaches for mars missions. Adv Sp Res 31:51–61

Drysdale AE, Rutkze CJ, Albright LD, LaDue RL (2004) The minimal cost of life in space. Adv Sp Res 34:1502–1508

Ducat DC, Way JC, Silver PA (2011) Engineering cyanobacteria to generate high-value products. Trends Biotechnol 29:95–103

Dykhuizen DE (1993) Chemostats used for studying natural selection and adaptive evolution. Methods Enzymol 224:613–631

Elena SF, Lenski RE (2003) Evolution experiments with microorganisms: the dynamics and genetic bases of adaptation. Nat Rev Genet 4:457–469

Ellis DI, Goodacre R (2012) Metabolomics-assisted synthetic biology. Curr Opin Biotechnol 23:22–28

ESA, IAA (2005) The impact of space activities upon society. ESA-BR-237, Noordwijk, The Netherlands

Ewing D (1995) The directed evolution of radiation resistance in E. coli. Biochem Biophys Res Commun 216:549–553

Fajardo-Cavazos P, Waters SM, Schuerger AC et al (2012) Evolution of Bacillus subtilis to enhanced growth at low pressure: up-regulated transcription of des-desKR, encoding the fatty acid desaturase system. Astrobiology 12:258–270

Ferrer M, Chernikova TN, Yakimov MM et al (2003) Chaperonins govern growth of Escherichia coli at low temperatures. Nat Biotechnol 21:1266–1267

Folcher M, Fussenegger M (2012) Synthetic biology advancing clinical applications. Curr Opin Chem Biol 16:345–354

Gao G, Tian B, Liu L et al (2003) Expression of *Deinococcus radiodurans* PprI enhances the radioresistance of *Escherichia coli*. DNA Repair 2(12):1419–1427

Giacomelli GA, Furfaro R, Kacira M et al (2012) Bio-regenerative life support system development for Lunar/Mars habitats. In: 42nd international conference on environmental systems, San Diego, California, 15–19 July

Gitelson I, Lisovsky G, MacElroy R (2003) Manmade closed ecological systems. Taylor and Francis, London

Gitelson J (1992) Biological life-support systems for mars mission. Adv Sp Res 12:167–192

Godia F, Albiol J, Montesinos J, Pérez J (2002) MELISSA: a loop of interconnected bioreactors to develop life support in space. J Biotechnol 99:319–330

Goldman RP, Travisano M (2011) Experimental evolution of ultraviolet radiation resistance in *Escherichia coli*. Evolution 65:3486–3498

Grace JM, Verseux C, Gentry D et al (2013) Elucidating microbial adaptation dynamics via autonomous exposure and sampling. In: AGU fall meeting abstracts, San Francisco, California, 9–13 Dec 2013, p 597

Graham JM (2004) The biological terraforming of mars: planetary ecosynthesis as ecological succession on a global scale. Astrobiology 4:168–195

Griffin M (2006) Science versus exploration: a false choice. Ad Astra 18:24

Hallenbeck PC (2012) Microbial technologies in advanced biofuels production. Springer, New York

Harris DR, Pollock SV, Wood EA et al (2009) Directed evolution of ionizing radiation resistance in *Escherichia coli*. J Bacteriol 191:5240–5252

Harrison JP, Gheeraert N, Tsigelnitskiy D, Cockell CS (2013) The limits for life under multiple extremes. Trends Microbiol 21:204–212

Hempel F, Bozarth AS, Lindenkamp N et al (2011) Microalgae as bioreactors for bioplastic production. Microb Cell Fact 10:81

Hendrickx L, De Wever H, Hermans V et al (2006) Microbial ecology of the closed artificial ecosystem MELiSSA (micro-ecological life support system alternative): reinventing and compartmentalizing the Earth's food and oxygen regeneration system for long-haul space exploration missions. Res Microbiol 157:77–86

Hendrickx L, Mergeay M (2007) From the deep sea to the stars: human life support through minimal communities. Curr Opin Microbiol 10:231–237

Henrikson R (2009) Earth food *Spirulina*: how this remarkable blue-green algae can transform your health and our planet, Revised edn. Ronore Enterprises Inc, Hana, Maui, Hawaii

Hepp A, Landis G, Kubiak C (1993) a chemical approach to carbon dioxide utilization on mars. In: Lewis JS, Matthews MS, Guerrieri ML (eds) Resources of near Earth space. The University of Arizona Press, Tucson

Hoffman SJ, Kaplan DI (1997) Human exploration of mars: the reference mission of the NASA mars exploration study team. Publication, NASA Special 6107

Horneck G, Facius R, Reichert M et al (2003) HUMEX, a study on the survivability and adaptation of humans to long-duration exploratory missions, part I: lunar missions. Adv Space Res 31:2389–2401

Horneck G, Facius R, Reichert M et al (2006) HUMEX, a study on the survivability and adaptation of humans to long-duration exploratory missions, part II: missions to mars. Adv Sp Res 38:752–759

Horneck G (2008) The microbial case for mars and its implication for human expeditions to mars. Acta Astronaut 63:1015–1024

Jonkers HM, Thijssen A, Muyzer G et al (2010) Application of bacteria as self-healing agent for the development of sustainable concrete. Ecol Eng 36:230–235

Kanervo E, Lehto K, Ståhle K et al (2005) Characterization of growth and photosynthesis of *Synechocystis* sp. PCC 6803 cultures under reduced atmospheric pressures and enhanced CO_2 levels. Int J Astrobiol 4:97–100

Khalil AS, Collins JJ (2010) Synthetic biology: applications come of age. Nat Rev Genet 11:367–379

Kim M, Zhang Z, Okano H et al (2012) Need-based activation of ammonium uptake in *Escherichia coli*. Mol Syst Biol 8:616

Kral TA, Altheide TS, Lueders AE, Schuerger AC (2011) Low pressure and desiccation effects on methanogens: implications for life on mars. Planet Space Sci 59:264–270

Kurahashi-Nakamura T, Tajika E (2006) Atmospheric collapse and transport of carbon dioxide into the subsurface on early Mars. Geophys Res Lett 33:L18205

Langhoff S, Cumbers J, Rothschild LJ et al (2011) What are the potential roles for synthetic biology in NASA's mission? NASA/CP-2011-216430. NASA Ames Research Center, Moffet Field, California, 30–31 July 2010

Lee SY (2012) Metabolic engineering and synthetic biology in strain development. ACS Synth Biol 1:491–492

Lehto K, Kanervo E, Stahle K, Lehto H (2007) Photosynthetic life support systems in the Martian conditions. In: Cockell C, Horneck G (eds) ROME: response of organisms to the martian environment (ESA AP-1299). ESA Communications, Noordwijk, The Netherlands, pp 151–160

Lehto KM, Lehto HJ, Kanervo EA (2006) Suitability of different photosynthetic organisms for an extraterrestrial biological life support system. Res Microbiol 157:69–76

Lobascio C, Lamantea M, Cotronei V et al (2007) Plant bioregenerative life supports: the Italian CAB project. J Plant Interact 2:125–134

Maggi F, Pallud C (2010) Space agriculture in micro- and hypo-gravity: a comparative study of soil hydraulics and biogeochemistry in a cropping unit on earth, mars, the moon and the space station. Planet Space Sci 58:1996–2007

Marlière P, Patrouix J, Döring V et al (2011) Chemical evolution of a bacterium's genome. Angew Chemie 50:7109–7114

Martin VJJ, Pitera DJ, Withers ST et al (2003) Engineering a mevalonate pathway in *Escherichia coli* for production of terpenoids. Nat Biotechnol 21:796–802

Massa GD, Emmerich JC, Morrow RC et al (2007) Plant-growth lighting for space life support: a review. Gravitational Sp Biol 19:19–30

Matheny JG (2007) Reducing the risk of human extinction. Risk Anal 27:1335–1344

McKay CP, Davis WL (1989) Planetary protection issues in advance of human exploration of mars. Adv Space Res 9:197–202

McKay CP, Marinova M (2001) The physics, biology, and environmental ethics of making mars habitable. Astrobiology 1:89–110

Menezes AA, Cumbers J, Hogan JA, Arkin AP (2014) Towards synthetic biological approaches to resource utilization on space missions. J R Soc Interface 12:20140715

Ming DW, Archer PD, Glavin DP et al (2014) Volatile and organic compositions of sedimentary rocks in Yellowknife Bay, Gale crater. Mars. Science 343:1245267

Möllers KB, Cannella D, Jørgensen H, Frigaard N-U (2014) Cyanobacterial biomass as carbohydrate and nutrient feedstock for bioethanol production by yeast fermentation. Biotechnol Biofuels 7:64

Montague M, McArthur GH, Cockell CS et al (2012) The role of synthetic biology for in situ resource utilization (ISRU). Astrobiology 12:1135–1142

Nelson M, Pechurkin NS, Allen JP et al (2010) Closed ecological systems, space life support and biospherics. In: Wang LK, Ivanov V, Tay J-H, Hung Y-T (eds) Environmental biotechnology. Humana Press, New York, pp 517–565

Nicholson WL, Fajardo-Cavazos P, Fedenko J et al (2010) Exploring the low-pressure growth limit: evolution of *Bacillus subtilis* in the laboratory to enhanced growth at 5 kilopascals. Appl Environ Microbiol 76:7559–7565

Nicholson WL, Krivushin K, Gilichinsky D, Schuerger AC (2013) Growth of *Carnobacterium* spp. from permafrost under low pressure, temperature, and anoxic atmosphere has implications for earth microbes on mars. PNAS 110:666–671

Niederholtmeyer H, Wolfstädter BT, Savage DF et al (2010) Engineering cyanobacteria to synthesize and export hydrophilic products. Appl Environ Microbiol 76:3462–3466

Olsson-Francis K, Cockell CS (2010a) Use of cyanobacteria for in-situ resource use in space applications. Planet Space Sci 58:1279–1285

Olsson-Francis K, Cockell CS (2010b) Experimental methods for studying microbial survival in extraterrestrial environments. J Microbiol Methods 80:1–13

Osanai T, Oikawa A, Numata K et al (2014) Pathway-level acceleration of glycogen catabolism by a response regulator in the cyanobacterium *Synechocystis* species PCC 6803. Plant Physiol 164:1831–1841

Patnaik R, Louie S, Gavrilovic V et al (2002) Genome shuffling of *Lactobacillus* for improved acid tolerance. Nat Biotechnol 20:707–712

Pagliaro M, Rossi M (2010) The future of glycerol, 2nd edn. The Royal Society of Chemistry, Cambridge, UK

Peralta-Yahya PP, Zhang F, del Cardayre SB, Keasling JD (2012) Microbial engineering for the production of advanced biofuels. Nature 488:320–328

Perani G (1997) Military technologies and commercial applications: public policies in NATO countries. Final report to the NATO Office for Information and Press, Centro Studi di Politica Internazionale, Rome, Italy

Poughon L, Farges B, Dussap CG et al (2009) Simulation of the MELiSSA closed loop system as a tool to define its integration strategy. Adv Sp Res 44:1392–1403

Presidential Commission for the Study of Bioethical Issues (2010) New directions: the ethics of synthetic biology and emerging technologies. Government Printing Office, Washington

Quintana N, Van der Kooy F, Van de Rhee MD et al (2011) Renewable energy from cyanobacteria: energy production optimization by metabolic pathway engineering. Appl Microbiol Biotechnol 91:471–490

Race M, Denning K, Bertka CM et al (2012) Astrobiology and society: building an interdisciplinary research community. Astrobiology 12:958–965

Radakovits R, Jinkerson RE, Darzins A, Posewitz MC (2010) Genetic engineering of algae for enhanced biofuel production. Eukaryot Cell 9:486–501

Raksajit W, Satchasataporn K, Lehto K et al (2012) Enhancement of hydrogen production by the filamentous non-heterocystous cyanobacterium *Arthrospira* sp. PCC 8005. Int J Hydrogen Energy 37:18791–18797

Ro D-K, Paradise EM, Ouellet M et al (2006) Production of the antimalarial drug precursor artemisinic acid in engineered yeast. Nature 440:940–943

Rothschild LJ (1990) Earth analogs for Martian life. Microbes in evaporites, a new model system for life on Mars. Icarus 88:246–260

Rothschild LJ (2010) A powerful toolkit for synthetic biology: Over 3.8 billion years of evolution. BioEssays 32:304–313

Ruder WC, Lu T, Collins JJ (2011) Synthetic biology moving into the clinic. Science 333:1248–1252

Schuerger AC, Ulrich R, Berry BJ, Nicholson WL (2013) Growth of *Serratia liquefaciens* under 7 mbar, 0 °C, and CO_2-enriched anoxic atmospheres. Astrobiology 13:115–131

Slade D, Radman M (2011) Oxidative stress resistance in *Deinococcus radiodurans*. Microbiol Mol Biol Rev 75:133–191

Spiller H, Latorre C, Hassan ME, Shanmugam KT (1986) Isolation and characterization of nitrogenase-derepressed mutant strains of cyanobacterium *Anabaena variabilis*. J Bacteriol 165:412–419

Subramanian G, Shanmugasundaram S (1986) Uninduced ammonia release by the nitrogen-fixing cyanobacterium *Anabaena*. FEMS Microbiol Lett 37:151–154

Thomas DJ, Boling J, Boston PJ et al (2006) Extremophiles for ecopoiesis: desirable traits for and survivability of pioneer Martian organisms. Gravitational Sp Biol 19:91–104

Tikhomirov AA, Ushakova SA, Kovaleva NP et al (2007) Biological life support systems for a mars mission planetary base: problems and prospects. Adv Sp Res 40:1741–1745

Tokano T (2005) Water on Mars and life. Springer, Berlin

Toprak E, Veres A, Yildiz S, Pedraza J (2013) Building a morbidostat: an automated continuous-culture device for studying bacterial drug resistance under dynamically sustained drug inhibition. Nat Protoc 8:555–567

Verseux C, Baqué M, Lehto K, de Vera J-PP, Rothschild LJ, Billi D (2015) Sustainable life support on mars—the potential roles of cyanobacteria. Int J Astrobiol. doi: 10.1017/S147355041500021X

Wang HH, Isaacs FJ, Carr PA et al (2009) Programming cells by multiplex genome engineering and accelerated evolution. Nature 460:894–898

Wassmann M, Moeller R, Reitz G, Rettberg P (2010) Adaptation of *Bacillus subtilis* cells to Archean-like UV climate: relevant hints of microbial evolution to remarkably increased radiation resistance. Astrobiology 10:605–615

Way JC, Silver PA, Howard RJ (2011) Sun-driven microbial synthesis of chemicals in space. Int J Astrobiol 10:359–364

Weber W, Fussenegger M (2011) Emerging biomedical applications of synthetic biology. Nat Rev Genet 13:21–35

Webster CR, Mahaffy PR, Atreya SK et al (2015) Mars methane detection and variability at Gale crater. Science 347:415–417

Zhang F, Rodriguez S, Keasling JD (2011) Metabolic engineering of microbial pathways for advanced biofuels production. Curr Opin Biotechnol 22:775–783

Zubrin R (1995) The economic viability of Mars colonization. J Br Interplanet Soc 48:407–414

Zubrin R, Wagner R (1996) The case for mars: the plan to settle the Red Planet and why we must. Free Press, New York

"First Species Whose Parent Is a Computer"—Synthetic Biology as Technoscience, Colonizing Futures, and the Problem of the Digital

Martin Müller

1 Introduction

Contemporary synthetic biology is a young field of research. Diverse initiatives, institutes, laboratories, scholarships, professorships, publications, and journals use the label, and it has attracted a lot of private and public funding. The shared interests of the synthetic biology stakeholders are to establish research infrastructure, to work on enabling technologies, to negotiate standards, and, finally, to build synthetic organisms and systems. The heterogeneous field expands very quickly on an institutional level and hence the risks and the technical potentials have only now been debated (e.g., Schmidt et al. 2009).

In this chapter I investigate synthetic biology as a cultural phenomenon, or to put it more precisely, as a phenomenon of *technoscientific culture* that emerges in contemporary societies (Weber 2010). When Craig Venter is asked by the Guardian if his project of the so-called *Digital Biological Converter*, a portable home gadget connectable to personal computers to synthesize proteins, viruses and living cells after downloading DNA sequences from the internet, is a serious proposal, he answers:

> Mine is not a fantasy look at the future, the goal isn't to imagine this stuff. We are the scientists actually doing this. [...] And we have the prototype. (Venter in Corbyn 2013)

M. Müller (✉)
Image Knowledge Gestaltung: An Interdisciplinary Laboratory, Cluster of Excellence, Humboldt University of Berlin, Berlin, Germany
e-mail: martin-mueller@culture.hu-berlin.de

© Springer International Publishing Switzerland 2016
K. Hagen et al. (eds.), *Ambivalences of Creating Life*, Ethics of Science and Technology Assessment 45, DOI 10.1007/978-3-319-21088-9_5

If we look at the works of George Church, Craig Venter, Drew Endy, and others[1] it becomes visible that synthetic biologists are actively engaging in contemporary challenges.[2] This self-understanding differs massively from a scientific discipline which wants to find facts and establish safe and sound representations of living phenomena. Bioengineers claim to create new living organisms from scratch, using genetically standardized parts and computer-based design: 'Living machines' which do not exist in 'nature' are supposed to serve human purposes. Beyond its actual (and limited) state of research, and beside critique (i.e. Kwok 2010) from within the field: some voices of synthetic biology offer bold claims of socio-technical scenarios, imagined objects, and future biotechnical experiments, which take place in society rather than behind laboratory doors. Leading institutions operate under the idea of a future that is in the making. The prestigious *Synthetic Biology Engineering Research Center* (Synberc) even carries the sentence 'Building the Future with Biology' in its logo. In popular and in scientific press, stakeholders and practitioners openly talk about 'vision', 'progress', 'breakthrough', and a 'new era of biology', where the 'creation of life' is promised to be a computational and engineering application (e.g., Carlson 2012; Collins 2012; Venter 2013). Craig Venter explains:

> We have now entered what I call "the digital age of biology", in which the once distinct domains of computer codes and those that program life are beginning to merge, where new synergies are emerging that will drive evolution in radical directions. (ibid, p. 1)

This chapter thematizes three intertwined topics for the discussion about the societal and philosophical dimensions of synthetic biology: the epistemic culture, the role of promises, and the 'problem of the digital'. The first part discusses the epistemic culture of synthetic biology by characterizing some of its *technoscientific specifica*. The second part discusses and critically analyzes the future-centered proposition and the production of promises within scientific and popular discourse. Here I critically thematize the (self-)representation of synthetic biology protagonists and their construction of socio-technical futures. The third part tries to open a new field of critical investigation on the discourse of synthetic biology by problematizing the ambivalent role of computer and information technology ('digital biology') in the design and material realization of living entities and systems. As a methodological and theoretical framework, I apply a loose conjunction of discourse analytical thinking, approaches from cultural and media theory as well as history and philosophy of technoscience. Therefore, my approach is rather more analytical and historical than empirical (Bublitz et al. 1999).

---

[1]Also outside academic and industrial framework of synthetic biology 'biopunks' and 'do-it-yourselfs-biologists' are actively engaging in contemporary challenges and call for a 'democratization' of biotechnology (Wohlsen 2011, p. 8).

[2]The academic field of synthetic biology is heterogeneous and multifaceted. Certainly there are researchers, who do not want to engage in contemporary challenges and whose approaches are not 'purely' technoscientific. For a general overview of the different approaches and research fields in synthetic biology: see Acevedo-Rocha, in this book.

2 Technoscientific Dimesions of Synthetic Biology[3]

Synthetic biology can indeed be seen as technoscience.[4] I will therefore start my chapter by giving a short insight into the concept and at the same time highlight some of the technoscientific characteristics to clarify the connection. The term technoscience has made an astonishing rise to prominence since the mid 1980s (Kastenhofer and Schwarz 2011). The compositum technoscience might provoke certain images: the conjunction of science and technology, or science falling entirely under the domain of technology. However, acclaimed scholars from science and technology studies, philosophy of science, and feminist studies have rejected such definitions. They have argued that more complex processes have been taking place: modern sciences have undergone a transformation regarding epistemology, practices, objects, vocabularies, and the role of public engagement (Haraway 1997; Weber 2003). The result is a new way of research where the 'grand narrative' of modern science is shifting from facts, truth, and representation of nature to technical design, hybridity, and problem solving from nano to a global scale (Weber 2003). Some philosophers, including Alfred Nordmann, dispute the existence of an epochal break (Nordmann et al. 2011), claiming that science and technoscience can rather be distinguished by their focus of attention: a modern science like evolutionary biology focuses on analyzing and representing the existing laws of nature, whereas a technoscience like synthetic biology focuses primarily on re-designing 'natural' entities and systems. In order to clarify this shift, it is important to look at conceptions of science and technology. Scientists typically aim at distinguishing effects of their theories and interventions from entities or phenomena of a "[...] given world or mind-independent reality" (Bensaude Vincent et al. 2011, p. 368). Bensaude Vincent and her co-authors further claim that the constitutive aspect of science is the "ontological presupposition": that the world is given and can be discovered by scientists and their experimental systems. Matters of fact are based on the idea of an eternal nature. In this sense

> [...] the world is typically taken to be composed of facts [...] – and a fact is 'that something is the case,' 'that a thing is so and so,' 'that this has been observed or measured' etc. (ibid, p. 370).

In this framework, science and technology have to be held apart. Scientists aim to provide fact-knowledge of natural phenomena that *can be adapted* by technological application. The interest of technology "[...] is to control the world, to intervene and change the 'natural' course of events" (Nordmann 2006, p. 8). The categorical and practical separation of the natural world and social world is the

---

[3]I would like to kindly thank Bernadette Bensaude Vincent, who introduced me to the technoscientific dimension of synthetic biology. My chapter draws widely on her comprehensive work and critical interventions on the discourse of synthetic biology.

[4]Synthetic biology is widely considered a technoscience, but unfortunately there has been very little theoretical work done to verify this assumption. An example is Schmidt et al. (2009) where the term technoscience is mentioned only once and there is no description of what it means.

constitutive ground of modern science. Pure science or basic science is only 'pure', when social or technical aspects do not contaminate facts. In technoscientific experimental settings it becomes rather difficult to identify and distinguish the contribution of a natural object and the contribution of human or technological apparatus within an experiment (Bensaude Vincent et al. 2011, p. 368). Among other research fields—such as nanotechnology, neuroscience, robotics, climate studies etc.—synthetic biology is characterized by its refusal of a given divide between nature and culture. Synthetic biology research does not distinguish between theoretical representation and technical intervention when, for example, the 'natural' functions of an engineered genome are studied.

In a broader sense, technoscientific epistemology practices aim at the appropriation and re-engineering of nature (Weber 2011b). The German philosopher and media theorist Jutta Weber argues

> [...] technoscience is no more mainly about representing the laws of nature and intervening in its processes but mostly about (re-)shaping new and hybrid worlds from a constructivist viewpoint. (2011a, p. 160)

Technoscience implicitly favors a different concept of time than the classical sciences. Classical science represents an idea of the future that is marked by increasing knowledge about nature: non-knowledge is turned into knowledge. In that sense, the world of the future is the evidence of nature represented as natural facts, and that's why the idea of discovery is so important in the self-conception of modern science. Science is unraveling, shedding light on, and lifting the veil from something that is already there: the timeless laws of an already given nature.

Technoscientists redesign nature with its own materials, but by mixing different elements they construct new objects. Therefore technoscience is interested in the future potentiality of nature, which is conceptualized as a pool of flexible resources that can be designed and transformed for human purposes; a field of materials, structures, functions, and their medialization, appropriation, and control. In this view "[...] life is becoming biomatter, waiting to be engineered" (Catts and Zurr 2014, p. 28). The German media historian Wolfgang Schäffner has pointed to the specific rhetoric involved when technoscientists talk about breaking 'nature' and 'matter' into its 'basic elements', putting them together again to build something new. When synthetic biologists start talking about 'bricks', 'circuits', and 'machines', it might seem as if they would be in the artificial world of architecture and design. Schäffner argues that the scientist as an observer and analyst of chemical, physical, and biological elements turns into a designer of something that didn't exist before (Schäffner 2010, p. 33). Tom Knight, synthetic biologist and computer engineer at MIT, calls that "the engineering of novel life". At an early point he described a turn in his own field as a revolution in the disciplinary conception of biology:

> Biology will never be the same. The remarkable scientific success of biology in describing, explaining, and manipulating natural systems is so well recognized as to be a cliche – but the engineering application of that scientific knowledge is just beginning. In the same way that electrical engineering grew from physics to become a separate discipline in the early

part of the last century, we see the growth of a new engineering discipline: one oriented to the intentional design, modeling, construction, debugging, and testing of artificial living systems. (Knight 2005, p. 1)

In the rhetoric of synthetic biologists the turn from fact-finding science to technoscientific epistemology is often phrased as a shift from 'reading' to 'writing'. The synthetic biologists Pengcheng Fu and Sven Panke claim:

[...] we are able to not only 'read' the genetic code to understand living systems but also 'write' the message for the creation of new life forms. All this fuels the need to frame these latest developments that promise to revolutionize our understanding of biology, blur the boundaries between the living and the engineered in a vital new bioengineering, and transform our daily relationship to the living world. (2009, p. 4)

One consequence of this "design turn" (Schäffner 2010, p. 33) is that the ontological and epistemological status of the research objects is called into question when, for example, the gene is not seen as an explanatory model of scientific discovery (see also Weber 2010). When genes are conceptualized as building tools for engineering, the focus of interest shifts from questions of explanation to the question of inherent potentiality and future application. Bensaude Vincent and her co-authors explain:

By becoming an object of technoscientific interest, an already familiar object becomes something new or something else. Indeed, its very nature changes in that it is no longer defined by what it is, but by its expected technical performance. Its structure, properties, and structure–property dependencies fade into the background, while potential functionalities acquired through dynamic modeling and re-engineering take center stage. This anticipatory performativity confers a strange temporal status to technoscientific objects that are simultaneously "already there" and "not yet realized". As such they function as proofs of concept that signify that a process or phenomenon has been demonstrated and at the same time refer to something that does not exist as yet but might come into being. (Bensaude Vincent et al. 2011, p. 374)

The result of technoscience practices is the design and production of objects that are neither purely natural nor technical. Technoscience epistemology negates the distinction between nature and culture. It is easy to draw a connection to the production of objects in synthetic biology when bioengineers speak of their constructions as 'living machines', or Craig Venter gives his *Mycoplasma laboratorium* the lovely name 'Synthia'. At a closer look technoscientific objects are unique cases. Every object demands individual inquiry.

Technoscientists in general and synthetic biologists in particular actively want to engage in social constellations. Bensaude Vincent argues that technoscientists do not retreat to a "[...] protected disciplinary sphere of facts as distinct from social values" (Bensaude Vincent et al. 2011, p. 369). Rather, technoscientists become agents who address and solve problems for society. Technoscientists have a different relation to public matters since the separation of science and society—and the laboratory and the public—have become blurred.[5]

---

[5]Under the term 'biocapital' there has been some critique of the commodification of biology. See Rose and Rose (2012, p. 12): "In the process the life sciences have been transformed into gigantic

The focus shifts from 'matters of fact' to what is called 'matters of concern': Technoscientists engage in complex problems that are not primarily scientific, but rather hybrid constellations of the social, natural, economical, juridical, etc.—with technical solutions (Latour 2008, see also 2009). According to Jutta Weber,

[t]his cultural turn is encouraged by technoscience's new epistemologies and ontologies which interpret our world as our product. (2011a, p. 161)

"Therefore," she explains further,

[...] many technoscientists and science managers invest increasingly in the popularization of technosciences [...] to demonstrate the usefulness of technosciences' endeavours for the public. (ibid.)

Therefore, matters of concern are indeed public matters, and because of their public (self-)representation,

[...] geneticists, nanotechnologists, brain researchers, or roboticists are perceived as technoscientists who mainly support, improve and perfect nature. (ibid.)

In this framework, the socio-technical future seems to be open and demands human imagination and intervention. The future lends itself to human invention—a collective future created by the construction of hybrid objects. In case of synthetic biology this means that science and society interact in a previously unforeseen manner: when synthetic biologists' 'living machines' leave the laboratory, whole ecospheres become experimental zones.

3 Colonizing Collective Futures: Synthetic Biology as Promise Cultures

The biologist and historian of science Donna Haraway argues that not only scientific fields are turning into technoscience research; our whole culture and society is affected by this transformation. For Haraway, the turn to technoscience and its effect on Western culture signifies

[...] mutation in historical narrative, similar to the mutations that mark the difference between the sense of time in European medieval chronicles and the secular, cumulative salvation histories of modernity. (1997, pp. 4–5)

Visions and promise-making are crucial elements that characterize synthetic biology as technoscience. Promises about potential achievements of synthetic biology influence how we imagine and construct futures. With Haraway, technoscience of synthetic biology is another example of a secular mode of salvation,

---

(Footnote 5 continued)

biotechnosciences, blurring the boundaries between science and technology, universities, entrepreneurial biotech companies and the major pharmaceutical companies, or 'Big Pharma'."

when salvation is understood as the promise and the realization of a perfect future world. For synthetic biology, nature and its phenomena are more or less flexible materials, adaptable forms, and fields of potentialities that can be used for re-engineering nature and living matter itself.

In this discourse of promise, the demarcation between the actual and the possible is not always clear (Bensaude Vincent 2013). The words of Harvard professor George Church, inventor of genetic technologies that are widely used by synthetic biologists, sound like rhetorics of the future, and are akin to rhetorics of science fiction:

> [...] we stand at the door of manipulating genomes in a way that reflects the progress of evolutionary history: starting with the simplest organisms and ending, most portentously, by being able to alter our own genetic makeup. Synthetic genomics has the potential to recapitulate the course of natural genomic evolution, with the difference that the course of synthetic genomics will be under our own conscious deliberation and control instead of being directed by the blind and opportunistic processes of natural selection. (Church and Regis 2012, pp. 12–13)

For those readers who are still unsure if 'we' already walked through that door he ensures:

> We are already remaking ourselves and our world, retracing the steps of the original synthesis – redesigning, recoding, and reinventing nature itself in the process. (ibid., p. 13)

This quote exemplifies how strongly the discourse on synthetic biology is marked by promises and utopian motives. It is not only the International Genetically Engineered Machine (iGEM) competition that rewards future driven projects. Also, some protagonists propagate prognoses and narratives about the future of 'life' and 'nature' to the wider scientific community and the public. Their agenda comprises claims with such 'humble' goals as the easy and inexpensive fabrication of medicine and materials by reprogrammed bacteria, the production of clean energy, the combat of the ecological crisis (Collins 2012; Highfield 2013), the development of powerful computer hardware using DNA, and a revolutionary bio-technologization of whole industries (Carlson 2012), the creation of novel genomes and species, and a new era of human control over evolution, initiated by 'digital design' (Venter 2013; Church and Regis 2012) and finally the possibility of humans colonizing other planets with the help of synthetic biology (see Cyprien Verseux et al., in this book).

The Austrian sociologist of science, Karin Knorr Cetina tries to approach the technoscientific transformation of Western culture by thinking about promises of future perfectibility. Her theory can be applied to the discourses of synthetic biologists who want to guide evolution in their own image, as George Church claims above. Knorr Cetina argues that we are living during a time of transition from the modern culture of humanism and enlightenment towards a posthumanist 'culture of life':

> One massive source of fantasies that fuel a culture of life and challenge traditional humanism is the biological sciences themselves. (2005, p. 78)

The reason for this transformation is a change within the configuration of scientific promise. The 'culture of humanism' promises political freedom and social perfectibility. This promise is articulated by the sciences, in particular the social sciences and humanities. There is a division of agency between the promise-maker (sciences) and the party realizing the promise (society). The responsibility of realizing the promise lies on the side of society (ibid.).

However, in the emerging technoscience culture the promise of perfectibility of society shifts towards the perfectibility of biological life, "what the biological sciences promise is the perfectibility of life." (ibid, p. 78). In this constellation, life sciences become a trustworthy agent to make societal and future promises. Synthetic biologists articulate a variety of promises that are centered on the idea of the technical perfectibility of life. However, contrary to the promise of humanism, Knorr Cetina claims that in the 'culture of life' the promise of perfectibility and responsibility to realize the promise falls on the side of technoscientists. In Knorr Cetinas words:

> fullfilment of the promise and the requirement of sincerity lie with the promise giver, all the promise receiver needs to contribute are plausible wants. (ibid, p. 80)

It is not difficult to find an example for the argument: The glowing trees of Cambridge, a project proposed by undergraduate students from the University of Cambridge, UK. Their vision is to construct a tree that glows in the dark as a substitute for electric streetlamps. Scientists, after finding a 'problem,' i.e. electric streetlamps are not ecological, propose a 'solution', i.e. 'let's build trees that glow in the dark'. The promise and the realization of this promise-solution were proposed by the project makers (Clark 2013; Reardon 2011).

The German sociologist of science Petra Lucht and her co-authors have stressed the image of a co-production of society and technoscience (2010). They argue that talking about the perfect future is a mode of producing individual and societal needs. The performance of the promise produces the desirability for technoscientific objects and projects, when synthetic biologists also talk of new desirable devices for consumer and popular culture (ibid.). Promises provoke the idea that coming innovations in synthetic biology don't need to adjust to the framework of current society. Instead the promise givers are marking a space and draw a picture of a biotechnical future that current society and individuals have to adjust to and prepare for already today (ibid.). Proposed synthetic biotechnical futures have an implicit impact on societal knowledge and practice. In short, it is not that synthetic biology has to be adjusted to contemporary society, but rather that contemporary society has to adjust to a future drawn by synthetic biologists.

In the promised future worlds of synthetic biology, artificial living entities have left the laboratory and the factory. Various hybrid entities will populate our world to perform useful tasks in everyday situations, which will lead to, for example, glowing trees lighting the streets of tomorrow. Here it is important to look at the cultural and historical implications of promise-making. Bernadette Bensaude Vincent argues—by referring to Barbara Adam and Steve Groves—that the epistemology and the practices of synthetic biology are implicitly informed by a modern conception of future as a "contested future" (2013, pp. 25–26). Promises are rhetorical devices to colonize and contest the future. In this sense, future is imagined as an open and empty space. The space has to be controlled by action. The future has to be made not only through actions, but also by imagination and crafted objects. These objects are supposed to populate the horizons of anticipated futures. In this conception the future is the problem and an opportunity for the present world: the future starts now and the future has to be made today (Adam and Groves 2007, p. 14). Synthetic biologists' "economy of promise" (Bensaude Vincent 2013; Joly 2010; Jones 2006; Rose and Rose 2012) practically invests in the management of possible futures by making novel objects and by suspending the idea to find an internal truth of nature. With their visions, synthetic biologists are becoming engineers of future societies. If 'life' and 'nature' could be programmed and controlled by so-called biological design, the future (one might claim) could also be programmed and controlled by science and engineering.

4 Phantasmatic Calculation: A Problematization of 'Digital Biology'

Evidently information science, cybernetics, and computer engineering play significant roles in the diverse field of synthetic biology. Many parts of synthetic biology research are intertwined with systems biology, the computational simulation and mathematical modeling of complex biological systems (see Fu and Panke 2009). Many synthetic biologists, like the pioneer Tom Knight, have a former career in information science and computer engineering. Daisy Ginsberg, Drew Endy and their co-authors explain:

> These self-styled pioneers of biological engineering aspire to redesign existing organisms using engineering principles like standardization; some even seek to construct completely novel biological entities. The field's engineering visions leads to parallels being drawn with the early days of computer technology, as researchers reimagine bits of DNA code as programmable parts, analogous to the components of computer software and hardware. (2014, p. x)

In the technoscientific promise-economy there is a trope of problematic claims when it comes to the topic of technical realization of designed living objects. Talking about synthetic biology with Der Spiegel in 2013, George Church expresses the following claim about the future of synthetic biology:

> Oh, life science will co-opt almost every other field of manufacturing. It's not limited to agriculture and medicine. We can even use biology in ways that biology never has evolved to be used. DNA molecules for example could be used as three-dimensional scaffolding for inorganic materials – with atomic precision. You can design almost any structure you want with a computer, then you push a button – and there it is, built-in DNA. (Church in Bethge and Grolle 2013)

What is the historical root of the connection between the anticipated success and the promise of feasibility of 'creating' living matter, and computer technology and digital design? The historian of science Lily E. Kay has shown that molecular biology underwent an epistemic transformation towards technoscience in the 1950s, leaving behind models of life explained by an epistemology of energy and mechanical thinking; instead, researchers began to conceptualize 'life itself' as information. Therefore, molecular biology understood itself as communication science, strongly affiliated with the emerging fields of cybernetics, information theory, and computer engineering. Molecular biologists "invented the genetic code", when they began to apply the vocabulary of these disciplines, using words like code, feedback, messages, codes, alphabet, instructions, texts, and programs (Kay 2000; see also Fox Keller 2002).

From that point on, organisms and molecules were viewed as information storage and retrieval systems. Kay stresses that the use of such language is ambivalent and problematic, as the line between ontological and metaphorical use has not been made clear by molecular biologists. These scriptural conceptions of 'life as text' are implicitly informed by the historic discourse of "The Book of Nature", where nature and the living world are basically written in a coded language, which can be deciphered by science (Kay 2000). Up until today, the aim of molecular biology is to read genomes by sequencing DNA in a linear and discrete code (Thacker 2004, 2006, 2009). According to Kay, the inscription of the code epistemology into the molecular levels of life accompanied by the readability of life promises new levels of control and feasibility. After 'reading' DNA became a standard application, 'writing' the 'text of life' for the 'creation' of new life forms seems to be the challenge for synthetic biology—as formulated above by Pencheng Fu and Sven Panke. Craig Venter explains in Wired:

> All the information needed to make a living, self-replicating cell is locked up within the spirals of DNA's double helix. As we read and interpret that software of life, we should be able to completely understand how cells work, then change and improve them by writing new cellular software. The software defines the manufacture of proteins that can be viewed as its hardware, the robots and chemical machines that run a cell. The software is vital because the cell's hardware wears out. Cells will die in minutes to days if they lack their genetic-information system. They will not evolve, they will not replicate, and they will not live. (Highfield 2013)

Crucial for synthetic biology's promise-economy are gene- and code-centered bio-cybernetic and even transhumanist figures of thought that (in)form new visions of life and nature as a field of potentials and even limitless treasures that can be programmed and controlled by computational procedures (Zakeriemail and Carremail 2015). Venter claims:

> DNA, as digitized information, is not only accumulating in computer databases but can now be transmitted as an electromagnetic wave at or near the speed of light, via a biological teleporter, to re-create proteins, viruses, and living cells at a remote location, perhaps changing forever how we view life. With this new understanding of life, and the recent advances in our ability to manipulate it, the door cracks open to reveal exciting new possibilities. As the Industrial Age is drawing to a close, we are witnessing the dawn of an era of biological design. Humankind is about to enter a new phase of evolution. (2013, p. 7)

Some bioengineers explicitly insinuate synthetic biology as a new universal biotechnology to materialize human wishes and projects in the form of living objects and systems. The *medium* of control over living matter seems to be the computational technology. It is remarkable how strong images of 'the computer' and 'the digital' Venter, Church and others use in their narratives of synthetic biology and the future of 'life'. One of the most famous examples was when in 2010, Venter and his colleagues presented

[...] the first synthetic cell, a cell made by starting with the digital code in the computer, building the chromosome [...] So this is the first self-replicating species that we've had on the planet whose parent is a computer. (Venter 2010)

After this point, it is deemed that evolutionary processes can be guided by human will as a project of engineering and computing. In that realm, life becomes a question of digital design.

In this narrative, the computer becomes an instrument of remarkable control and enables the rational production of living entities, structures, and systems. For Venter, his so-called "digital biology" is fulfilling the ambivalent promise of early modern science, when he is pointing to Bacon's 'Nova Atlantis'. Venter argues:

The fusion of the digital world of the machine and that of biology would open up remarkable possibilities for creating novel species and guiding future evolution. We had reached the remarkable point of being at the beginning of "effecting all things possible," and could genuinely achieve what Francis Bacon described as establishing dominion over nature. (2013, p. 78)

In that sense Venter and Church resume a figure of thought from the computer and cybernetics discourse of the late 1930s. Back then the computer was mathematically conceptualized as a 'universal machine': universal, because it was supposed to imitate every other *calculation* machine. Synthetic Biology even radicalizes this utopian motif and expands it into the domain of 'life itself'. Here, the computer is imagined as a universal machine that is able to simulate every possible form of life as code language. Even more utopian: the computer should control the materialization of living entities as designed objects, as 'living machines'. The universality of this approach—the digital 'creation of life' and limitless plasticity of 'living matter'—lies at the heart of synthetic biologists' techno-utopianism.

References

Adam B, Groves C (2007) Future matters. Brill, London

Bensaude Vincent B (2013) Between the possible and the actual: philosophical perspectives on the design of synthetic organisms. Futures 48:23–31

Bensaude Vincent B, Loeve S, Nordmann A, Schwarz A (2011) Matters of interest: the objects of research in science and technoscience. J Gen Phil Sci 42:365–383

Bethge P, Grolle J (2013) Interview with George Church: Can Neanderthals be brought back from the dead? Spiegel-Online. http://www.spiegel.de/international/zeitgeist/george-church-explains-how-dna-will-be-construction-material-of-the-future-a-877634.html. Accessed 14 June 2015

Bublitz H, Bührmann AD, Hanke C, Seier A (eds) (1999) Das Wuchern der Diskurse. Perspektiven der Diskursanalyse Foucaults, Campus, Frankfurt/Main

Carlson R (2012) Biology is technology. The promise, peril, and new business of engineering life. Harvard University Press, Cambridge, Massachusetts

Catts O, Zurr I (2014) Countering the engineering mindset: the conflict of art and synthetic biology. In: Ginsberg AD, Calvert J, Schyfter P, Elfick A, Endy D (eds) Synthetic aesthetics: investigating synthetic biology's designs on nature. MIT Press, Cambridge, Massachusetts

Clark L (2013) Glowing trees could pave the way for solving world problems with biology. Wired UK. http://www.wired.co.uk/news/archive/2013-05/9/glowing-plants-kickstarter. Accessed 14 June 2015

Church GM, Regis E (2012) Regenesis. How synthetic biology will reinvent nature and ourselves. Basic Books, New York

Collins J (2012) Synthetic biology: bits and pieces come to life. Nature 7387:8–10

Corbyn Z (2013) Craig Venter: 'this isn't a fantasy look at the future. We are doing the future'. In: The guardian, 12 Oct, p 41

Fox Keller E (2002) Making sense of life. Explaining biological development with models, metaphors, and machines. Harvard University Press, Cambridge, Massachusetts

Fu P, Panke S (eds) (2009) Systems biology and synthetic biology. Wiley, Hoboken

Ginsberg AD, Calvert J, Schyfter P, Elfick A, Endy D (eds) (2014) Synthetic Aesthetics: investigating synthetic biology's designs on nature. MIT Press, Cambridge, Massachusetts

Haraway DJ (1997) Modest_Witness@Second_Millennium. FemaleMan©_Meets_OncoMouse™. Feminism and technoscience. Routledge, New York

Highfield R (2013) J Craig Venter sequenced the human genome. Now he wants to convert DNA into a digital signal. Interview, Wired UK. http://www.wired.co.uk/magazine/archive/2013/11/features/j-craig-venter-interview. Accessed 14 June 2015

Joly P-B (2010) On the economics of techno-scientific promises. In: Akrich M, Barthe Y, Muniesa F, Mustar P (eds) Débordements. Mélanges offerts à Michel Callon. Presse des Mines, Paris

Jones R (2006) Economy of promises. Nature Nanotech 3:65–66

Kastenhofer K, Schwarz A (2011) Editorial: probing technoscience. Poiesis Prax 8(2–3):61–65

Kay LE (2000) Who wrote the book of life? A history of the genetic code. Stanford University Press, Stanford

Knight TF (2005) Engineering novel life. Mol Syst Biol 1:E1

Knorr Cetina K (2005) The rise of a culture of life. EMBO Rep 6:76–80

Kwok R (2010) Five hard truths for synthetic biology. Nature 463:288–290

Latour B (2008) What is the style of matters of concern? Two lectures in empirical philosophy. Assen, Van Gorcum

Latour B (2009) A cautious Prometheus? A few steps toward a philosophy of design. In: Hackne F, Glynne J, Minto V (eds) Proceedings of the 2008 annual international conference of the design history society.Falmouth, 3–6 Sept 2009, e-books, Universal Publishers, pp 2–10

Lucht P, Erlemann M, Ruiz B (eds) (2010) Technologisierung gesellschaftlicher Zukünfte. Nanotechnlogien in wissenschaftlicher, politischer und öffentlicher Praxis. Centaurus, Freiburg

Nordmann A (2006) Collapse of distance: epistemic strategies of science and technoscience. Dan Yearb Philos 41:7–34

Nordman A, Radder H, Schiemann G (eds) (2011) Science transformed: debating claims of an epochal break. University of Pittsburgh Press, Pittsburgh

Reardon S (2011) Visions of synthetic biology. Science 333:1242–1243

Rose H, Rose SPR (2012) Genes, cells, and brains. The Promethean promises of the new biology. Verso, London

Schäffner W (2010) The design turn. Eine wissenschaftliche Revolution im Geiste der Gestaltung. In: Claudia Mareis (eds) Entwerfen - Wissen - Produzieren. Designforschung im Anwendungskontext. Transcript, Bielefeld

Schmidt M, Kelle A, Ganguli-Mitra A, de Vriend H (eds) (2009) Synthetic biology. The technoscience and its societal consequences. Springer, Dordrecht

Thacker E (2004) Biomedia. University of Minnesota Press, Minneapolis

Thacker E (2006) The global genome. Biotechnology, politics, and culture. MIT Press, Cambridge, Massachusetts

Thacker E (2009) "De anima": on life and the living. In: Bock von Wülfingen, Bettina and Ute Frietsch (eds) Epistemologie und Differenz. Zur Reproduktion des Wissens in den Wissenschaften. Transcript, Bielefeld

Venter JC (2010) Synthetic life. Transcript Press conference. https://www.ted.com/talks/craig_venter_unveils_synthetic_life/transcript. Accessed 14 June 2015

Venter JC (2013) Life at the speed of light from the double helix to the dawn of digital life. Viking, New York

Weber J (2003) Umkämpfte Bedeutungen. Naturkonzepte im Zeitalter der Technoscience, Campus, Frankfurt/Main

Weber J (2010) Making worlds: epistemological, ontological and political dimensions of technoscience. Poiesis Prax 7(1–2):17–36

Weber J (2011a) Technoscience as popular culture. On pleasure, consumer technologies and the economy of attention. In: Nordman A, Radder H, Schiemann G (eds) Science transformed: debating claims of an epochal break. University of Pittsburgh Press, Pittsburgh

Weber J (2011b) Die kontrollierte Simulation der Unkontrollierbarkeit—Kontroll- und Wissensformen in der Technowissenschaftskultur. In: Bublitz H, Kaldrack I, Röhle T, Winkler H (eds) Unsichtbare Hände. Automatismen in Medien-, Technik- und Diskursgeschichte. Fink, Paderborn

Wohlsen M (2011) Biopunk: DIY scientists hack the software of life. Pinguin, New York

Zakeriemail B, Carremail PA (2015) The limits of synthetic biology. Trend Biotechnol 33(2):57–58

"Some Kind of Genetic Engineering… Only One Step Further"—Public Perceptions of Synthetic Biology in Austria

Walburg Steurer

1 Introduction

> I have an idea for how we could define synthetic biology. Namely, if we could define it, if we could accept the definition that it is some kind of genetic engineering… only one step further, a considerable step further.[1]

This proposition was made by a participant in a citizen panel (CP) conducted in November 2012 in Vienna, Austria. The quote points to a central issue related to synthetic biology and its public perception: the embedding of the research field within the discursive frame of genetic engineering. The drawing of parallels between synthetic biology and genetic engineering has for several years been discussed in the scientific literature (Kronberger et al. 2009, 2012; Pauwels 2009; Torgersen and Hampel 2012; Torgersen and Schmidt 2013). Some stakeholders and policy makers deem the parallel problematic because they fear a repetition of the controversy over genetically modified crops at the end of the 1990s[2] (Kaiser 2012; Kronberger 2012; Tait 2009, 2012; Torgersen 2009; Torgersen and Hampel 2012). This concern can be reinforced by empirical studies showing a persistent low public support for genetically modified crops in Europe (Gaskell et al. 2010) and suspicion towards genetic engineering in general (Rehbinder et al. 2009, p. 152). On the other hand, an understanding of synthetic biology as being a continuation of classic

---

[1]Female participant in CP 1, adults aged 50+, Vienna. Citizen panels were conducted in German and transcripts quoted in this chapter have been translated to English by the author.

[2]The anti-GMO movement in Europe at the end of the 1990s was triggered by two events: (1) the import of GM crops—not labeled as such—from the U.S. to Europe, (2) the outbreak of the BSE scandal. Within this context, Austria was one of the first countries where anti-GMO movements emerged (Seifert 2002, 2003).

W. Steurer (✉)
Department of Political Science, University of Vienna, Vienna, Austria
e-mail: walburg.steurer@univie.ac.at

© Springer International Publishing Switzerland 2016
K. Hagen et al. (eds.), *Ambivalences of Creating Life*, Ethics of Science and Technology Assessment 45, DOI 10.1007/978-3-319-21088-9_6

genetic engineering could downplay the risks, unknowns and depth of interventions of synthetic biology, which may exceed those of traditional biotechnology (Engelhard 2010, 2011).

For synthetic biology the issue of framing is challenging because concrete applications are rare, expected benefits rely on promissory visions of the future, and long-term impacts are unpredictable. Furthermore, even between experts consensus about the exact definition of synthetic biology seems difficult to find (Calvert and Martin 2009; Kitney and Freemont 2012; SCENIHR et al. 2014). In one commonly quoted definition, synthetic biology is referred to as "(A) the design and construction of new biological parts, devices, and systems and (B) the re-design of existing, natural biological systems for useful purposes".[3] Synthetic biology is described as an interdisciplinary research field involving knowledge and practices of biology, chemistry, physics, engineering, and computer science (Nature Biotechnology 2009) with possible applications ranging from the production of drugs and vaccines (Ruder et al. 2011), via biosensors for the detection of toxins, to the production of biofuels and biodegradable plastics (Kitney and Freemont 2012; Schmidt 2012). One of its pioneers, Drew Endy, named it the "engineering of biology" (Endy 2005). Critical voices, such as the NGOs Friends of the Earth and ETC Group, even name it "extreme genetic engineering" (ETC Group 2007; Friends of the Earth et al. 2012). These manifold conceptualizations leave room for the question whether synthetic biology is something new (Andrianantoandro et al. 2006; Ball 2004; Benner and Sismour 2005; Endy 2005), or a progression of traditional genetic engineering (De Lorenzo and Danchin 2008; De Vriend 2006). Consequently, it will be interesting to see how synthetic biology will be perceived, framed, and discussed outside of expert circles.

Throughout the last decades, and since the advent of the anti-GMO movement in particular, the introduction of new technologies has been accompanied by governance strategies that foster dialogue with stakeholders and the public (Marris and Rose 2010; Tait 2009). Synthetic biology is no exception to this. Every year national and international expert panels, ethics boards, and governmental advisory commissions publish reports and recommendations which underline the importance of public dialogue in emerging research fields like synthetic biology (e.g. European Group on Ethics 2009; Nuffield Council on Bioethics 2012; OECD Royal Society 2010; Presidential Commission for the Study of Bioethical Issues 2010). On the other hand, parts of the scientific community fear that public engagement could trigger technophobic discourses that hamper innovation (Graur 2007; Stirling 2012). Others question the ability of "ordinary" and "non-scientifically trained" (McHughen 2007) citizens to voice informed opinions on complex technological issues. Still others argue that public engagement is used as strategy to prevent conflict, restore trust in authorities, and would serve the primary goal of science promotion (Torgersen 2009; Stirling 2012; Wynne 2006). Indeed, within science and technology studies discussions about aims, impacts, power relations, and inclusiveness of engagement

---

[3]www.syntheticbiology.org. Accessed 25 Mar 2015.

experiments have a long tradition (Bogner 2012; Delgado et al. 2011; Irwin et al. 2013; Marris and Rose 2010; Stirling 2008; Tait 2009; Wickson et al. 2010; Wynne 2006; see also Seitz, this volume), and social scientists are critically reflecting their own roles within engagement experiments, (ELSI) research programs, or framing in general (Calvert and Martin 2009; Marris and Rose 2010; Mohr and Raman 2012; Stilgoe et al. 2014; Stirling 2008, 2012; Torgersen 2009).

So far, public engagement projects with the aim of involving citizens and/or stakeholders on questions of governance and incorporating their recommendations into policy making have been rare in the field of synthetic biology (e.g. BBSRC and EPSRC 2010; Royal Academy of Engineering 2009; but cf. Rerimassie, this volume). In the literature we rather find empirical studies focusing on public perceptions per se. Large-scale surveys have been carried out on both sides of the Atlantic (Gaskell et al. 2010; Hart Research Associates 2008, 2009, 2010, 2013; Kahan et al. 2009), and a considerable number of studies have used a combination of qualitative and quantitative methods to investigate the public perception of synthetic biology in a variety of countries (Dragojlovic and Einsiedel 2012; Hart Research Associates 2008, 2009; Kronberger et al. 2009, 2012; Navid and Einsiedel 2012; Pauwels 2009, 2013; Schmidt et al. 2008). The results of these studies show that recurring issues that matter to respondents are: long-term impacts, side effects, economic interests, intellectual property rights, distributional justice, notions of "life" and morality of constructing "artificial life", safety, security and regulation of synthetic biology.

Furthermore, qualitative studies focused on the framing of synthetic biology by experts, stakeholders or the media and the influence of certain framings on public perceptions (Ancillotti and Eriksson, this volume, Cserer and Seiringer 2009; Gschmeidler and Seiringer 2012; Kronberger et al. 2009, 2012; Lehmkuhl 2011; Pauwels 2009; Pearson et al. 2011). In addition, the importance of past experiences and the drawing of parallels to other research fields, such as biotechnology, nanotechnology, computer science, cloning or stem cell research (Pauwels 2009, 2013; Tait 2009; Torgersen 2009; Torgersen and Hampel 2012; Torgersen and Schmidt 2013), and the role of metaphors like e.g. "playing God", "creating life" (Dabrock 2009; Dragojlovic and Einsiedel 2012; Eichinger, this volume; Falkner this volume; van den Belt 2009), "Frankenstein" (Ball 2010; Gschmeidler and Seiringer 2012; van den Belt 2009) or "living machines" (Deplazes and Huppenbauer 2009) have been investigated in empirical studies and theoretical papers.

What contribution can the Austrian CP study add to the scientific discussion? First, qualitative studies provide the means for exploring public perceptions in-depth and contextualize quantitative data, such as those from Eurobarometer, where Austrians were attested of being particularly cautious about synthetic biology (Gaskell et al. 2010), or from a multi-country comparative survey conducted by Pardo et al. (2009), where Austria ranged at the lower end of 15 investigated societies regarding acceptability of producing biopharmaceuticals in genetically modified animals and plants (Pardo et al. 2009; Rehbinder et al. 2009). Second, by taking qualitative work on public perceptions of synthetic biology conducted in Austria (Kronberger et al. 2009, 2012), in the UK (BBSRC and EPSRC 2010), and in the US (Pauwels 2009, 2013) into account it can be seen if and how attitudes towards synthetic biology have changed over time and vary between societies. It can especially be investigated if the anchoring of synthetic biology within biotechnology—as stated by Kronberger et al. (2009, 2012) and by Pauwels (2009, 2013)—is still persistent, or if new frames and comparators have emerged. Summarizing, the Austrian CP study which will be presented in this chapter explores questions thrown up by quantitative data, paying attention to discourses and framings of synthetic biology outside of expert and media circles, and to changes in attitudes towards synthetic biology.

2 Methods

2.1 Citizen Panel Methodology

To investigate how members of the Austrian public encounter synthetic biology, citizen panels (CPs) were conducted in this study with participants from a variety of socio-demographic backgrounds.[4] The method of CPs as mode of political participation was developed in the 1970s. CPs were introduced as an innovative method for giving voice to the public and for incorporating its inputs into policy decisions. The idea behind CPs as described by Crosby and colleagues is to put a

> "group of the public in dialogue with public officials so that the officials get the reactions of 'the people themselves' on a particular subject, rather than simply getting the views of those who are lobbying from a particular point of view or interest" (Crosby et al. 1986, p. 171).

However, in the scientific literature, the definition of CPs varies across projects and authors, and a clear distinction from other methods of public engagement seems at times difficult to discern. Rowe and Frewer have criticized the unclear and sometimes contradictory nomenclature of public engagement mechanisms in general (Rowe and Frewer 2005), and their critique also holds true for the CP method. While some authors use the term "citizen panel" synonymously to "consensus conference" (Brown 2006; Guston 1999; Lin 2011), "citizen jury", "planning cell" (Brown 2006; Crosby et al. 1986; Lin 2011), "citizens' review panel" (Fiorino 1990) or "deliberative poll" (Brown 2006), others define each of these mechanisms as different methods for public engagement (Abelson et al. 2006; Nanz et al. 2010; Rowe and Frewer 2005; Sheedy et al. 2008; United Nations Department of Economic and Social Affairs 2011).[5]

---

[4]The CPs were conducted in the framework of work provided to the Austrian Research Promotion Agency (FFG). The citizen panel study was coordinated and supervised by Herbert Gottweis.

[5]Also the design of CPs is characterized heterogeneously in the scientific literature, which is due to the association of the term with completely different engagement mechanisms (Rowe and Frewer 2005). Accordingly, designs of CPs range from discussions within small groups of participants assumed to represent a specific community (Abelson et al. 2006; Guston 1999; Sheedy et al. 2008), to projects involving several hundred participants constituting a statistically representative sample and having more the form of surveys (Abelson et al. 2003; Nanz et al. 2010). Meetings are

The CP conception of the Austrian study follows a characterization given by Abelson et al. (2006), which is very close to the method of focus groups (Barbour 2008; Bloor et al. 2001; Krueger and Casey 2009; Liamputtong 2011). For Abelson et al. (2006), CPs are composed of small groups of citizens who discuss a predetermined issue in a face-to-face meeting. An expert provides participants with balanced and accessible information on the subject. In addition to being invited to discuss and deliberate on the issue, participants are asked to formulate recommendations based on their deliberations (Abelson et al. 2006, p. 15). Effectively, a CP is here a special kind of workshop with the public, which is composed of information and discussion phases with the aim of bringing different perspectives, ideas and opinions to the fore. As such, CPs are particularly suitable for exploring public understandings of synthetic biology, and for generating inputs for policy making.

2.2　Sample Design

For participant selection a purposive, non-representative sampling approach was chosen (Barbour 2014) as the intent was not to produce statistically representative data mirroring the perceptions of the general Austrian population, but rather to get access to different ways of understanding and debating synthetic biology by including the perspectives and experiences of a diverse set of societal groups. The sample was supposed to reflect diversity in terms of age, gender and living area. Men and women involving a mixture of academic grades and professional backgrounds were invited, and participants were divided into two age groups: adults aged 18–49, and adults aged 50+. This separation was chosen on the basis of the hypothesis that in older age groups past experiences with protest movements during the 1970s and 1980s in Austria could have an influence on public perceptions of synthetic biology and its governance.[6]

Participants were recruited during October and November 2012 by snowball sampling and with the help of online advertisements.[7] A total of eight CPs were subsequently conducted—with half of the CPs taking place in Innsbruck, in western

---

(Footnote 5 continued)

in some cases organized as singular events and in others as a sequence of meetings over a longer time period, with a selected standing group of participants.

[6]In this connection it is important to know about two events that are considered particularly significant for Austria's political culture and popular understanding of protest and democracy. First, during the 1970s a protest movement formed against the activation of the nuclear power plant "Zwentendorf"—ever since, Austria has been nuclear-free in electricity production. Second, in the mid-1980s a protest movement and mass-occupation of the wetland "Hainburger Au" hindered the construction of a hydroelectric power plant in the nature reserve. As a consequence of protester's demonstrations, the natural ecosystem of the "Hainburger Au" has been left untouched until today and Austria's national energy policy deeply influenced by the event (Seifert 2002).

[7]Participant recruitment was carried out by Ursula Gottweis, Walburg Steurer, and Viktoria Veith.

Austria, and half in Vienna, in the eastern part of Austria, so as to account for different regional areas. In each city two groups with 18- to 49-year-old adults and two groups with adults aged 50+ were organized. The number of participants within individual CPs varied from five to twelve people. The overall number of participants was 67. Ages ranged from 18 to 78, with a mean age of 43 years. Overall, there was a small surplus of citizens with higher educational backgrounds due to some participants not showing up on short notice (for details see Table 1).

2.3 Data Collection

All CPs took place in November and December 2012.[8] The discussions were audio recorded with the informed consent of the participants and afterwards transcribed in order to facilitate analysis. Participants were assured that their personal data will be treated confidentially and their statements remain anonymous. Each CP lasted for about two hours and was led by two trained moderators who provided balanced information about synthetic biology and ensured that every participant had equal opportunity to speak. For the moderation of the CPs, a semi-structured topic guide, composed of five thematic units, was followed. The same topic guide was used in every CP for comparability of results. Each thematic unit was divided into two alternating phases: (1) information phases in which participants received information about objectives, strategies and fields of application of synthetic biology, and (2) discussion phases in which participants were invited to bring in their perspectives and opinions on synthetic biology, to discuss challenges and opportunities within the group and to formulate recommendations on synthetic biology governance.[9]

The CPs started with an introduction by the moderators and the disclosure of the topic to be discussed. In order to avoid that participants inform themselves beforehand, when inviting them, they had been told that the CPs would be about the role of science and technology in general, and about a novel research field to be disclosed during the CP in particular. Therefore—and as a warm-up exercise—in the first thematic unit participants were invited to discuss the impacts of science and technology on their everyday lives.

Within the second unit, participants were provided with basic information about functioning of cells, genome, and genetic blueprint, and the ways in which synthetic biology uses, (re-)constructs, and (re)designs them by combining the knowledge and practices of biology, chemistry, physics, engineering and information

---

[8]The CPs were organized and conducted by Ursula Gottweis and Walburg Steurer.

[9]Regarding the composition of the thematic units and selection of example cases the topic guide was inspired by those used in the UK "Synthetic Biology Dialogue" by BBSRC and EPSRC (2010) and in the public dialogue organized by the Royal Academy of Engineering (2009). Furthermore, case selection was inspired by a focus group study conducted by a group of researchers from the Chair of Ethics at the Friedrich-Alexander University Erlangen-Nuremberg, which is yet to be published.

Table 1 Characteristics of sample

		Number (n)
Number of citizen panels		8
Number of participants		67
Gender	Male	38
	Female	29
Living area	Innsbruck	36
	Vienna	31
Nationality[a]	Austria	33
	Other	26
Age	18–49	36
	50+	31
	Average age of 18–49 year olds	30
	Average age of 50+ year olds	62
	Age range of total sample	18–78
	Average age of total sample	43
Educational level[b]	Basic education	9
	Vocational education	11
	Secondary education	23
	Tertiary education	36
Employment status[b]	Student	27
	Employed	24
	Freelance	8
	Unpaid work	3
	Unemployed	3
	Retired	19

[a]Not specified by 8 participants
[b]Multiple answers possible

technology. Further, it was explained in how far synthetic biology is different from traditional genetic engineering. This first information phase was followed by a discussion about participants' understandings and interpretations of the term "synthetic biology" and about its ethical, legal, social and economic implications.

The third thematic unit was dedicated to possible applications and products of synthetic biology. The moderators presented examples from three different fields of application: (1) Medicine: Synthesis of artemisinic acid—a precursor substance for the anti-malarial drug "Artemisinin"—in redesigned yeast. Traditionally, the substance is extracted from the sweet wormwood tree (Artemisia annua) cultivated primarily in China, Vietnam, Kenya, Tanzania, Uganda, Madagaskar and India. Stakeholders promise that with the help of synthetic biology production costs could be reduced and access to the drug for less developed countries be guaranteed (Collins 2012; Hommel 2008; Keasling 2009; Ro et al. 2006; Weber and Fussenegger 2009; Westfall et al. 2012). (2) Agriculture: Construction of modified organisms (bacteria, viruses, or insects) for plant pest control (Gilbert and Gill 2010; Jin et al. 2013; Thomas et al. 2000; Weber and Fussenegger 2009). The main arguments advanced by researchers for supporting this kind of research is that the use of pesticides and potential impacts on human health could be reduced through this biological alternative, and that non-target species would remain untouched by the artificially constructed organisms (Jin et al. 2013; Weber and Fussenegger 2009). (3) Environment: Bio-fuels from redesigned algae as an alternative for fossil fuels and biofuels from crops (Georgianna and Mayfield 2012; Gimpel et al. 2013; Service 2011; Wang et al. 2012). Each presentation was followed by a discussion about the application of synthetic biology in that specific case, and possible challenges and opportunities. Participants were invited to reflect as well about positive as about negative implications.

In the fourth section, questions were asked concerning the governance of synthetic biology. Focus was drawn to the role of researchers, policy makers, and funding bodies. Participants were asked to give recommendations on how the field should be regulated, who should regulate it, and how supervision could be guaranteed. Furthermore, they were asked about requirements that should be met for research funding and about conditions for synthetic biology products to enter the market.

Finally, participants were invited to imagine a future where synthetic biology would be part of their everyday lives, and to describe their imaginations, expectations, and feelings, such as hopes, fears and concerns.

2.4 Data Analysis

For data analysis, a mixed methods approach combining structured content analysis (Kuckartz 2012; Mayring 2008) and interpretive frame analysis was chosen (Fischer 2003; Schön and Rein 1994). Structured content analysis allows for combining inductive and deductive approaches in category development, and is furthermore suited for coding manifest as well as latent contents within texts (Kuckartz 2012). Consequently, in the first step of the analysis, key issues were identified and a category system developed. This was based on the thematic structure of the topic guide and on prior knowledge gained from the scientific literature. After coding about 20 % of the data material, categories and codes were revised and new ones formulated inductively as they emerged from the empirical data. The analysis focused on manifest contents as well as on in-depth structures and latent contents within the transcripts. Frame analysis allows to analyse underlying frames that shape discourses (Fischer 2003; Schön and Rein 1994). By conducting frame analysis special attention was put on the framing of synthetic biology by the CP participants and on comparators chosen for making synthetic biology graspable. Throughout the whole project, the use of the qualitative data analysis software Atlas.ti facilitated the management, storage and organization of the data (Friese 2012).

3 Results

3.1 Something Old, Something New—Making the Unknown Tangible by Drawing Parallels

The majority of the participants was not familiar with the term "synthetic biology", even though they knew about the practice of (re-)constructing organisms in laboratories with the help of modern technologies. Thus, while the practice itself was known, participants did not associate it with the term "synthetic biology".

Generally, the term "synthetic biology" evoked surprise and puzzlement. This was due to the combination of the two words "synthetic" and "biology", which were understood as being opposed to each other. In order to make sense of the "contradictory" concept, participants on the one hand looked at each of the two words separately, and on the other, at the relation between them. They often concluded that the term "synthetic biology" was a contradiction in itself: "synthetic" as something artificial, unnatural, technical or man-made, and "biology" as something natural, living, and detached from human power. Participants put it as follows:

> That's a contradiction. Synthetic biology is a contradiction. 'Bio' is a Greek word, as far as I remember from school, and it means, means 'life', but *synthetic life*, I am not sure, if you can call this life at all.[10]

> Participant (P) 10: It's a paradox, when I only see these two words, 'synthetic' and 'biology'. I associate 'biology' automatically with a natural product and 'synthetic' is just its opposite.
> P11: Well, but what *is* nature? [...] *Nature* is itself only a construct made by man, hum, something he invented, created somehow.[11]

Statements of this type led to discussions about the definition of "life" as such. While for some the main concern lay in drawing a demarcation line between "dead matter" and "life", "artificial" and "natural", or "animate" and "inanimate", others classified the word "synthetic" as simply not being appropriate to describe neither the "material" from which parts and systems are constructed nor the organisms resulting thereof. Single components as well as "life forms" constructed by synthetic biology were perceived as being built from living substance, not from dead material or scratch—as the word "synthetic" would suggest—even though the constructed parts, systems and organisms do not exist in nature and presumably would never have been generated by it. In the words of a participant:

> There must already have been something living within it, living organisms can't be built from dead matter.[12]

---

[10]Male participant in CP 8, adults aged 50+, Innsbruck.

[11]Conversation between two male participants in CP 6, adults aged 18–49, Innsbruck.

[12]Male participant in CP 1, adults aged 50+, Vienna.

Furthermore, imaginations and meanings of the attribute "synthetic" from other contexts and shared discourses were mobilized. It made people think of synthetic foodstuff, such as "imitation cheese" and "imitation ham",[13] energy drinks and E-numbers. The word "synthetic" was associated with something "lab-grown", "unhealthy" and "faked". To sum it up, the term "synthetic biology" caused irritation and evoked rather negative yuck feelings and imaginations. Noteworthy in this context is the following comment:

> Err, but in general, for me, this term, I mean synthetic biology, sounds, err, is negatively loaded, err, um, err, because it is, err, I think there are so many dangers that could come up within science. And for me this sounds a bit like, like certain science fiction novels I have read, and they seldom had a happy end.[14]

Beyond reflections about the literal sense of the term "synthetic biology", participants also tried to make sense of the practices and consequences of synthetic biology by drawing parallels. This can be interpreted as strategy to cope with the unknown and uncertain: by comparing the abstract with the concrete, the former becomes tangible and understandable. In the CPs especially the imagined challenges and opportunities of synthetic biology were compared to experiences with scientific innovations from throughout the history of humankind, as the next quotes show:

> There we are in a similar situation as Marie Curie was, who did research into uranium, had uranium all over her body, and died from it. But, can we put into question that it was a breakthrough? Didn't it generate fundamental knowledge for contemporary science?[15]

> A little bit it reminds me of the time when the steam engine, the steam locomotive was invented, and then also people were against it, and for heaven's sake, devil's work, and dangerous, you die when you move so fast. Ah, it's the *uncertainty*.[16]

> But you could see it also with drugs, that many drugs, starting with Contergan,[17] till I don't know, many drugs have been released, that afterwards had completely different side-effects, we should treat it with caution, the whole thing.[18]

While the examples cited above refer to historical events, parallels were also drawn to more recent phenomena and empirical values from neighboring research fields, such as nanotechnology, information technology, pre-implantation genetic diagnostics, stem cell research or cloning. However, the most common reference made was to genetic engineering—with a tendency to equate synthetic biology and

---

[13]These terms are known in German under the buzzwords "*Analog-Käse*" and "*Mogel-Schinken*", which had been at the centre of heated public debates throughout the previous five years (Die Welt 2009).

[14]Male participant in CP 2, adults aged 18–49, Vienna.

[15]Male participant in CP 6, adults aged 18–49, Innsbruck.

[16]Male participant in CP 7, adults aged 50+, Innsbruck.

[17]Contergan was the trade name of a drug containing thalidomide, which was freely available in pharmacies in Western Germany from 1957 to 1961. It was, amongst others, used against morning sickness in pregnant women, and caused severe damage to children, most notably with regard to limb development.

[18]Female participant in CP 5, adults aged 18–49, Innsbruck.

genetic engineering or to understand synthetic biology as the obvious and logical progression of genetic engineering. This becomes clear from the citation in the introduction to this chapter as well as from the following excerpt:

> But on the other side, the risks and opportunities that emerge out of this new, or maybe not so new technology are old hat. It was thirty years ago, if you think about what lies ahead, for example with genetic engineering, I don't see any difference with my lay knowledge. If you do it this way, or that way, it's all, there are incredible opportunities, but there is also an incredible amount of things that could fall on us.[19]

In summary, participants tried to make sense of synthetic biology and its implications by drawing parallels to other technologies and past experiences. The fields associated with the (re-)construction of organisms were preferably inscribed within the discursive frame of "genetic engineering" or "genetic modification". Furthermore, the term "synthetic biology" caused irritation and evoked rather negative feelings and expectations. Therefore, the next section will focus on concrete dangers and challenges brought to the fore by CP participants and on the hopes and opportunities they perceived within "this new, or maybe not so new"[20] research field.

3.2 Something Good, Something Bad—"An Ambivalent Thing"

In order to further investigate risks and opportunities of synthetic biology, participants were introduced to three examples for its (future) application and asked to discuss positive and negative aspects for each example.

The first example presented was the application of synthetic biology for the production of artemisinic acid in modified yeast. Other than the high hopes and promissory future scenarios raised by scientists and stakeholders, attitudes towards the synthetic Artemisinin were divided in the CPs. While some participants showed enthusiasm, others were more cautious and pointed to unknown risks and economic interests. This ambivalence becomes apparent in the following conversation:

> P11: I think, finally something happens, because down there [in Africa], where people are really poor, in an economic sense, they could really be helped, with this drug at a cheap rate. So, I really appreciate and support it, and I say 'it's a good thing'. But the question is always who really takes profit. Those, who receive the treatment, or again a big company, or companies? But we will never be able to prevent this.
> P9: No, but principally, if an effective drug against Malaria can be produced, we can only appreciate it.[21]

---

[19]Male participant in CP 1, adults aged 50+, Vienna.

[20]Male participant in CP 1, adults aged 50+, Vienna.

[21]Conversation between two male participants in CP 8, adults aged 50+, Innsbruck.

After initial fascination, people were primarily concerned about the interests of (pharmaceutical) companies and industries behind the research. Throughout all CPs, the topics of monopolies, intellectual property rights and distributional justice constituted issues of discussion. One participant cited the example of drugs against AIDS to underpin his concern that access to novel products and scientific achievements would remain a privilege of the rich. He argued that multinational companies would hold patents and prevent large-scale supply with drugs at an affordable price for the poorest countries.

Other comments focused on the future of the farmers who cultivate the medicinal plant from which the artemisinic acid is traditionally extracted:

Will they in the future have a means of existence?[22]

Furthermore, in two CPs, participants voiced suspicions about the selection of the Artemisinin-case for the CPs. In their perception the Artemisinin-example could be misused as a door-opener argument for the application of synthetic biology in other fields:

You have chosen an example that is effective as good publicity for synthetic biology. Because against Malaria, we know it, something has to be done, because hundreds of thousands of people are dying of it, maybe more.[23]

Beyond economic interests, uncontrollability and uncertainty mattered to participants. In particular, they were concerned about long-term risks and side effects. Participants pointed out that research was still at an early stage and worried that consumers could be used as "human guinea pigs". Furthermore, participants questioned what would happen if the modified yeast would "escape" from laboratories and crossbreed with natural organisms. It was argued that unintended evolution and mutations could be the consequence and the sensitive balance of natural ecosystems be damaged. Several participants invoked Goethe's ballad "The Sorcerer's Apprentice" or Shelly's "Frankenstein" to epitomize their visions of a future with synthetic biology.

Concerns about uncontrollability were even more pronounced in the second example case: the construction of bacteria, viruses or insects for plant pest control. Again long-term impacts and unforeseeable side effects were perceived as critical. Participants pointed to the risk of crossbreeding between natural and artificially constructed organisms and misuse of synthetic biology for terrorist purposes and warfare:

P6: I would have a bad feeling with this thing, that this, that they could somehow mutate and become killer viruses [generalized laughing, talking across each other], yes
P1: extremely
P3: that's really how it is
P6: and to me, to my mind there are always and immediately coming weapons
P1: chemical

---

[22]Female participant in CP 5, adults aged 18–49, Innsbruck.

[23]Male participant in CP 8, adults aged 50+, Innsbruck.

P3: Anthrax

P6: yes, chemical, [talking across each other] I really wouldn't need that

P3: No, me neither, I don't want that

P1: Me too, I am rather, yes, quite skeptical, against that. I would also be scared that someone could say (..), yes, because, we always believe, man always nicely believes, that he can control everything, but

P2: we often saw that

P1: that it went wrong.[24]

Taken together, also in the agricultural field participants' attitudes towards the application of synthetic biology were marked by perceptions of risks and distrust in authorities and scientists to overlook the field. Compared to the medical field, risk perceptions and fears were much more pronounced in agriculture, as participants imagined the use of synthetic biology for medical purposes in laboratories or as drugs within the human body as less problematic than field release of novel organisms in open environments. Their reasoning was that in the first case controllability could to some degree be possible due to research taking place within confined spaces, while uncontrollable evolution and mutations were perceived as being the logic consequence of field release of novel organisms into natural ecosystems. In addition, it is notable that parallels to past experiences with genetically modified crops were again drawn:

I don't know, I am rather skeptical about interventions within the natural ecosystem. I think it's a completely different thing if you do it only, only in the medical field. I think in the medical field I am much more tolerant and I think, there you can try much more, but when man impinges on nature, which is not a human being, but algae or insects, I simply have a bad feeling, and I don't think that we can control everything like we suppose to do, because also when you just use it within confined areas - they did it as well with genetically modified maize, which was spread by the winds, and, I don't know, I simply have a bad feeling.[25]

A second example further illustrates the embedding of synthetic biology within the discursive frame of genetic engineering and the mobilization of respective imaginaries:

P1: [...] There's again a danger connected to it. How should it be possible to test it? It's difficult to say, because there exist a huge variety of bacteria and viruses, and, and all these things are so huge. I think you can't test it, you simply have to apply it (.) with force, stop, punctum.

P11: That would mean almost additionally to genetic manipulation, right? If there is some vermin in maize, and now if [...] I would culture a virus, which fights that particular vermin, and then we do genetic engineering

P1: all inclusive

P11: [laughs cynically] yes. But I think a layperson lacks the overview. How can this really be done within boundaries?

---

[24]Conversation between two male (P1, P2), and two female (P3, P6) participants in CP 5, adults aged 18–49, Innsbruck.

[25]Female participant in CP 4, adults aged 18–49, Vienna.

P1: Right. Is science able to assess the risks? Or how much time do such trials take, to have at least a minimum of security? That's alarming, isn't it?[26]

Overall, in both example cases—medicine and agriculture—distrust against authorities, scientists and sponsors of synthetic biology seem to be at the heart of participants' skepticism. Distrust became visible not only when participants questioned the possibility to control or overlook the field, but also when they communicated their suspicion that those interested in the research would obscure their "real" aims and interests—thus, participants perceived a fundamental lack of transparency. A similar distrust had already become apparent with the suspicion that the Artemisinin-example could be used as door-opener, but it became even more apparent when participants put into question the arguments that by using synthetic biology, arable farm land could be saved for food production and thereby the growing world population be fed.

Along this line of distrust, in the third example case—the production of biofuels from redesigned algae as an alternative to fossil fuels or biofuels from crops—participants criticized a perceived instrumentalization of the issue of "hunger" or of the argument of "ensuring food supply for the growing world population" for the promotion of synthetic biology. Further, participants were suspicious that other energy sources would intentionally remain unexplored and research left without funding due to economic interests. The next quote illustrates this generalized distrust:

I don't think that bio-algae are the solution, don't think that this will become commonplace, impossible for me, honestly, to be perfectly honest. I worked in the automobile industry for twenty years, I know that there was research going on in the past; with steam you can power cars, in principle that doesn't cost anything, but those in power are against it, they hinder that these things enter the market, because everything would collapse, it's determined by money, and power, and avarice, but that's the world.[27]

Within the pattern of distrust and skepticism moral questions played a critical role, as they again displayed participants' pessimistic and suspicious attitude towards synthetic biology. Participants were not only critical about the construction of "life" as such, but worried also that optimization and purposeful selection could easily be drawn into extremes. They argued that synthetic biology would begin with the (re-)construction of bacteria, insects, or algae, but in the future could be used for eugenic purposes with the aim of constructing the "perfect human". While the question of how far man is allowed to go and the reproach of transgressing nature were mentioned several times, interestingly the metaphor of "playing God" was only used twice. This may be due to a perception of synthetic biology as not being something completely new, but another form or a progression of classic genetic engineering, and as such something which has become "normal" or perceived as being within the realm of humans' mighty.

---

[26]Conversation between a female (P1) and a male (P11) participant in CP 8, adults aged 50+, Innsbruck.

[27]Female participant in CP 3, adults aged 50+, Vienna.

Summarizing the inputs, the application of synthetic biology was perceived as "an ambivalent thing" with hopes, fears, and moral concerns openly voiced in the CPs. In addition, distrust could be identified as a central attitude towards synthetic biology. An excerpt brings this exemplarily to the point:

P9: But it's an ambivalent thing. On the one side there are many risks and dangers, on the other science and technology have made an increase in life expectancy possible. [...] I would say 'boon and bane', right, both.
[...]
P6: But in the end it's always man who is responsible, will it be good, or will it be bad.
P2: For sure.
P6: So we should try to control him [man].
[Laughing]
P5: But maybe it could be like with the atomic bomb. Someone too invented that, I think, and only afterwards he realized how deadly it was. Many might not even know what will be the end product.[28]

This ambivalent image raises questions about how participants cope with the perceived uncertainty and intransparency, and which attitudes they developed out of their distrust and skepticism. The answers to these questions will be provided in the next section.

3.3 Being "Just a Lay Person"—Between Resignation and Self-Activation

The attitudes that participants developed as a result of their distrust were—like distrust itself—mostly not expressed directly, but became discernible as implicit attitudes within discussions. Two major attitudes—understood as coping strategies —could be identified: resignation and self-activation. Resignation comprises on the one hand the perception of oneself as being "just a lay person" who lacks the overview and is obliged to believe what scientists—perceived as "insiders"—and the media say, and on the other hand, the feeling of being powerless, to have no influence on the progress of science. Concerning the latter, participants voiced first, the fear that research would continue anyway, independently of peoples' demands, wishes and opinions, and second, the feeling of being powerless against (economic) interests:

P5: Well, science will always move on, regardless of whether you are for or against it, whether it's forbidden or not, it will go on. But I'm not doing well with that, so.
P1: In the course of years, technology will maybe have progressed so far, that a normal, a mortal individual won't be able to manage it, right?[29]

---

[28]Conversation between a female (P5) and three male (P2, P6, P9) participants in CP 8, adults aged 50+, Innsbruck.
[29]Conversation between two male participants in CP 7, adults aged 50+, Innsbruck.

The fact that this drug [Artemisinin] does already exist, demonstrates that regardless of what we are discussing here, it has already been done.[30]

[...] my estimation is realistic, it's not preventable. All that's imaginable, all that's researchable, will be researched, if not now, then in 10 years, it's a stream that is in a state of flux. They will try to regulate it as good as possible, but they will fail.[31]

However, distrust and the feeling of powerlessness did not only result in resignation, but in some participants turned into its opposite: an attitude of self-activation. Self-activation in this context means a form of emancipation of the individual, who becomes aware of his/her power as critical and self-reflexive consumer. Accordingly, participants defined it as crucial, first, to inform themselves very well, for example by reading food labels attentively, in order to be empowered as individuals to decide which products to buy. Second, it was seen as essential to rethink one's consumer behavior on a more general level. This included self-criticism of being members of a "throw-away-society" which incites a run for the cheapest products regardless of the conditions of production, as well as very concrete suggestions like the reduction of individual car use, which would make the need for alternative energy sources—and therefore also research into biofuels from redesigned algae—less urgent. Thus, self-activation emerged as a defense reaction against a general distrust in scientists, industry, funders and regulators, and against a perceived non-transparency, manipulation by media and advertisement, and an imbalance of power between "insiders" and the "lay public".

Finally, the call for individual responsibility was further expanded into a call for societal responsibility, especially for the next generations. This brings us to the question of how synthetic biology should be regulated.

3.4 The Big "If"—Setting the Conditions

The overall impression gained from data analysis is that support for synthetic biology is always conditional. Thus, especially—but not only—when CP participants were asked about preconditions for the application of synthetic biology and recommendations for its governance, there appeared a big "if". This big "if" could be identified as a recurring pattern within different contexts.

The first big "if" concerned information and transparency and was therefore closely related to the distrust identified before. In the preceding section, the importance of informing oneself and taking the role of the responsible consumer was addressed within the pattern of self-activation: the labeling of products in a clear and visible manner had high priority for most participants as it symbolizes the guarantee for being oneself the person who decides what to consume. Thus,

---

[30]Male participant in CP 6, adults aged 18–49, Innsbruck.

[31]Male participant in CP1, adults aged 50+, Vienna.

information and transparency were established as preconditions for autonomy and for the restoration of trust in authorities. An excerpt illustrates this well:

> We can't stop it from progressing anyway. It's happening in the background anyway, all that research does already exist, I don't know for how many years, but research continues, it will come anyway. But there must be someone who communicates that in a hundred per cent transparent way to the people, what exactly happens there, what are the negative sides, what is positive about it. So, I really would like to be informed very well, then I can decide for myself, if I want to buy it or not, that's the point.[32]

The quote is interesting in two dimensions: On the one hand, it highlights a shift from resignation to self-activation, on the other, it shows that knowledge—gained through information—symbolizes and enables a form of power as it facilitates the emancipation of the individual. Hence, it can be concluded that transparency and information, first, help to restore the balance of power between informed "insiders" and "lay people", and second, to restore trust in authorities.

The balancing of risks and benefits constituted a second big "if". Participants underlined that uncertainties should be disclosed from the outset. An excerpt illustrates the importance of knowing challenges and opportunities of synthetic biology—thus, again, the importance of transparency—and of having the possibility to choose:

> Well, if I could say from the beginning 'these are the opportunities and those are the risks', put all I know on the table, then I could maybe better form my opinion than if I always have the feeling that we are manipulated, only the advantages, only the advantages. And about the risks I have to think on my own.[33]

Within the context of balancing risks and benefits, the exploration of alternatives was perceived as critical. The big "if" here refers to missing alternatives and a perception of synthetic biology as being the lesser evil. Thus, if no alternatives are available, synthetic biology becomes a viable practice. The following conversation taken from a discussion about biofuels from redesigned algae shows this well:

> P2: The question is, so it seems to me, if we want to continue pumping up fossil fuel from the soil, or if we want to extract it from algae.
> P1: I believe that there must be an *alternative*. I simply cannot imagine that there doesn't exist *anything*.
> P2: Me neither, I don't find any of the two possibilities cool, but if you have to take a decision,
> P1: yes, then algae would in any case be the lesser evil. Now, I have to admit, that it *is* no bad idea.
> P2: That's a technical question.
> P2: Well, it's no bad idea, but (..)[34]

---

[32]Male participant in CP 3, adults aged 50+, Vienna.

[33]Female participant in CP 1, adults aged 50+, Vienna.

[34]Conversation between a female (P1) and a male (P2) participant in CP 6, adults aged 18–49, Innsbruck.

Missing alternatives were a decisive factor for the support of synthetic biology particularly within the medical field—the field of application that received most support within the CPs. Thus, if the question is about life and death, the ways how drugs are produced play only a minor, if any role at all, as the next quotes demonstrate:

P4: If I were affected by a certain disease, I would be happy if drugs were available, regardless of whether they are produced with the help of synthetic biology, or not, I would not care about that.
P1: Yes, if they could help you, right?
P2: Year, that's
P1: that's clear, yes
P4: Then I would be happy, if something would exist, then.[35]

P3: I think, and this is very interesting, when it comes to physical health, our attitudes change completely [affirmation from other participants] regarding those things. So, I think if I suffered from Malaria and had no money and could get this drug, at a cheap rate
P1: then you would take it
P3: cheaper, I would buy it
[Several participants speak simultaneously, incomprehensible]
P3: it's something completely different
P6: of course.[36]

The cost factor was, thus, decisive for the support of synthetic biology. This was not only true for the medical field, but also for other fields of applications, such as agriculture:

Sounds promising, especially in regard of fuels, if it can really be produced at a cheaper rate.[37]

Third, prevention and containment were named as pivotal requirements. Participants recommended that long-term studies and reliable tests on side effects should be carried out before synthetic biology products enter the market and before modified organisms are released into natural ecosystems. Furthermore, it was requested that agents for drugs should one-to-one correspond to agents extracted from natural sources, and research and application of synthetic biology should remain within confined spaces or closed areas. The scientific literature differentiates between "biosecurity" as prevention of intended harm (e.g. bioterrorism), and "biosafety" as prevention of unintended harm (e.g. natural disasters) (Kelle 2009); these concepts could also be identified within the CP discussions, even though participants did not use these terminologies. To give a few examples, participants' recommendations sounded as follows:

---

[35]Conversation between a female (P2) and two male (P1, P4) participants in CP 7, adults aged 50+, Innsbruck.

[36]Conversation between a male (P1) and two female (P3, P6) participants in CP 5, adults aged 18–49, Innsbruck.

[37]Male participant in CP 5, adults aged 18–49, Innsbruck.

Well, yes. I imagine that this could be a great thing, if it were safe, whereas safety is a two-edged sword. What might seem safe to one person might seem unsafe to the other. But there are lots of things that should be tested, and checks that should be implemented.[38]

P11: I could imagine that, if research would really engage this intensively, so that they could also manage to control the side effects.
P6: Year, year.
P11: If they could say 'it will only do this, and, and, nothing else'.[39]

I think as long as it's within confined spaces or something, like it's with this drug [Artemisinin], I don't know, I wouldn't have any problems with it.[40]

I think it would be good to know how things work, just in case. It should only be used within confined spaces. So, my fear is rather, that this falls in the hands of the wrong decision makers, who could lead us into a world wherein we would not want to live.[41]

Finally, participants were asked how and by whom these "ifs" could be met, i.e. how the field should be regulated, who should regulate it, and who should be responsible for its oversight. A broad variety of recommendations were given, but opinions were divided. While some participants argued that research and application of synthetic biology should be regulated by each state individually, others were more inclined to regulation at international level. Alternatively, it was proposed that an independent regulatory and supervisory body should be created—be it at national or at international level. Still others argued that existing regulations on biotechnology would be sufficient for regulating synthetic biology.

On the other hand, there were participants who questioned the possibility of regulation in general. They argued that both, policy makers as well as scientists would be corrupted and that only in a utopian world regulation and control could be possible. Participants explained that when money comes into play regulation and control would be an illusion, because on the one hand there would be an entanglement between economy and politics, and on the other between economy and scientific research:

Even though this might sound radical, when economy gets into the game it always becomes a little bit corrupted, and at that moment morals do not play a role anymore, then it says 'profit or not'; and in a perfect world you could separate it, there you could say 'here is the research and everything happens for the common run of mankind, and there is the market', but it's a healthcare industry, and not everything is love, peace and harmony.[42]

The citation entails two central messages: first, it sets the fourth big "if", in that it implicitly points to distributional justice and equal access to benefits arising out of synthetic biology as preconditions for research in synthetic biology. Second, it contains the assumption that the entanglement between research and economy and between research and politics also means that researchers cannot be trusted.

---

[38]Male participant in CP 8, adults aged 50+, Innsbruck.
[39]Conversation between two male participants in CP8, adults aged 50+, Innsbruck.
[40]Female participant in CP 2, adults aged 18–49, Vienna.
[41]Female participant in CP 4, adults aged 18–49, Vienna.
[42]Female participant in CP 4, adults aged 18–49, Vienna.

As such, the option of having the conscience of researchers as guiding principle or regulatory mechanism—as proposed by some participants—is also implicitly rejected. In some CPs this option was even explicitly rejected, for example, when participants argued that conscience of researchers would not suffice for regulation because it would depend on each subject's point of view and personal standards for integrity could vary between researchers.

Summarizing, CP participants had very clear ideas about the preconditions that should be given in order to guarantee responsible research in synthetic biology and safe applications, even though there was no consensus about who exactly should regulate and control the field. Most notably, while there were manifold and diverging opinions regarding the latter question, there was one shared opinion: that citizens should be given more voice within synthetic biology governance.

4 Discussion and Conclusion

Incorporating citizens' views by, for example, taking the openly stated opinions and recommendations—as well as the implicit attitudes—of the Austrian CP participants seriously could make research and governance of synthetic biology more socially robust. Interestingly, in the present study, differences due to age groups, gender, or educational and residential backgrounds were almost not found. Only the attitudes towards the application of modified organisms in agriculture and the investigation of alternatives to fossil fuels and biofuels from crops seemed more affirmative in the Innsbruck CPs. Overall, similarities between groups outweighed differences, which is contrary to our original hypothesis, which was that past experiences with protest movements during the 1970s and 1980s could play a role in elder generation's discussions.

The analysis showed that CP participants' awareness of synthetic biology was rather low when they were first confronted with the term. This low level of awareness was no big surprise, as it corresponds to Eurobarometer data from 2010, where 83 % of respondents across the EU member states declared that they had not heard about synthetic biology yet (Gaskell et al. 2010, p. 30). Interestingly, however, despite not knowing the term "synthetic biology", CP participants were well aware of the practices that are used in synthetic biology and of research going on in that field. Therefore, the CP results raise the question of whether respondents interviewed for the Eurobarometer survey were not aware of synthetic biology research or whether they simply were not familiar with the term.

Overall, the results of the Austrian CPs did—with regard to manifest contents— not differ significantly from the results of the "Synthetic Biology Dialogue" set up in the UK by BBSRC and EPSRC (2010), and the focus group studies conducted by Kronberger et al. (2009, 2012) in Austria and by Pauwels (2009, 2013) in the US. Issues identified as critical within those studies, and in the scientific literature more generally, did also come up in the Austrian CPs: risk-benefit-tradeoff, biosecurity, biosafety, economic interests and intellectual property rights, equal access, definitions of life and moral questions concerning the construction of artificial life, lack of information and transparency. Also, the difficulties in distinguishing synthetic biology from traditional genetic engineering and the embedding of practices of (re-)constructing organisms or their parts within the discursive frame of "genetic engineering" were clearly visible in the CPs.

However, what differentiates the Austrian CPs from the other three engagement experiments is that distrust seems to be much more pronounced. Distrust became first and foremost discernible when looking at the in-depth structure of the textual material. It became manifest in a sceptical and rather pessimistic fundamental attitude, with participants underlining the ambivalent character of synthetic biology and voicing suspicions about the "real" interests behind research. This threw up the question how participants cope with their distrust, and led to the subsequent identification of two main coping strategies. While on the one hand, resignation could be identified as one possible coping strategy, on the other hand, people tended to call upon individual responsibility. For example, participants suggested reconsidering their own life style in the context of their appeal to rethink consumer behaviour. Nonetheless, CP participants had the clear understanding that emancipation is only possible when "insiders"—understood as the synthetic biology community, regulators, and the media—provide information to the public. Comparison with the BBSRC/EPSRC study shows that participants' feelings of powerlessness to understand the science or to have any influence on scientists point to similar attitudes of resignation (BBSRC and EPSRC 2010, p. 41), whereas the call for self-activation seems to be rather specific to the Austrian CPs.

Furthermore, while participants acknowledged positive sides of synthetic biology as well and voiced high hopes, support for synthetic biology was always conditional. Within the pattern of setting conditions, four big "ifs" could be identified as being essential for acceptance of synthetic biology. These were: (1) information and transparency, (2) the balancing of risks and benefits and the investigation of alternatives, (3) the application of synthetic biology only within confined and controlled spaces and after thorough testing (biosecurity and biosafety), and (4) equal access to products and benefits. The big "ifs" were often uttered implicitly and seem to rely on unconscious constructions that are influenced by past experiences and empirical values from other fields. Thus, the construction of the big "ifs" is related to the attitude of drawing parallels. Drawing parallels can be interpreted as a strategy to make the uncertain and unknown tangible and understandable—a process in which past experiences and empirical values offer a repertoire of imaginaries for developing visions for the future. Drawing parallels and referring to past experiences was also identified in the UK public dialogue (BBSRC and EPSRC 2010), and in Austrian (Kronberger et al. 2009, 2012) and US focus groups (Pauwels 2009, 2013).

The imaginative repertoire which was most prominently mobilized when drawing parallels was that of genetic engineering. It can thus be concluded that the fields associated with the (re-)construction of organisms are occupied by the discursive frame of "genetic engineering" and "genetic modification". This perception of synthetic biology as not being something completely new, but another form or a progression of traditional biotechnology may also explain why the metaphor of "playing God"—which the author had expected to come up frequently as it is often taken up by the media—was pronounced only twice. It seems that the manipulation and construction of living organisms was rather perceived as something man had done for years and therefore not as a skill solely ascribed to God or nature. It is noteworthy in this context that also in the BBSRC and EPSRC (2010) and the Kronberger et al. (2009, 2012) studies the "playing God" metaphor seemed not to matter in the first place. However, this paralleling between synthetic biology and genetic engineering, and the perception of synthetic biology as being within humans' mighty does not mean that synthetic biology is perceived as something positive—rather it is understood as being even worse than genetic engineering in that it goes "one step further".[43]

Acknowledgements The empirical data used in this study were generated in the framework of work conducted for the Austrian Research Promotion Agency (FFG) as partner of ERASynBio. Special thanks go to all participants in the CPs in Vienna and Innsbruck for their interest and willingness to discuss the issue of synthetic biology. I am especially grateful to my former supervisor Herbert Gottweis, who sadly passed away on March 31, 2014. His supervision and advice during data collection and analysis, and feedback on earlier drafts of this chapter were very valuable for me. I am indebted to Ursula Gottweis for her collaboration during script development, participant recruitment, and moderation of the CPs in Vienna and Innsbruck. Special thanks go to the organizers and participants at the International Summer School "Analyzing the Societal Dimensions of Synthetic Biology", in Berlin, September 15–19, 2014, and to Ingrid Metzler, Katharina T. Paul, and Johannes Starkbaum from the University of Vienna for their inspiring and helpful feedback.

References

Abelson J, Eyles J, McLeod CB, Collins P, McMullan C, Forest P-G (2003) Does deliberation make a difference? Results from a citizens panel study of health goals priority setting. Health Policy 66(1):95–106

Abelson J, Gauvin F-P, MacKinnon MP, Watling J (2006) Primer on public involvement. Document prepared for the Health Council of Canada

Andrianantoandro E, Basu S, Karig DK, Weiss R (2006) Synthetic biology: new engineering rules for an emerging discipline. Mol Syst Biol 2(2006):0028

Ball P (2004) Synthetic biology: starting from scratch. Nature 431(7009):624–626

Ball P (2010) Making life: a comment on 'Playing god in Frankenstein's footsteps: synthetic biology and the meaning of life' by Henk van den Belt (2009). Nanoethics 4(2):129–132. doi:10.1007/s11569-010-0091-x

Barbour R (2008) Doing focus groups. Sage, London

Barbour R (2014) Introducing qualitative research: a students guide, 2nd edn. Sage, London

BBSRC, EPSRC (2010) Synthetic biology dialogue. http://www.bbsrc.ac.uk/web/FILES/Reviews/1006-synthetic-biology-dialogue.pdf. Accessed 17 June 2015

Benner SA, Sismour AM (2005) Synthetic biology. Nat Rev Genet 6(7):533–543

---

[43]Female participant in CP 1, adults aged 50+, Vienna.

Bloor M, Frankland J, Thomas M, Robson K (2001) Focus groups in social research. Sage, London

Bogner A (2012) The paradox of participation experiments. Sci Technol Hum Values 37(5):506–527. doi:10.1177/0162243911430398

Brown MB (2006) Survey article: citizen panels and the concept of representation. J Polit Philos 14(2):203–225

Calvert J, Martin P (2009) The role of social scientists in synthetic biology. EMBO Rep 10 (3):201–204

Collins J (2012) Synthetic biology: bits and pieces come to life. Nature 483(7387):S8–S10

Crosby N, Kelly JM, Schaefer P (1986) Citizens panels: a new approach to citizen participation. Pub Adm Rev 46(2):170–178. doi:10.2307/976169

Cserer A, Seiringer A (2009) Pictures of synthetic biology: a reflective discussion of the representation of synthetic biology (SB) in the German-language media and by SB experts. Syst Synth Biol 3(1–4):27–35

Dabrock P (2009) Playing God? Synthetic biology as a theological and ethical challenge. Syst Synth Biol 3(1–4):47–54. doi:10.1007/s11693-009-9028-5

De Lorenzo V, Danchin A (2008) Synthetic biology: discovering new worlds and new words. EMBO Rep 9(9):822–827

De Vriend H (2006) Constructing life. Early social reflections on the emerging field of synthetic biology. Working document 97, The Hague

Delgado A, Lein Kjølberg K, Wickson F (2011) Public engagement coming of age: from theory to practice in STS encounters with nanotechnology. Pub Underst Sci 20(6):826–845. doi:10. 1177/0963662510363054

Deplazes A, Huppenbauer M (2009) Synthetic organisms and living machines: positioning the products of synthetic biology at the borderline between living and non-living matter. Syst Synth Biol 3(1):55–63

Die Welt (2009) Nach Analog-Käse nun der Mogel-Schinken. http://www.welt.de/wissenschaft/article4049225/Nach-Analog-Kaese-nun-der-Mogel-Schinken.html. Accessed 25 Mar 2015

Dragojlovic N, Einsiedel E (2012) Playing God or just unnatural? Religious beliefs and approval of synthetic biology. Pub Underst Sci doi:10.1177/0963662512445011

Endy D (2005) Foundations for engineering biology. Nature 438(7067):449–453

Engelhard M (2010) Biosicherheit in der Synthetischen Biologie. Die Unterschiede zur Gentechnik erfordern neue Sicherheitsstandards. Die Politische Meinung 493:17–22

Engelhard M (2011) Die Synthetische Biologie geht weit über die klassische Gentechnik hinaus. In: Dabrock P, Bölker M, Braun M, Ried J (eds) Was ist Leben – im Zeitalter seiner technischen Machbarkeit? Beiträge zur Ethik der Synthetischen Biologie. Karl Alber, Freiburg, pp 43–60

ETC Group (2007) Extreme genetic engineering: an introduction to synthetic biology. http://www.etcgroup.org/sites/www.etcgroup.org/files/publication/602/01/synbioreportweb.pdf. Accessed 21 Aug 2013

European Group on Ethics (2009) Opinion no. 25—ethics of synthetic biology

Fiorino DJ (1990) Citizen participation and environmental risk: a survey of institutional mechanisms. Sci Technol Hum Values 15(2):226–243. doi:10.2307/689860

Fischer F (2003) Reframing public policy: discursive politics and deliberative practices. Oxford University Press, Oxford

Friends of the Earth, International Center for Technology Assessment, ETC Group (2012) The principles for the oversight of synthetic biology. http://www.etcgroup.org/sites/www.etcgroup. org/files/The-Principles-for-the-Oversight-of-Synthetic-Biology-FINAL.pdf. Accessed 3 Aug 2015

Friese S (2012) Qualitative data analysis with ATLAS.ti. Sage Publications, London

Gaskell G, Stares S, Allansdottir A, Allum N, Castro P, Esmer Y et al (2010) Europeans and biotechnology in 2010. Winds of change? Eurobarometer

Georgianna DR, Mayfield SP (2012) Exploiting diversity and synthetic biology for the production of algal biofuels. Nature 488(7411):329–335

Gilbert LI, Gill SS (eds) (2010) Insect control: biological and synthetic agents. Elsevier, London

Gimpel JA, Specht EA, Georgianna DR, Mayfield SP (2013) Advances in microalgae engineering and synthetic biology applications for biofuel production. Current Opin Chem Biol 17(3):489–495. doi:10.1016/j.cbpa.2013.03.038

Graur D (2007) Public control could be a nightmare for researchers. Nature 450(7173):1156

Gschmeidler B, Seiringer A (2012) "Knight in shining armour" or "Frankenstein's creation"? The coverage of synthetic biology in German-language media. Pub Underst Sci 21(2):163–173. doi:10.1177/0963662511403876

Guston DH (1999) Evaluating the first U.S. consensus conference: the impact of the citizens' panel on telecommunications and the future of democracy. Sci Technol Hum Values 24(4):451–482. doi:10.1177/016224399902400402

Hart Research Associates (2008) Awareness of and attitudes toward nanotechnology and synthetic biology: a report of findings. http://www.nanotechproject.org/process/assets/files/7040/final-synbioreport.pdf. Accessed 3 Aug 2015

Hart Research Associates (2009) Nanotechnology, synthetic biology, and public opinion. http://www.wilsoncenter.org/sites/default/files/nano_synbio.pdf. Accessed 17 June 2015

Hart Research Associates (2010) Awareness and impressions of synthetic biology: a report of findings. http://www.synbioproject.org/library/publications/archive/6456. Accessed 16 Sept 2013

Hart Research Associates (2013) Awareness and impressions of synthetic biology: a report of findings. http://www.wilsoncenter.org/sites/default/files/synbiosurvey2013_0.pdf. Accessed 3 Aug 2015

Hommel M (2008) The future of artemisinins: natural, synthetic or recombinant? J Biol 7(10):38

Irwin A, Jensen TE, Jones KE (2013) The good, the bad and the perfect: criticizing engagement practice. Soc Stud Sci 43(1):118–135. doi:10.1177/0306312712462461

Jin L, Walker AS, Fu G, Harvey-Samuel T, Dafa'alla T, Miles A, Marubbi T, Granville D, Humphrey-Jones N, O'Connell S, Morrison NI, Alphey L (2013) Engineered female-specific lethality for control of pest lepidoptera. ACS Synth Biol 2:160–166

Kahan DM, Braman D, Mandel GN (2009) Risk and Culture: Is Synthetic Biology Different? Harvard law school program on risk regulation research paper no 09-2; Yale Law School, Public law working paper no 190

Kaiser M (2012) Commentary: looking for conflict and finding none? Pub Underst Sci 21(2):188–194. doi:10.1177/0963662511434433

Keasling J (2009) Synthetic biology in pursuit of inexpensive, effective, anti-malarial drugs. BioSocieties 4(2–3):275–282. doi:10.1017/S1745855209990147

Kelle A (2009) Synthetic biology and biosecurity. EMBO Rep 10(S1):S23–S27

Kitney R, Freemont P (2012) Synthetic biology—the state of play. FEBS Lett 586(15):2029–2036

Kronberger N (2012) Synthetic biology: taking a look at a field in the making. Pub Underst Sci 21(2):130–133. doi:10.1177/0963662511426381

Kronberger N, Holtz P, Kerbe W, Strasser E, Wagner W (2009) communicating synthetic biology: from the lab via the media to the broader public. Syst Synth Biol 3(1–4):19–26. doi:10.1007/s11693-009-9031-x

Kronberger N, Holtz P, Wagner W (2012) Consequences of media information uptake and deliberation: focus groups' symbolic coping with synthetic biology. Pub Underst Sci 21(2):174–187. doi:10.1177/0963662511400331

Krueger RA, Casey MA (2009) Focus groups: a practical guide for applied research, 4th edn. Sage Publications, Los Angeles

Kuckartz U (2012) Qualitative Inhaltsanalyse: Methoden, Praxis, Computerunterstützung. Beltz Juventa, Weinheim

Lehmkuhl M (2011) Die Repräsentation der synthetischen Biologie in der deutschen Presse. Abschlussbericht einer Inhaltsanalyse von 23 deutschen Pressetiteln. Deutscher Ethikrat, Berlin. http://www.ethikrat.org/dateien/pdf/lehmkuhl-studie-synthetische-biologie.pdf. Accessed 15 June 2015

Liamputtong P (2011) Focus group methodology: principle and practice. Sage, London

Lin AT (2011) Technology assessment 2.0: revamping our approach to emerging technologies. Brooklin Law Rev 76(4):1–62

Marris C, Rose N (2010) Open engagement: exploring public participation in the biosciences. PLoS Biol 8(11):e1000549

Mayring P (2008) Qualitative Inhaltsanalyse: Grundlagen und Techniken, 10th edn. Beltz, Weinheim

McHughen A (2007) Public perceptions of biotechnology. Biotechnol J 2(9):1105–1111. doi:10. 1002/biot.200700071

Mohr A, Raman S (2012) Representing the public in public engagement: the case of the 2008 UK stem cell dialogue. PLoS Biol 10(11):e1001418

Nanz P, Fritsche M, Isaak A, Hofmann M, Lüdemann M (2010) Verfahren und Methoden der Bürgerbeteiligung. In: Bertelsmann Stiftung (ed) Politik beleben, Bürger beteiligen: Charakteristika neuer Beteiligungsmodelle. Gütersloh, pp 6–49

Nature Biotechnology (2009) What's in a name? Nature Biotechnol 27(12):1071–1073

Navid EL, Einsiedel EF (2012) Synthetic biology in the science café: what have we learned about public engagement? J Sci Commun 11(4):1–9

Nuffield Council on Bioethics (2012) Emerging biotechnologies: technology, choice and the public good

OECD Royal Society (2010) Symposium on opportunities and challenges in the emerging field of synthetic biology. Synthesis report

Pardo R, Engelhard M, Hagen K, Jørgensen RB, Rehbinder E, Schnieke A, Szmulewicz M, Thiele F (2009) The role of means and goals in technology acceptance. A differentiated landscape of public perceptions of pharming. EMBO Rep 10(10):1069–1075. doi:10.1038/embor.2009.208

Pauwels E (2009) Review of quantitative and qualitative studies on U.S. public perceptions of synthetic biology. Syst Synth Biol 3(1–4):37–46. doi:10.1007/s11693-009-9035-6

Pauwels E (2013) Public understanding of synthetic biology. Bioscience 63(2):79–89. doi:10. 1525/bio.2013.63.2.4

Pearson B, Snell S, Bye-Nagel K, Tonidandel S, Heyer L, Campbell AM (2011) Word selection affects perceptions of synthetic biology. J Biol Eng 5(1):9

Presidential Commission for the Study of Bioethical Issues (2010) New directions: the ethics of synthetic biology and emerging technologies. Washington, DC

Rehbinder E, Engelhard M, Hagen K, Jørgensen RB, Pardo-Avellaneda R, Schnieke A, Thiele F (2009) Pharming: promises and risks of biopharmaceuticals derived from genetically modified plants and animals, vol 35. Ethics of science and technology assessment. Springer, Berlin

Ro D-K, Paradise EM, Ouellet M, Fisher KJ, Newman KL, Ndungu JM, Ho KA, Eachus RA, Ham TS, Kirby J, Chang MCY, Withers ST, Shiba Y, Sarpong R, Keasling JD (2006) Production of the antimalarial drug precursor artemisinic acid in engineered yeast. Nature 440(7086):940–943. doi:http://www.nature.com/nature/journal/v440/n7086/suppinfo/nature04640_S1.html

Rowe G, Frewer LJ (2005) A Typology of public engagement mechanisms. Sci Technol Hum Values 30(2):251–290. doi:10.1177/0162243904271724

Royal Academy of Engineering (2009) Synthetic biology: public dialogue on synthetic biology. www.raeng.org.uk/synbiodialogue. Accessed 02 Apr 2015

Ruder WC, Lu T, Collins JJ (2011) Synthetic biology moving into the clinic. Science 333 (6047):1248–1252. doi:10.1126/science.1206843

SCENIHR, SCCS, SCHER (2014) Opinion on synthetic biology I, Definition. http://ec.europa.eu/health/scientific_committees/emerging/docs/scenihr_o_044.pdf. Accessed 17 Dec 2014

Schmidt M (ed) (2012) Synthetic biology: industrial and environmental applications. Wiley, Weinheim

Schmidt M, Torgersen H, Ganguli-Mitra A, Kelle A, Deplazes A, Biller-Andorno N (2008) SYNBIOSAFE e-conference: online community discussion on the societal aspects of synthetic biology. Syst Synth Biol 2(1–2):7–17. doi:10.1007/s11693-008-9019-y

Schön DA, Rein M (1994) Frame reflection: toward the resolution of intractable policy controversies. Basic Books, New York

Seifert F (2002) Gentechnik - Öffentlichkeit - Demokratie: Der österreichische Gentechnik-Konflikt im internationalen Kontext. Profil-Verlag, München, Wien

Seifert F (2003) Demokratietheoretische Überlegungen zum österreichischen Gentechnik-Konflikt. SWS-Rundschau (1/2003):106–128

Service RF (2011) Algae's second try. Science 333(6047):1238–1239. doi:10.1126/science.333. 6047.1238

Sheedy A, MacKinnon MP, Pitre S, Watling J (2008) Handbook on citizen engagement: beyond consultation. Canadian Policy Research Networks Inc. Ottawa. http://cprn.org/documents/ 49583_EN.pdf. Accessed 23 Aug 2013

Stilgoe J, Lock SJ, Wilsdon J (2014) Why should we promote public engagement with science? Pub Underst Sci 23(1):4–15. doi:10.1177/0963662513518154

Stirling A (2008) "Opening up" and "closing down": power, participation, and pluralism in the social appraisal of technology. Sci Technol Hum Values 33(2):262–294. doi:10.1177/ 0162243907311265

Stirling A (2012) Opening up the politics of knowledge and power in bioscience. PLoS Biol 10(1): e1001233

Tait J (2009) Upstream engagement and the governance of science. EMBO Rep 10(S1):S18–S22

Tait J (2012) Adaptive governance of synthetic biology. EMBO Rep 13(7):579

Thomas DD, Donnelly CA, Wood RJ, Alphey LS (2000) Insect population control using a dominant, repressible, lethal genetic system. Science 287(5462):2474–2476. doi:10.1126/ science.287.5462.2474

Torgersen H (2009) Synthetic biology in society: learning from past experience? Syst Synth Biol 3 (1–4):9–17. doi:10.1007/s11693-009-9030-y

Torgersen H, Hampel J (2012) Calling controversy: assessing synthetic biology's conflict potential. Pub Underst Sci 21(2):134–148. doi:10.1177/0963662510389266

Torgersen H, Schmidt M (2013) Frames and comparators: how might a debate on synthetic biology evolve? Futures 48(100):44–54. doi:10.1016/j.futures.2013.02.002

United Nations Department of Economic and Social Affairs (2011) Guidelines on citizens' engagement for development management and public governance. http://unpan1.un.org/ intradoc/groups/public/documents/un-dpadm/unpan045265.pdf. Accessed 23 Aug 2013

van den Belt H (2009) Playing God in Frankenstein's footsteps: synthetic biology and the meaning of life. Nanoethics 3(3):257–268. doi:10.1007/s11569-009-0079-6

Wang B, Wang J, Zhang W, Meldrum DR (2012) Application of synthetic biology in cyanobacteria and algae. Front Microbiol 3 doi:10.3389/fmicb.2012.00344

Weber W, Fussenegger M (2009) The impact of synthetic biology on drug discovery. Drug Discov Today 14(19–20):956–963. doi:10.1016/j.drudis.2009.06.010

Westfall PJ, Pitera DJ, Lenihan JR, Eng D, Woolard FX, Regentin R, Horning T, Tsuruta H, Melis DJ, Owens A, Fickes S, Diola D, Benjamin KR, Keasling JD, Leavell MD, McPhee DJ, Renninger NS, Newman JD, Paddon CJ (2012) Production of amorphadiene in yeast, and its conversion to dihydroartemisinic acid, precursor to the antimalarial agent artemisinin. Proc Natl Acad Sci 109(3):E111–E118. doi:10.1073/pnas.1110740109

Wickson F, Delgado A, Kjolberg KL (2010) Who or what is 'the public'? Nat Nano 5(11):757– 758

Wynne B (2006) Public engagement as a means of restoring public trust in science—hitting the notes, but missing the music? Community Genet 9(3):211–220

Synthetic Biology in the Press

Media Portrayal in Sweden and Italy

Mirko Ancillotti and Stefan Eriksson

1 Introduction

1.1 The Role of the Public

Synthetic biology is an emerging field, still in its infancy in light of difficulties in defining it (Arkin et al. 2009), legal disputes about intellectual ownership (Nelson 2014), and calls for regulation before it is given free reign (Schmidt 2008; Kelle 2013). The Global Network of Science Academies has recently issued a positional statement in which they acknowledge the need for specific regulation, encourage the dissemination of guidelines, and call for assuming scientific responsibility. A point of special interest is the importance assigned to the public: science outreach and public engagement are heavily promoted (Global Network of Science Academies 2014).

It is a trend that technology assessment should involve the public. This seems particularly true in the case of synthetic biology; given the potential the field holds to affect everyone's life. When scientists reach out to society, however, there is a risk for spinning (i.e. giving a biased view of) anticipated results or future applications. Andrew D. Ellington, Professor of Molecular Biosciences, has observed that "synthetic biology's key utility is to excite engineers, undergraduates and funding agencies" (Arkin et al. 2009). This-coupled with some ethicists focussing on anticipative or even speculative ethics (of what might come to be) and media focussing on drama—can put the public and policy makers at a disadvantage

M. Ancillotti (✉) · S. Eriksson
Centre for Research Ethics and Bioethics, Uppsala University, Uppsala, Sweden
e-mail: Mirko.Ancillotti@crb.uu.se

S. Eriksson
e-mail: Stefan.Eriksson@crb.uu.se

© Springer International Publishing Switzerland 2016
K. Hagen et al. (eds.), *Ambivalences of Creating Life*, Ethics of Science
and Technology Assessment 45, DOI 10.1007/978-3-319-21088-9_7

141

regarding their ability to properly assess and its possible applications. Therefore, the way the public is addressed and the way synthetic biology is made addressable are key factors for public involvement.

Some of the rather problematic, albeit central, issues in technology assessment are then how and to which extent the public should be involved in it. Two main impulses can be recognized: on the one hand a tendency to rely on experts' analyses, on the other the drive for public deliberation, i.e. to include the views of the public and social interests in the determination of the path of science and technology (Hennen 2013). Ideally, public engagement facilitates that particularly sensitive scientific research and fields develop in accordance with public interest and in a way that makes sense of common moral intuitions.

The strive for public engagement is not immune from criticism and, as remarked by Richard A.L. Jones, there will always be resistance to public engagement influencing the process of setting priorities (Jones 2014). The reasons for resistance can be many. Jones pinpoints three political reasons. First, it can be assumed that politics and science are separate spheres and that the scientists providing advice are reliable while external opinions are expressions of non-objective and biased positions. Second, the idea of the engaged public influencing policy is contrary to representative democracy, since they are not answerable to Parliament. Third and foremost, the free market might be considered a better way to aggregate public preferences about new technologies (Jones 2014).

There is also resistance rooted in the individual integrity of citizens. There might be a risk that both governments and lobbying groups, in their attempt to influence and ameliorate things, become "oppressors" of others and make them feel obliged to embrace a specific view (Hansson 2008). Sociological studies have pointed to the fact that public engagement can lead to a hasty acceptance and justification of new technologies or research programs (Irwin 2001; Árnason 2012).

Nonetheless, there has been an increase in the last 10–15 years of calls for more public engagement. This tendency can be spotted among scholars (e.g., Hennen 2013; Wareham and Nardini 2013), professional societies (e.g., Global Network of Science Academies 2014), funding programmes (e.g., Horizon 2020), and governmental organizations (e.g., Synthetic Biology Roadmap Coordination Group (SBRCG) 2012).[1]

Science outreach is mostly beneficial for scientists (Bentley and Kyvik 2011). The reason for this is that science outreach not only fills a perceived knowledge gap or enhances citizens' scientific literacy, but legitimates the research. Adrian Mackenzie recognizes two ways in which synthetic biology can be furthered by appeals to the public (Mackenzie 2013). First, by scientists announcing that their research is "momentous and vital" (exemplified by J. Craig Venter), second, by scientists including in "doing" science the task of rendering it more accessible or

---

[1]For a deeper analysis of the role of public engagement in the assessment of synthetic biology see Seitz, this volume.

interesting to the public. The difference lies not merely in a more or less hyperbolic communication style, but in the extent to which and the way one conceives of science as a social enterprise.

The investigation of the ethical and social implications of synthetic biology can benefit from empirical data about the public perception of the field. There are basically two ways to conceive of this task; either to investigate and try to measure the public reception, for example by conducting group interviews (e.g., see Steurer in this volume) or by analyzing how information is presented to the public. The present work is based on the latter strategy and studies the media portrayal of synthetic biology.

1.2 The Role of the Media

The media can be considered the primary arena in the selection both of which issues to bring forth and of the form for bringing them to the attention of the public, decision makers, and interest groups (Nisbet et al. 2003). According to Dorothy Nelkin, media do not only frame issues to be served as news to the recipients, they also frame social relationships and shape the public consciousness on science events (Nelkin 2001). According to some authors, these are indeed the *effects* of media communicating science and technology to a wide audience, but this should not be confused with media's primary function, which is to set agendas and bring issues to the public attention (McCombs and Shaw 1972; Nisbet et al. 2003). Media do have social impact but an undetermined one, as they attract the attention of a non-committed, fragmented, busy audience that is looking for entertainment (Dunwoody 1987). Reporters usually work under deadline pressure and deal with complex issues. Much of the content of their stories depends on the way their sources provide the information to them (Kruvand 2012). Let's take the case of press releases; journalists should use them to attain knowledge about a certain scientific development and communicate it to news consumers, but they are also consciously used by researchers to attract media and funding bodies' attention to positive results of their research (Yavchitz et al. 2012).

The media should not be given a role that they do not and should not have; to wit, to educate citizens, but they can surely be helpful in attracting the attention of a broad audience on important scientific and technological issues, such as synthetic biology. In this they can give a more or less adequate picture of the issue and its consequences and thus shape future deliberations. It is therefore interesting to investigate how an issue such as synthetic biology has been portrayed and from where depictions find their substance.

The present chapter presents empirical data on the relationship between the media (the daily press) and synthetic biology and investigates how the public of two countries, Italy and Sweden, are faced with this new field.

2 Aims and Research Questions

The overall aim of the study is to investigate how the media have been portraying synthetic biology to the public in the light of the idea that mass media contribute not only to informing the public but can also contribute to shape ideas about the issues they write about.

In order to understand how the media in the examined countries have been portraying synthetic biology, and—as a consequence of that—what news consumers have been told about it, the following research questions were formulated:

- What were the reasons for coverage?
- What figures of speech were recurrent and what were the most used framing words?
- Are there notable differences in how synthetic biology was covered in Sweden and Italy?
- How was synthetic biology described?
- What were the featured risks and benefits?
- Was public engagement promoted?
- Does the press coverage mirror the contents of the academic debate?

3 Materials

Three major Swedish and Italian newspapers were analyzed. The press was chosen over other kinds of media as print media are easily accessible and the tools and methods for analyzing them are consolidated. In addition, the newspapers that were chosen can be considered newspapers of record, which means that they are not tabloid, they are not only entertaining, and although their readers are not particularly committed to deal with issues of science and technology, they can still be considered more critical than the recipients of many other media. Adopting a different terminology, the selected newspapers can be defined in their countries as *elite media*; they are those kinds of media capable of setting the frameworks into which other media operate (Chomsky 1997). Also, the audience of a mainstream and traditional medium such as newspapers is quite broad and probably more representative of the lay public than the audience of alternative and new media. The audience of, e.g., a scientific blog need to put in more effort to get and to stay in touch with its preferred media outlets and thus it represents an already attentive public.[2]

---

[2]The lay public is here used to describe people, including scientists, who are no experts in the field. An attentive public is "the part of the general community already interested in (and reasonably well-informed about) science and scientific activities" (Burns et al. 2003).

This analysis concerns two European countries that are quite different in terms of cultural roots, social dynamics, media freedom,[3] and probably also regarding attitudes towards science and technology.[4] What they have in common is that they have not, or have only marginally, been considered in previous studies on synthetic biology public reception, and neither Sweden nor Italy has proceeded with a structured involvement or engagement of its citizens with regard to synthetic biology. Thus, the media coverage of synthetic biology in these countries has probably not been influenced by any political agenda promoting science outreach or public engagement. This is in contrast with, for example, the UK, where there are important governmental programs to inform and engage the public (Bhattachary et al. 2010; SBRCG 2012), or with Germany, where there is a considerable amount of bottom up public participation on biotechnology, since research in life sciences is deemed particularly ethically sensitive (Gloede and Hennen 2002; Peters et al. 2007; Hansen 2010).

The three largest (by circulation) paid-for newspapers in Sweden and Italy from January 1, 2009 to December 31, 2013 were considered. The data about print media circulation were obtained from TS Mediefakta for Sweden[5] and Accertamenti Diffusione Stampa for Italy.[6] The Swedish newspapers are *Dagens Nyheter*, *Svenska Dagbladet*, and *Göteborgs-Posten* (all in Swedish) and the Italian are *Corriere della Sera*, *la Repubblica*, and *il Sole 24 Ore* (all in Italian). Both printed and online versions were considered.

4 Methods

The present work was designed and conducted as a qualitative content analysis following Mayring (2000). This research method was preferred mainly for the reason of completeness and because it is suitable for answering several different kinds of research questions (Bryman 2006).

Articles were retrospectively collected using the media databases *Mediearkivet* and *PressText* and through the archives of each newspaper. Articles were found using search terms, which were selected deductively from the scientific literature (Cserer and Seiringer 2009; Gschmeidler and Seiringer 2012; Pauwels et al. 2012) and inductively (Table 1). In the search were also included the names of renowned

---

[3]Freedom—from 2009 to 2013—ranks Italy as a nation whose press is defined as "partially free" and Sweden as "free" (Freedom House 2009, 2010, 2011, 2012, 2013). The same conclusions are supported by the yearly index from Reporters Without Borders: http://en.rsf.org/. Accessed 05 Jun 2015.

[4]The 2013 Eurobarometer on the extent to which European citizens feel well informed about developments in science and technology ranked Sweden very high, with a score of 61 %, and Italy very low, with a score of 29 % (European Commission 2013).

[5]TS Mediefakta: http://www.ts.se. Accessed 05 Jun 2015.

[6]Diffusione Stampa: http://www.adsnotizie.it/. Accessed 05 Jun 2015.

Table 1 Search terms, ordered by the best hit rate

Search terms	
Terms	Scientists
[synthetic biology]	Venter Craig (JCVI, US)
[artificial OR synthetic] life	Keasling Jay (UC Berkeley. US)
[artificial OR synthetic] bacterium	Church George (Harvard University, US)
[artificial OR synthetic] DNA	Luisi Pier Luigi (Roma Tre, Italy)
[artificial OR synthetic] cell	Endy Drew (Stanford University, US)
[artificial OR synthetic] protein	Collins James (Harvard University, US)
artemisinin	Stano Pasquale (Roma Tre, Italy)
designer AND organism	Benner Steven (FfAME, US)
[artificial OR synthetic] virus	Chen Bor-Sen (NTHU, Taiwan)
bioterrorism	Forster Anthony (Uppsala University, Sweden)
iGEM	Fussenegger Martin (ETH Zürich, Switzerland)
biobrick	Knight Tom (MIT, USA)
minimal [organism OR genome]	Larsson Christer (Chalmers, Sweden)
bioengineer	Mansy Sheref (University of Trento, Italy)
[artificial OR synthetic] gene	Nielsen Jens (Chalmers, Sweden)
[artificial OR synthetic] genome	Silver Pamela (Harvard University, US)
XNA	Smolke Christina (Stanford University, US)
biosafety	Weber Wilfried (Freiburg University, Germany)
biosecurity	Weiss Ron (MIT, US)

international scientists, selected deductively from the literature and inductively, including three prominent scientists active in Sweden and three in Italy (Table 1) (Oldham et al. 2012; Mackenzie 2013).

Inclusion criteria were the following: all newspaper articles that included a search term and had even a slight connection to synthetic biology as a subject were considered. The absence of the term "synthetic biology" did not represent a discriminating factor. The relevance of the stories found through the search was instead rated on the basis of how much the content engaged with synthetic biology as it has been defined by Benner and Sismour (2005), the Royal Academy of Engineering (2009), and the Global Network of Science Academies (2014).

Each article was read twice. After the first reading articles were formally coded, and after the second reading they were content coded, as detailed below. Both deductive and inductive development categories were applied. Thus, new issues (primarily new topics and metaphors) that recurred in many articles were added to a set of pre-identified categories.

Articles were formally coded with regard to their date of publication, length, and media type. Word counting was performed using QSR International's NVivo 10 software, and three categories were applied: short (0–299 words), medium (300–999 words), and long (1000 words or more). Three main article types were identified: news, feature articles, and—as one type—editorials, columns, and opinion pieces (ECOs).

Content wise, articles were divided into three categories according to the extent to which synthetic biology was central to them: weak, medium, or strong. Articles barely mentioning synthetic biology or giving it just a few words were subsumed under weak connection stories. Medium and strong categories were those in which synthetic biology was given moderate or extensive space, respectively.

For each article it was considered whether the reason for coverage was a specific event, such as a conference or a scientific publication, or whether the story was about an issue or a certain argument. Articles were also put into four categories according to the narrative: thematic (T), episodic (E), thematic with episodic discussions (TE), and episodic with thematic discussions (ET). The thematic narrative approaches issues in a general context and focuses on long-term outcomes; the episodic refers to case studies, events, and focuses on concrete outcomes; thematic with episodic discussions were the articles whose narratives were mainly thematic, but involved episodic parts; episodic with thematic discussions were the articles whose narratives were mainly episodic, but involved thematic parts (Iyengar 1991; Morgan 2002).

Another aspect investigated concerned the clarity with which synthetic biology was described and whether its characterization was clear enough to enable recipients to distinguish it from other biotechnologies. From this analysis newspaper articles with a weak connection to synthetic biology were excluded. Four categories were assigned: clear, not clear, misleading, and missing (when it was not possible to assess the clarity of the description; in most cases, due to its absence).

A further topic for investigation concerned the language. QSR International's NVivo 10 software was used for producing a frequency word list from all of the articles. A set of framing key words was then selected from the most frequently recurring words employed by the journalists to describe synthetic biology. Besides the framing key words, the type of metaphors used was noted. In this context, the overall tone of the article was also assessed, with the intention of describing the general normative impression given by the articles. The following labels were assigned to each article: positive, neutral, negative, skeptical, or cautious.

Other aspects investigated were the topics of the articles, as well as risks and benefits thought to be related to these. Several topics might have occurred in a single article, thus they are not mutually exclusive. In order to describe the way specific topics were communicated, an evaluative label was assigned to how each topic was portrayed: as positive, neutral negative, or absent. Only articles that had a medium or strong connection to synthetic biology were considered in this particular analysis.

The last aspect investigated concerned the number of times calls for oversight or public interest or public engagement were mentioned (even if only in passing) in the articles.

5 Results

Between 2009 and 2013 the Swedish newspapers considered in this study covered synthetic biology in 36 articles and the Italian ones in 95 articles (131 in total). The percentage of articles weakly connected to synthetic biology was in both countries about 17 % (Table 2). With 65 articles (50 % of the total), 2010 was the most

Table 2 Articles on synthetic biology in Sweden and Italy from 2009 to 2013: strength of connection with synthetic biology

	Weak		Medium		Strong		Total #
	(#)	(%)	(#)	(%)	(#)	(%)	
Italy							
Corriere della Sera	8	27	7	23	15	50	30
Repubblica	2	6	11	30	23	64	36
Sole 24 Ore	7	24	11	38	11	38	29
Tot	17	18	28	30	49	52	95
Sweden							
Dagens Nyheter	3	16	7	37	9	47	19
Svenska Dagbladet	3	30	3	30	4	40	10
Göteborgens-Posten	0	0	2	29	5	71	7
Tot	6	17	12	33	18	50	36
Italy and Sweden	23	18	41	31	67	51	131

productive year (Fig. 1). It is noteworthy that 33 out of the 65 stories were issued between May 20th and May 22nd, 2010. They represent 31 % of all the articles that had medium or strong connection to synthetic biology, and all of them relate to the Venter group's *Science* publication (Gibson et al. 2010) detailing their success in transplanting a synthetic genome into a recipient cell. The length of the articles in Sweden was distributed as follows: 22 % short, 64 % medium, and 14 % long. In Italy the result was comparable: 16 % short, 71 % medium, and 13 % long. The article types were distributed as follows: in Sweden 19 % news, 53 % feature articles, and 28 % editorials, columns, and opinion pieces (ECOs), while Italy had 29 % news, 51 % feature articles, and 20 % ECOs.

Driving reasons for coverage in both countries were events (publications, press releases, conferences, etc.) rather than issues (synthetic biology itself or issues related to it), with a score of 67 % in Sweden and 69 % in Italy. Concerning the

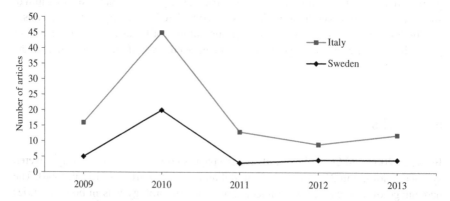

Fig. 1 Press coverage of synthetic biology in Sweden and Italy from 2009 to 2013

narrative, Sweden had 31 % thematic, 8 % episodic, 39 % thematic with episodic discussions, and 22 % episodic with thematic discussions articles, while Italy had 20 % thematic, 33 % episodic, 24 % thematic with episodic discussions, and 23 % episodic with thematic discussions articles.

With regard to the clarity with which synthetic biology was described, in Sweden, 33 % of the descriptions were categorized as clear, 6 % as not clear, 3 % as misleading, and 58 % as missing. In Italy, 34 % were clear, 18 % not clear, 3 % misleading, and 45 % missing. The term "synthetic biology" appeared in 25 % of the Swedish articles and in 46 % of the Italian articles.

The language of Italian articles copiously resorted to models and metaphors to explain issues, but also to express opinions. Although this is an aspect hardly quantifiable, it can be said that approximately 60 % of the stories with medium or strong connection to synthetic biology made heavy use of such stylistic devices. The most common were metaphors of computers and software, creativity, construction, and machines. Swedish stories adopted a more sober language in communicating synthetic biology; stylistic devices were found in about the 25 % of the stories with medium and strong connection to synthetic biology. In Sweden the most commonly used metaphors were those of religion and design. Expressions like "made/created artificial life" were equally pervasive in Sweden and Italy. One of the preferred expressions in Sweden was "milestone in the history of biology/science" with reference to the "creation" of the first synthetic cell. Indeed, there was lot of coverage and great value given to this event. Nonetheless, one of the major preoccupations in many of the Swedish and Italian articles was to not overestimate its scientific or ethical importance.

In the Swedish stories the most used framing key words were *bacteria, creation,* and *artificial,* while in the Italian stories they were *bacteria, synthetic,* and *artificial* (see Fig. 2). It is noteworthy that in both countries the most recurring term was *Venter* with about 50 % more occurrences than the otherwise most used framing key words. Venter was not considered a framing key word because it is not a term used to describe synthetic biology, but the finding nevertheless gives an idea of what journalists wrote about the field.

The overall normative tone of the articles was neutral to positive in both countries (differences between the two countries were negligible): 33 % were positive, 4 % negative, 52 % neutral, 5 % skeptical, and 6 % cautious. In this count weakly connected articles were also included because it was deemed useful to know in which context slight references to synthetic biology emerged as well.

Table 3 shows the ten most discussed topics in the Swedish and Italian newspaper articles from 2009 to 2013.

The major benefits envisioned in the articles of both countries concern *the environment* (depollution, bioremediation): Italy 24 % and Sweden 13 % of articles; *production of energy* (biofuels): Italy 20 % and Sweden 22 %; *healthcare related improvements* (vaccines, pharmaceutical products): Italy 3 % and Sweden 22 %; and lastly *economic*: Italy 4 % and Sweden 8 %. These percentages refer to openly positive considerations of what synthetic biology is expected to contribute, not merely to the fact that synthetic biology may or can find an application in these fields.

Fig. 2 Specific framing key words mentioned in Italian and Swedish newspaper articles from 2009 to 2013; frequencies are expressed as a percentage of the highest word recurrence for each country, 64 for Sweden and 190 for Italy, respectively

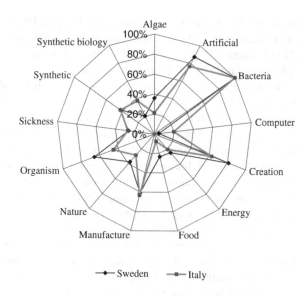

Table 3 The ten most discussed topics in the Italian and Swedish newspaper articles (total n = 95 and 36, respectively) from 2009 to 2013, differentiated by their normative tone. Note than one article might include several topics

Topic	Positive		Negative		Neutral		Mention		Total
	Italy	Sweden	Italy	Sweden	Italy	Sweden	Italy	Sweden	It and Sw
Healthcare	3	8	1	0	1	1	37	10	61
Environment	23	5	1	1	10	5	6	2	53
Energy/biofuels	19	8	1	0	5	6	10	3	52
Ethics	1	0	3	0	19	10	7	3	43
Economy	4	3	2	1	10	1	8	9	38
Biohazard	1	0	0	0	10	3	4	1	19
Religion	0	0	0	0	10	2	4	1	17
Food/GMO	4	0	0	0	1	0	5	7	17
Agriculture	6	0	0	0	5	3	2	0	16
Research Ethics	0	0	2	0	9	0	3	1	15

The major risks mentioned were *biohazard* (accidental release of pathogens): Italy 15 % and Sweden 11 % of articles; and *bioterrorism*: Italy 4 % and Sweden 5 %. It is noteworthy that none of the articles considering these risks displayed a negative stance towards synthetic biology because of them; the risks were mentioned just as possible issues.

Calls for oversight (direct, or a report of it, or just a mention of this issue) were seen in 13 % of the articles while a mention of or promotion of public engagement occurred in 7 % of the articles.

6 Discussion

Swedish and Italian press coverage of synthetic biology in the period from January 2009 to December 2013 differed only sporadically in terms of articles' length, types, and the strength of connection to synthetic biology. However, the countries' press coverage markedly differed in quantity: Italian newspapers covered the theme approximately three times more than Swedish newspapers throughout the period. One possible reason could have been different financial resources. As the three largest newspapers by circulation of each country were considered this can hardly be the sole explanation for such a marked difference in coverage. More likely, the difference is caused by editorial interests or agenda-settings. The high level of social trust typical of Nordic countries may also have contributed to the particularly low presence of synthetic biology in Swedish newspaper articles (Delhey and Newton 2005). This high level of trust, which can be extended to science and technology as well, might have induced Swedish reporters and editors to not pay greater attention to synthetic biology because it was not perceived as particularly dangerous or controversial.

It is currently not possible to properly assess whether the press coverage of SB in Sweden and Italy was substantially different from other countries or not. This is due to a lack of basis for direct comparison. The few existing studies cover different periods and countries, and are differently designed. Pauwels et al. (2012) studied press coverage in the US and some European countries (including Italy, but not Sweden) from 2008 to 2011. They found no significant difference between Italy and other countries. Although the experimental design of the study was different from ours, this does suggest that the Swedish press may have under-covered synthetic biology. Overall, synthetic biology has hitherto not found much media interest; only sporadically it has gained resonance. The main drivers for the attention of the media in both countries were prominent events. Craig Venter undoubtedly represented the major catalyst of media attention; he was mentioned in 68 % of the articles that were medium or strongly connected to synthetic biology. These two aspects, event driven coverage and the massive attention on one scientist (often controversially described) embodying the field, suggest that the role of synthetic biology in media coverage is, according to a trend in science popularization described by Burnham (1987), that of a media commodity, easily replaceable with the next big thing from the life sciences.

One of the aims of this study was to discuss the results of the analysis in the light of the idea that mass media contribute not only to informing the public but also to the shaping of ideas about a number of issues (Scheufele 1999; Valkenburg et al. 1999; Kronberger et al. 2012). Media frame issues and thus influence the opinions of the public by underscoring specific facts or values and providing interpretation schemes (Nelson et al. 1997; Scheufele 1999). The most frequent framing key words give an idea of the way synthetic biology was depicted to the public: one main subject are bacteria, on which—or starting from which—some creational or manufacturing procedures are performed, and the bacteria are described either as artificial or as synthetic. The framing key words and, more specifically, the description of synthetic biology found in most articles were appropriate to depict to the public what synthetic biology is about without excessive hypes.

It can be considered an established fact that media resort to figurative language in describing scientific and technological contents and this has also been seen in studies about two of the most recent emerging issues, nanotechnologies and synthetic biology (Scheufele and Lewenstein 2005; Hellsten and Nerlich 2011; Gschmeidler and Seiringer 2012). The present study reveals a marked difference between the Italian and Swedish recourse to metaphors, where Italian articles were rich in figures of speech and Swedish articles mostly preferred basic representations. This might be due to the different rhetoric of the two countries or to different ways in which media interpret their role of informing the public about scientific developments.

The pervasiveness of the expression "creation of artificial life", however, was common to the two countries. This is not surprising: on the one hand it represents a *leitmotiv* of media communication of biotechnologies, on the other hand this is the language (design and creation of life, computer as parents, etc.) adopted by many influential scientists communicating with the media. Craig Venter used it in his highly storied press release and speech announcing the "creation" of the first synthetic cell (J. Craig Venter Institute 2010; Venter 2010). What the media did in such cases was simply to echo these words, which make good copy. The obvious point here is that the way the same words are meant and used by scientists on the one hand and by the media and the public on the other hand, can vary (Pauwels 2013).

The relation between Venter and the media appears to be a mutually beneficial relationship. On one side there are the media, interested in new appealing stories, and on the other there is a scientist-entrepreneur (as Venter is often presented) who values publicity. As Marjorie Kruvand has commented on Arthur L. Caplan, whom the media have turned into a sort of bioethics' "Dr. Soundbite"; such experts shape the news by providing stories with comments and context (Kruvand 2012). Venter is synthetic biology's "Dr. Soundbite"; the language he uses and the aspects that he draws attention to heavily contribute to the way media communicate synthetic biology.

We found that media presented synthetic biology as a field with high potential, which is in line with other studies (Gschmeidler and Seiringer 2012; Pauwels et al. 2012). Almost no emphasis on the component of novelty was found. This may be due to the fact that synthetic biology is not perceived as very different from other biotechnologies (Kronberger et al. 2012). That synthetic biology is not clearly distinguished from other fields does not appear to be related to poor media descriptions. It is more likely an effect of the fact that many elements that are interesting about synthetic biology and its applications coincide with the elements that feature in other biotechnologies. Another point of convergence, as pointed out by Gschmeidler and Seiringer (2012), is the vocabulary, as many key words used are part of the basic biotech jargon.

The most frequent topics to which synthetic biology was related in our study were healthcare, the environment, and energy production. These topics are public issues of great interest, and effective communication strategies "[...] necessitate connecting a scientific topic to something the public already values or prioritizes, conveying personal relevance" (Nisbet and Scheufele 2009, p. 1774). However, in the way they are framed and presented in relation to synthetic biology, they actually represent expectations about the positive potential of synthetic biology. As observed by Kronberger et al. (2009), there is a tendency for journalists writing about synthetic biology to focus on its practical applications. In doing so, they run the risk of conveying to readers the notion that synthetic biology is already fulfilling, or will soon fulfill these practical expectations, which is far from reality. Indeed, in relation to human health and to the environment, we saw merely four and one negatively inclined articles, respectively. The remaining articles in our study were neutral or positive regarding bioremediation or new drugs/vaccines, such as the semi-synthetic production of the anti-malaria drug artemisinin. In addition, the tone about ethical and religious issues was neutral, in most cases only mentioning the fact that synthetic biology, as a field or through its applications, may create certain moral tensions. So we can see that the major benefits envisioned in the articles of both countries overlap with the most treated topics; they were emphasized both in quantity (recurrence) and weight (positively presented).

The possible risks in relation to synthetic biology, mainly biohazards and bioterrorism, were only mentioned a few times in the articles, receiving much less consideration, both in quantity and weight (neutrally presented) than the positive topics. In marked difference to the media portrayal, academic studies and policy reports usually display a balanced view and consideration of both benefits and risks involved with synthetic biology. Similarly, the involvement of the public or the need for an oversight of synthetic biology research was very rarely mentioned in the newspapers, although in academic and other settings the debates about the ethical, societal, and legal dimensions of synthetic biology often raise the necessity of involving the public. This tendency can be spotted among social scientists and philosophers (Hennen 2013; Wareham and Nardini 2013), but also professional societies (Global Network of Science Academies 2014), funding programmes (e.g., Horizon 2020), and governmental organizations (SBRCG 2012).

Lastly, the influence of the sources (individual scientists, press releases and papers, etc.) of the articles on how they are shaped must be considered strong. As previously mentioned, Sweden and Italy are two countries rather different in cultural and social terms as well in terms of media freedom and public consideration of science and technology. The marked similarities of language, selection of topics, and risks and benefits envisioned suggest that the common sources of Swedish and Italian stories are responsible for this common framing. Essentially, Swedish and Italian readers have been told the same things about synthetic biology. This signifies at least two things; first, that reporters didn't filter or process substantially the information received from their sources, and second, that those who are the sources can heavily influence the framing of how synthetic biology is communicated to the public and, as a consequence, to a certain degree, the public perception and shaping of ideas about synthetic biology.

7 Conclusions

Synthetic biology has until now not gained a big media resonance in Sweden and Italy. There are not many articles, drivers for attention are mainly events, and the overall impression is that the field is treated as a media commodity. The portrayal of synthetic biology offered to the public is very positive and is that of a biotechnology holding great potential to improve our life at many levels, with only minor risks that relates to malicious external agents or accidental events. While Swedish and Italian newspapers were generally adequate in their choice of language when describing synthetic biology, they were rather unbalanced in the choice and presentation of topics evoked by it.

The differences between the countries were principally quantitative, where Italian papers devoted much more space to synthetic biology. This may be explained by considering the different financial resources of the newspapers and the different attitudes towards science (Swedish trust). The similarities in contents and forms seem to be strongly related to a marked dependence on the way scientists frame their accomplishments and the lack of critical scrutiny on the behalf of the media.

References

Arkin A, Berry D, Church C et al (2009) What's in a name? Nat Biotech 27(12):1071–1073

Árnason V (2012) Scientific citizenship in a democratic society. Public Underst Sci 22(8):927–940

Benner SA, Sismour AM (2005) Synthetic biology. Nat Rev Genet 6(7):533–543

Bentley P, Kyvik S (2011) Academic staff and public communication: a survey of popular science publishing across 13 countries. Public Underst Sci 20(1):48–63

Bhattachary D, Calitz JP, Hunter A (2010) Synthetic biology dialogue. TNS-BMRB Report

Bryman A (2006) Integrating quantitative and qualitative research: how is it done? Qual Res 6 (1):97–113

Burnham JC (1987) How superstition won and science lost: popularizing science and health in the United States. Rutgers University Press, New Brunswick

Burns TW, O'Connor DJ, Stocklmayer SM (2003) Science communication: a contemporary definition. Public Underst Sci 12(2):183–202

Chomsky N (1997) What makes mainstream media mainstream. Z Magazine. October

Cserer A, Seiringer A (2009) Pictures of synthetic biology. Syst Synth Biol 3(1–4):27–35

Delhey J, Newton K (2005) Predicting cross-national levels of trust: global pattern or nordic exceptionalism. Eur Sociol Rev 21(4):311–327

Dunwoody S (1987) Scientists, journalists, and the news. Chem Eng News 65(46):47–49

European Commission (2013) Eurobarometer Responsible Research and Innovation, Science and Technology. Special Eurobarometer 401. Brussels: European Commission

Freedom House (2009) Freedom of the Press 2009. https://freedomhouse.org/sites/default/files/ FOTP%202009%20Full%20Release%20Booklet.pdf. Accessed 05 Jun 2015

Freedom House (2010) Freedom of the Press 2010. https://freedomhouse.org/sites/default/files/ FOTP2010–Final%20Booklet_5May.pdf. Accessed 05 Jun 2015

Freedom House (2011) Freedom of the Press 2011. https://freedomhouse.org/sites/default/files/ FOTP%202011%20Full%20Release%20Booklet.pdf. Accessed 05 Jun 2015

Freedom House (2012) Freedom of the Press 2012. https://freedomhouse.org/sites/default/files/Booklet%20for%20Website_0.pdf. Accessed 05 Jun 2015

Freedom House (2013). Freedom of the Press 2013. https://freedomhouse.org/sites/default/files/FOTP%202013%20Booklet%20Final%20Complete%20-%20Web.pdf. Accessed 05 Jun 2015

Gibson DG, Glass GI, Lartigue C et al (2010) Creation of a bacterial cell controlled by a chemically synthesized genome. Science 329(5987):52–56

Gloede F, Hennen L (2002) A difference that makes a difference? Participatory technology assessment in Germany. In: Joss S, Bellucci S (eds) Participatory technology assessment. European Perspectives. Centre for the Study of Democracy, London, pp 92–107

Global Network of Science Academies (2014) IAP Statement on Realising Global Potential in Synthetic Biology: Scientific Opportunities and Good Governance. IAP Report

Gschmeidler B, Seiringer A (2012) "Knight in shining armour" or "Frankenstein's creation"? The coverage of synthetic biology in German-language media. Public Underst Sci 21(2):163–173

Hansen J (2010) Biotechnology and public engagement in Europe. Palgrave Macmillan, Basingstoke

Hansson MG (2008) The private sphere. An emotional territory and its agent. Springer, Dordrecht

Hellsten I, Nerlich B (2011) Synthetic biology: building the language for a new science brick by metaphorical brick. New Genet Soc 30(4):375–397

Hennen L (2013) Parliamentary technology assessment in Europe and the role of public participation. In: O'Doherty K, Einsiedl E (eds) Public engagement and emerging technologies. UBC Press, Vancouver

Irwin A (2001) Constructing the scientific citizen: science and democracy in the biosciences. Public Underst Sci 10(1):1–18

Iyengar S (1991) Is anyone responsible? How television frames political issues. University of Chicago Publisher, Chicago

J. Craig Venter Institute (2010) First self-replicating synthetic bacterial cell. [Press release] Retrieved from: http://www.jcvi.org/cms/press/press-releases/full-text/article/first-self-replicating-synthetic-bacterial-cell-constructed-by-j-craig-venter-institute-researcher/home/. Accessed 05 Jun 2015

Jones RAL (2014) Reflecting on public engagement and science policy. Public Underst Sci 23 (1):27–31

Kelle A (2013) Beyond patchwork precaution in the dual-use governance of synthetic biology. Sci Eng Ethics 19(3):1121–1139

Kronberger N, Holtz P, Kerbe W et al (2009) Communicating Synthetic biology: from the lab via the media to the broader public. Syst Synth Biol 3(1–4):19–26

Kronberger N, Holtz P, Wagner W (2012) Consequences of media information uptake and deliberation: focus groups' symbolic coping with synthetic biology. Public Underst Sci 21 (2):174–187

Kruvand M (2012) Dr. Soundbite: the making of an expert source in science and medical stories. Sci Commun 34(5):566–591

Mackenzie A (2013) From validating to objecting: public appeals in synthetic biology. Sci Cult 22 (4):476–496

McCombs ME, Shaw DL (1972) The agenda-setting function of mass media. Public Opin Q 36 (2):176–187

Mayring P (2000) Qualitative content analysis. FQS 1(2) Art. 20. http://nbn-resolving.de/urn:nbn:de:0114-fqs0002204. Accessed 05 Jun 2015

Morgan D (2002) A content analysis of media coverage of health care and the uninsured 2002. Frame Works Institute, Washington, DC

Nelkin D (2001) Beyond risk: reporting about genetics in the Post-Asilomar Press. Perspect Biol Med 44(2):199–207

Nelson B (2014) Cultural divide. Nature 509(7499):152–154

Nelson TE, Clawson RA, Oxley ZM (1997) Media framing of a civil liberties conflict and its effect on tolerance. Am Polit Sci Rev 91(3):567–583

Nisbet MC, Brossard D, Kroepsch A (2003) Framing science: the stem cell controversy in an age of press/politics. IJPP 8(2):36–70

Nisbet MC, Scheufele DA (2009) What's next for science communication? Promising directions and lingering distractions. Am J Bot 96(10):1767–1778

Oldham P, Hall S, Burton G (2012) Synthetic biology: mapping the scientific landscape. PLoS ONE 7(4):e34368. doi:10.1371/journal.pone.0034368

Pauwels E (2013) Communication: mind the metaphor. Nature 500(7464):523–524

Pauwels E, Lovell A, Rouge E (2012) Trends in American and European Press coverage of synthetic biology. Synbio 4 (Synthetic Biology Project). Wilson Center

Peters HP, Lang JT, Sawicka M et al (2007) Culture and technological innovation: impact of institutional trust and appreciation of nature on attitudes towards food biotechnology in the USA and Germany. Int J Public Opin Res 19(2):191–220

Royal Academy of Engineering (2009) Synthetic biology: scope, applications and implications. The Royal Academy of Engineering, London

Scheufele DA (1999) Framing as a theory of media effects. J Commun 49(1):103–122

Scheufele DA, Lewenstein BV (2005) The public and nanotechnology: how citizens make sense of emerging technologies. J Nanopart Res 7(6):659–667

Schmidt M (2008) Diffusion of synthetic biology: a challenge to biosafety. Syst Synth Biol 2 (1–2):1–6

Synthetic Biology Roadmap Coordination Group (SBRCG) (2012) A synthetic biology roadmap for the UK. Research Councils UK. Technology Strategy Board, Swindon

Valkenburg PM, Semetko HA, Vreese CHD (1999) The effects of news frames on readers' thoughts and recall. Commun Res 26(5):550–569

Venter CJ (2010) Craig venter: watch me unveil "synthetic life". [Video file] Retrieved from: http://www.ted.com/talks/craig_venter_unveils_synthetic_life#t-38502. Accessed 05 Jun 2015

Wareham C, Nardini C (2013) Policy on synthetic biology: deliberation, probability, and the precautionary paradox. Bioethics 29(2):118–125

Yavchitz A, Boutron I, Bafeta A et al (2012) Misrepresentation of randomized controlled trials in press releases and news coverage: a cohort study. PLOS Medicine 9(9):e1001308. doi:10.1371/journal.pmed.1001308

Let's Talk About… Synthetic Biology—Emerging Technologies and the Public

Stefanie B. Seitz

1 Introduction

Synthetic biology is a relatively young discipline that emerged at the interfaces between molecular biology, biotechnology, organic chemistry, engineering, and informatics/systems biology. It now looks back on about 15 years of rapid development from first initiatives by a few pioneers to a sizeable discipline. Its meetings (especially the BioBricks Foundation Synthetic Biology Conference Series SBx.0) attract hundreds of participants (Way et al. 2014).

The term "synthetic biology" appeared as early as 1912 in a paper by Stéphane Leduc ("La Biologie Synthétique") which did, however, not refer to what it is now. Rather, it may have been the Nobel Prize winning work of Werner Arber, Daniel Nathans, and Hamilton Smith on restriction enzymes in 1978 that "has led us in the new era of synthetic biology" (Szybalski and Skalka 1978, p. 181). In 2000, Eric Kool was one of the first who used the term synthetic biology in a rather "contemporary" understanding (Benner and Sismour 2005). However, a universal and agreed definition of synthetic biology does not yet exists, as is often the case with new and emerging fields of science and technology, and it may actually be useful if the definition remains somewhat open to debate. For example, "boundary objects" that lack clear definition (Guston 2001) may facilitate interdisciplinary work as it is

S.B. Seitz (✉)
KIT, Institute for Technology Assessment and Systems Analysis (ITAS), Karlstraße 11, 76133 Karlsruhe, Germany
e-mail: stefanie.seitz@kit.edu

© Springer International Publishing Switzerland 2016
K. Hagen et al. (eds.), *Ambivalences of Creating Life*, Ethics of Science and Technology Assessment 45, DOI 10.1007/978-3-319-21088-9_8

practiced in synthetic biology. Also, as an "umbrella term" (Rip and Voß 2013) synthetic biology may allow strategic science governance.[1]

One definition of synthetic biology which is often referred to came from the EU NEST High Level Expert Group:

> Synthetic biology is the engineering of biology: the synthesis of complex, biologically based (or inspired) systems which display functions that do not exist in nature. This engineering perspective may be applied at all levels of the hierarchy of biological structures – from individual molecules to whole cells, tissues and organisms. In essence, synthetic biology will enable the design of 'biological systems' in a rational and systematic way. (European Commission 2005, p. 7)

Synthetic biology thus extends beyond mere genetic engineering which is about the *manipulation* of natural organisms and moves on to the *creation* of novel, synthetic life. Thereby the focus lays on *novelty* and *synthetic* in the sense that forms of life are designed from the scratch and made to merely serve their human creators and nothing else—no extravagant metabolism, no lavish variability—just mineralized functional machines made from organic material, so far the wish and vision of the scientists. The focus on novelty has been criticised on the basis that contemporary synthetic biology approaches still use and recombine only what can be found in nature and only along the laws of nature (cf. Fussenegger 2014). Nevertheless, it still holds true that the development of methodology and the knowledge associated with synthetic biology, especially the recent advances in DNA synthesis, the standardization of functional genetic components (the BioBrick approach[2] is a prime example here), and the development of increasingly complex algorithms for the simulation of biological systems (Systems Biology[3]) have increased scientists' ability to design and build robust and predictable biological systems using engineering design principles[4] (Cameron et al. 2014).

Applications based on the principles of synthetic biology are sometimes expected to contribute to the solution of many of the world's most significant challenges. The main fields of application for synthetic biology are [for an elaborated overview see Benner and Sismour (2005) or König et al. (2013)]:

---

[1] In the literature "boundary objects" (Star and Griesemer 1989; Guston 2001) or "umbrella terms" (Rip and Voß 2013) describe terms that are not precisely defined. Therefore, they are vague enough to be used by quite diverse groups or disciplines and allow for interpretations. At the same time they are sufficiently defined to keep some kind of global identity and guarantee a shared understanding.

[2] More information can be found at http://biobricks.org. Accessed 14 Aug 2014.

[3] Systems biology is an interdisciplinary field that combines quantitative data generation and mathematical modelling with biomedical laboratory experiments to elucidate general principles governing the properties of complex systems that give rise to biological functions.

[4] This becomes even more evident in the interpretation of synthetic biology for military use as illustrated in the Official US Department of Defence Science Blog "Armed with Science", where a programme manager of the Defence Advanced Research Projects Agency, Justin Gallivan, states: "By making these systems more robust, stable and safe, BRICS seeks to harness the full range of capabilities at the intersection of engineering and biology". http://science.dodlive.mil/2014/08/12/the-future-of-synthetic-biology-applications. Accessed 14 Aug 2014.

- Energy supply: synthetic biology aims to develop approaches for the production of fuels like biodiesel or bioethanol that could replace petrol and hydrogen (see also Wünschiers in this volume).
- Disease control: synthetic biology aims to contribute to the development of systems for drug discovery, diagnosis of disease as well as production of drugs, active components, and vaccines.
- Environmental protection: synthetic biology aims to supply systems that function as biosensors, chemicals as renewable resources, or even can be used to remediate polluted sites with specially adapted decomposing organisms.
- Military and Space applications are under development as well (see also Verseux et al. in this volume).

Despite brisk research activities only a relatively modest—in terms of its own rhetoric and visions—number of applications are available for economic exploitation to date, and most of the possible applications of synthetic biology are still the promises of some of its protagonists (König et al. 2013). And if and when these applications are feasible is difficult or impossible to predict (cf. Grunwald 2012). However, in the health sector some prominent examples can be found: Sanofi-Aventis, for example, uses reprogrammed yeast for the production of the malaria drug Artemisinin and the Swiss start-up firm Bioversys offers a synthetic biology-based approach for drug discovery that has helped to discover new molecules in order to overcome the antibiotic resistance of several pathogens (cf. Fussenegger 2014). Many other applications are in a testing stage and supposed to reach market maturity in the coming years.

But synthetic biology is not only about applications—much of the research effort in synthetic biology is aimed at the understanding of fundamentals of life: how metabolism or signalling processes function within organisms and how processes are intertwined on a systems level. Top down or in vivo approaches aim to minimize the genome of existing organisms in order to create a tailor-made minimal organism or cell. Therefore, natural structures and processes are reconstructed in an engineering tradition. The bottom up or in vitro approach uses chemical precursors for a de novo synthesis of a minimal cell which then can either replicate nature or is orthogonal to it. Furthermore, this approach aims to build a non-natural genome, e.g. by extending the set of bases or the genetic code (for an overview see Boldt et al. (2009) and also Acevedo-Rocha in this volume).

Thus, synthetic biology is yet mainly basic lab research about fundamental questions of life that seems quite far away from actual applications, except for the above mentioned examples. But with a growing number of applications, consumers and patients will directly face synthetic biology-based uses in their everyday life. Society therefore needs to negotiate on synthetic biology in terms of safety, security, ethics, and justice. In particular, policy makers and stakeholders will have to tackle regulation concerning synthetic biology by anticipating its novelty, complexity, and uncertainties.

In this chapter I will argue why an emerging science as synthetic biology is an issue for Technology Assessment (TA). The assessment of these kinds of technologies challenges TA because it has to overcome knowledge deficits. I will point out that TA addresses this by analyzing public perceptions and engaging into public dialogue. As an example I will give an impression on the status quo of public perception and public debate in Germany. Finally, I will argue that engaging the public has further functions and effects beside contribute to knowledge production to inform governance decisions and will discuss its consequences.

2 Why Technology Assessment for Synthetic Biology?

Technology Assessment in general evaluates all kinds of technologies as well as scientific and science-based technological developments regarding their intended and unintended impacts and consequences for society. In its evaluations, TA aims to reveal the complex interrelationships between technology, science, and society, provide knowledge for responsible design and the societal embedding of new technologies and develop strategies for sustainable development. Problem-oriented research—which refers to the societal need for advice and thus for policy advice and participation in the public debate—is part of the mission of TA. Knowledge production in TA is therefore in close relation to the value dimension (e.g. in the form of ethical analyses or sustainability assessments), has a prospective component, and aims to provide "knowledge for action". Moreover, TA aims to develop strategies to deal with the uncertainties of knowledge and is highly interdisciplinary as well as transdisciplinary through the involvement of non-scientific actors like stakeholders, decision-makers, and citizens (Grunwald 2010).

One special field of TA assesses the so-called "new and emerging science and technologies" (NEST). This umbrella term first appeared within the sixth EU framework programme.[5] The function of TA in the field of NEST is the early detection of scientific and technological developments which may significantly change society in the mid- or long-term. Among the NESTs, synthetic biology is one of the prime examples of the so-called "techno-sciences" in which the traditional boundaries between (knowledge-oriented) natural science and (application-oriented) engineering dissolve and basic scientific research ab initio is placed in a context of utilization (Grunwald 2012, p. 10).

Thus, synthetic biology as such an emerging techno-science is a well suited issue for TA. The early onset of accompanying research and thus the early detection of risks and societal unwanted developments can enable the responsible actors to start interventions in time—although this is a demanding task. And indeed, a substantial amount of work about societal implications and risk analyses, which have been a topic of (scientific) debate since the very beginning of synthetic biology, underlines the importance of TA activities. Moreover, several publications have suggested the

---

[5]More information can be found at http://cordis.europa.eu/fp6/nest.htm. Accessed 14 Aug 2014.

anticipation of possible negative implications (e.g. Boldt et al. 2009; Dana et al. 2012; ETC Group 2007; Parens et al. 2009) which are among others:

Biosafety: Concerns about environmental, health, and safety issues with respect to the intended or accidental release of synthetic organisms into nature[6] (close to debate on genetically modified organisms) and doubts about the appropriateness of risk assessment methods relating to the divergence to well-known biological systems.

- Biosecurity: The aim to provide an easy-to-use toolbox to create synthetic organisms has raised concerns regarding its dual-use potential (e.g. concerning bio-terrorism).
- Ethical problems that arise from conceptual questions: Notions of 'playing God' and 'messing with life' could change the understanding of 'life', 'nature', and 'human as creator' (see also Eichinger in this volume).
- Social justice: Competition between traditional production traits and those using synthetic organisms for recourses, markets, etc. as well as questions of intellectual property.

In the past years TA scholars started to bring the critical aspects of synthetic biology to the attention of policy makers (cf. EPTA 2011; Pei et al. 2012; van Doren and Heyen 2014) and TA activities like studies (e.g. Sauter 2011; TA-SWISS 2012) and projects (e.g. Albrecht 2014) were initiated.

3 Addressing the Challenges of TA for Synthetic Biology by Including the Public

The aim of TA activities is to accompany research and development in order to avoid socially negative but instead to promote desirable developments. It became clear quite early that TA efforts could be running into a dilemma, the "Control Dilemma"—sometimes also named after its originator David Collingridge (1980, p. 11): In an early stage of a development (of a certain technology) there is too little knowledge about possible technical and social implication to legitimize (governance) interventions on one hand. On the other hand, when first implications become detectable, the development may have progressed so far that correcting interventions are hard to enforce against the established paths.

Evidently this dilemma applies for synthetic biology. There are plenty of visions, promises, hopes, and also fears, but at the same time little testable data and few cases are available. Thus, TA for synthetic biology faces the specific difficulties of assessing NEST: the lack of knowledge—and even worse—evidence concerning its

---

[6]In the case of synthetic biology, it is essential to consider the precautionary principle: Once "synthetic organisms" as products of the "synthetic biology technology" have been released, they cannot be retrieved or even traced.

consequences for society and therefore a lack of scientific legitimating for interventions. Consequently, Grundwald (2012) asked which direction can TA research take and how can it be relevant to the decision-making process while it is by no means clear to what intervention and design can refer to (ibid., p. 10).

TA scholars have taken on this challenge in a number of different ways (for an elaborated overview see Kollek and Döring 2012): One approach is to focus on the conditions of constitution of science and its interplay with society, as in the "Science Assessment" concept put forward by Gill (1994) and Böschen (2005). Other concepts rather focus on the earliest possible onset of TA: "Constructive TA" as described by Schot and Rip (1996) seeks to shape the design, development, and implementation of a technology, and "Real-Time TA" (Guston and Sarewitz 2002) aims to accompany technology development from the outset and integrate societal issues as well as policy and governance aspects at an early stage. This was explicitly tried in the Berkeley Human Practices Lab of SynBERC (Rabinow and Bennett 2009). In a different type of approach, Grunwald (2012) pleads for "Hermeneutical TA" that analyses and reflects visions and narratives.

All these TA approaches towards an early assessment of NESTs have one thing in common: they targets public perceptions and attitudes as well as public and political debate rather than the technology itself. The rationale behind is the aim to provide an evidence-based set of arguments for early intervention in order to bridge the lack of scientific evidence (with respect to toxicological studies, risk assessments, etc.). As Patrick Sturgis pointed out:

> If our ambition is for science policy to enable and encourage technological choices that maximise the public good, who better to ask about which areas of science should and should not be pursued and supported than the public themselves? (Sturgis 2014, p. 39)

These approaches involve the public (here citizens and relevant stakeholders) in more or less direct way and thereby the whole set of method of empirical social studies are used. In the next section I will give an impression on the status quo of public perception and public debate in Germany in order to demonstrate how these methods can be employed.

4 Excursus: Synthetic Biology and the Public—Status Quo in Germany

In order to analyse public perception of a technology and the state of public controversies polls as quantitative methods are always a good starting point and thereby reveal sometimes astonishing facts. For example the past technology-related Eurobarometer polls revealed that—despite the persisting prejudice among life scientists—the population of Western industrialized nations and Germany in particular does *not* oppose technological developments which was confirmed in other quantitative and qualitative research. According to the most recent Eurobarometer (2013) survey, 54 % of the Germans are interested in technological developments, 38 % do regularly inform themselves about it, and in total 76 % agree that science and technology have a positive impact on society—and this is even below the EU27[7] mean.

But does this mean they are also positive about synthetic biology? Well, knowledge about synthetic biology is low: Only 18 % of the polled Germans knew the term (Eurobarometer 2010). Thus, like the majority of the Europeans, most Germans are not familiar with this technology (Gaskell et al. 2010, p. 29) and this is not much different in the US.[8] Interestingly, Germany shows the highest disagreement on the support (approval) of synthetic biology among the EU27 (ibid., p. 34). The low public awareness of synthetic biology indicates that there is no broad public debate on this issue—in contrast to genetically modified organisms: Here only 18 % had *not* heard of it.

Beside quantitative methods qualitative are used to gain more insight into the public perceptions and attitudes, for example different kinds of individual and group interviews including panel discussion.[9] However, for Germany data is rare—only a transcript of a public conference by the German Ethic Council in 2011 was found. This document showed that synthetic biology is yet mainly recognized by scientists and stakeholders—only a few citizens were present, and they hardly participated in the discussions (GEC 2011). However, some of the recent projects on synthetic biology (e.g. the EU project synenergene[10]) are also focusing on public awareness and their results will give further insights in the future.

Another source of information is media content analysis. Although this allow a rather indirect view, it still helps to estimating the thematic input for public debate in a certain area[11] and drawing conclusions on *framing* (correlation between the characterization of a topic by the media and its understanding by an audience), *priming* (media content influences viewpoints of the audience in general), and *agenda setting* (correlation between occurrence/frequency of a topic in the media and the importance attributed by the audience) (Scheufele and Tewksbury 2007, p. 11).

---

[7]EU27 includes the member states of the European Union as of June 2013.

[8]More information available at: http://www.synbioproject.org/process/assets/files/6655/synbio survey2013.pdf. Accessed 14 Aug 2014.

[9]Steurer (in this volume) presents the results of this kind of method—citizens panels and focus groups—for analyses on public debate/perceptions in Austria.

[10]For more information see: http://www.synenergene.eu/. Accessed 14 Aug 2014.

[11]That applies only if on assumes that media consumers take up the presented topic as well as the framing used by the journalists (Brüggemann 2014). Here, it is worth mentioning that the assessment of media effects on the recipient side regarding the consequences for opinion-forming or decision-making is accordingly demanding and maybe even impossible (Bonfadelli and Friemel 2011).

One of the most accessible sources for analysing the state of public debate in Germany is a media content analysis of 23 German press[12] titles by Lehmkuhl (2011). He drew the following conclusions: Journalists had problems to distinguish synthetic biology from genetic engineering, and the overall media coverage of synthetic biology in Germany was low, although slightly increasing until 2010 (the end of Lehmkuhl's study; see Fig. 1). The majority of the articles (46 %) described and explained the synthetic biology approach—only 21 % discussed the pros and cons of this technology. The latter was only the case in the context of focused reporting in response to public relation activities of the group around Craig Venter and the regional iGEM[13] contributions. Lehmkuhl identified four major themes: The first theme is "progress" which reflects the expectations that synthetic biology could help to solve some of the 'grand challenges' of mankind. The second is "runaway" which indicates unsubstantiated unease and fatalism towards scientific-technological progress. The third was called "Pandora's box/Devil's bargain" and broach the issue of inability to control unwanted side effects of the technology. And the fourth is "race" which interprets science as a sports competition in the sense of who is first and who wins the prize.

Based on this analysis, Lehmkuhl drew some conclusions about the state of the public debate in Germany which he held was only existent between the actors of the so-called periphery (referring to the deliberative model of the public sphere by Habermas (1992) and Gerhards et al. (1998)). It was thus a debate in "authority-free" space dominated by scientists and civil society organizations; in the Habermas model an indication that this public debate was not (yet) of political relevance. From a journalistic point of view, there were no prominent speakers for synthetic biology, which is a sign for the absence of a discourse that is declared relevant by journalists. The debate centred on the interpretation of what synthetic biology is and what it can/cannot do—but there was no real arguing (in a sense of statement and objection) in it. The main issues that were brought into the debate by non-scientist speakers were bio-terrorism, uncontrollable propagation, and the "monopolization of life" which alludes to the risks of economic utilization of synthetic biology.

In order to get an impression of the development since 2011, I investigated the German press from 2011 to 2014 using Lehmkuhl's methods.[14] My investigations revealed that his findings concerning the contents and framings are still valid.

---

[12]Although the print media is still the number one information resource, there is a shift towards a more mixed usage of resources including TV and the Internet (Eurobarometer 2013). Access to the latter is much more time-consuming and cost-intensive and their analysis much more elaborate; scientific approaches are still in a rather experimental state (Mitchelstein and Boczkowski 2010).

[13]The international Genetically Engineered Machine (iGEM) competition is a worldwide synthetic biology competition for young talents (university and high school students) that aims at raising attention for the field. More information at www.igem.org. Accessed 14 Aug 2014.

[14]A quantitative and qualitative analysis for the period between August 31, 2011 and December 31, 2014 was done in the framework of the EA Summer School as described in Lehmkuhl (2011, p. 20): Using the GENIOS database and the German search term "synthetische Biologie".

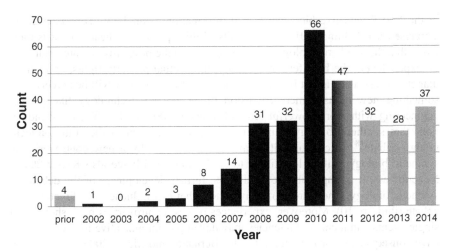

Fig. 1 Media coverage in German print media. Data until August 31, 2011 and method for media analysis were taken from Lehmkuhl (2011, p. 32) and are represented in *black bars*. For own data GENIOS database was used, shown in *grey bars*

However, the quantity of media coverage showed an interesting development (see Fig. 1): after 2010, the number of articles decreased to the level of 2008/2009, and rose again slightly in 2014. Thus, the curve shape may resemble in some aspects the hype-disappointment cycle which is known as "Gartner Hype Cycle"[15] Hence, synthetic biology definitely experienced its peak in media attention in 2010 when Craig Venter claimed to produce the first "synthetic cell".[16] Notably many reports assessing the potentials of synthetic biology were published by the end of 2009 which brought many visions down to earth but may have create media attention.

Since this curve is based on media coverage data, one cannot transfer the conclusions to technical applications of synthetic biology. Nevertheless, the data may give a week indication that the knowledge about synthetic biology and synthetic biology as a scientific and technological field was settled and is producing increasingly applicable knowledge and fewer visions (cf. Haslinger et al. 2014 for the case of nanotechnology).

Beside, some scholars are using themes taken up by the media as indicators for the future direction of the public debate[17] or relate to previous debates on subjects

---

[15]The Gartner Hype Cycle represents a "graphic representation of the maturity and adoption of technologies and applications, and how they are potentially relevant to solving real business problems and exploiting new opportunities". More information can be found at http://www.gartner.com/technology/research/methodologies/hype-cycle.jsp and http://www.spiegel.de/netzwelt/tech/aufmerksamkeits-kurven-die-hype-zyklen-neuer-technologien-a-443717.html. Accessed 14 Aug 2014.

[16]Ancillotti and Eriksson (this volume) show for Swedish and Italian press, which also covered synthetic biology more extensively in 2010, that this event was indeed the main topic in that year.

[17]For limitations of this approach see footnote 11.

of the NEST field like bio-, nano-, and information technology (Torgersen 2009; Torgersen and Schmidt 2013). The results of this approach indicate that it is most likely that the synthetic biology debate will take up the problems that are dealt with on expert level, like biosafety, biosecurity, intellectual property rights, and particular ethical aspects. Here it is crucial whether synthetic biology will be perceived as something new or as a mere extension of an existing technology like genetic engineering with more powerful tools (Torgersen 2009, p. 13). There are indications that synthetic biology is perceived as a prolongation of the "old" biotechnology debate[18] which ironically may be the reason why a new controversy on synthetic biology will be less likely to arise (ibid., p. 14; see also Kaiser 2012). Nevertheless, it is impossible to predict the further development of the debate on synthetic biology. There are too many factors that can influence the direction of a debate (ibid., p. 10). Moreover, experiences of past controversies have shown that single events and local contingencies may develop communicative impact and thus have consequences for the whole debate (Bernauer and Meins 2003).

5 Motivations for Public Engagement in the Assessment of Synthetic Biology

The German case study above illustrates how methods involving the public are used by TA in the first way: to produce knowledge but also as legitimation for the recommendations they conclude out of their findings. Thus, asking the question why to include public engagement in TA on has to keep in mind that the motivations to use such deliberative or participative methods can be multiple. Stirling (2008) defined the three most common types of motivations: First, the motivation can be *substantive* which means that choices concerning the nature and trajectory of innovation can be co-produced with publics in ways that authentically embody diverse sources of social knowledge, values, and meanings. Here, citizens are considered as "experts for the everyday" that complement professional knowledge with their own pragmatic expertise and life experience (Hahn et al. 2014). Transdisciplinary generated the conditions to think outside the box and thus eventually discover aspects that has been overlooked by the experts in the field.

Second, the motivation can be *normative* which means that dialogue is the right thing to do for reasons of democracy, equity, and justice. Along with this mind set come high expectation toward the effects of public engagement, e.g. better decisions, responsible innovations and seemingly even better people. Here participation would aim at eliciting further engagement of the participants and the initiation of a broad public debate with consequences for decision-making.

---

[18]Still there are no indications that this is the case: David Shukman "Will synthetic biology become a GM-style battleground?" BBC News, July 21, 2014. http://www.bbc.com/news/science-environment-23274175. Accessed 14 Aug 2014.

And third, in some cases the motivation is an *instrumental* one which means that the organizing entity aims a certain outcome e.g. creating acceptance or avoidance/solution of conflicts. Here, public engagement exercises aiming at educating the "ignorant public" (cf. Bauer et al. 2007) about the technology in order to avoid controversies and gain rational public debate (e.g. Philp et al. 2013; ter Meulen 2014) as well as public acceptance.

Each motivation for advocating public debate and promoting public engagement may be reasonable on its own merits, but they are partially conflicting. This fact might be part of the explanation for the divergent opinions in the scientific debate on participatory TA: the expectation and/or motivation of the scholar regarding the capacity or impact of participation influence their level of support for public engagement. It would be good if scholars revealed these motivations and rationales (e.g. Bogner 2012; Torgersen and Schmidt 2013; Walk 2013) and thereby helped to classify the evaluations and conclusions of public engagement events.

The majority of TA approaches mentioned in third section of this chapter—especially those in the tradition of social constructivist studies of science and technology (STS)—include notions of participation and engagement by lay people. In line with this it is also important to distinguish between the different notions of participation: talking about the inclusion of views and expertise of non-scientific actors into knowledge production and decision-making processes is *not* the same as citizens' participation within the political system in a sense of direct democracy (cf. Walk 2013). Although the growing popularity of citizens' participation within the political system appears in parallel to the trend of more participation in the scientific context, this should not be confused with participatory or deliberative approaches in TA that aim on knowledge production in the first way (Hennen 2012). Nevertheless, it is not easy to draw the line—scholars promoting public engagement with science "share a normative commitment to the idea of democratic science policy, and have argued that public engagement can be part of this" (Stilgoe et al. 2014, p. 5). And thus, the motivations for using approaches involving the public—in whatsoever format—can be substantive, normative and/or instrumental at the same time. But among the proponents of public engagement it is widely agreed that the public need to be engaged "up-stream" (Wilsdon and Willis 2004) without agreement about to how this should be done (Torgersen and Schmidt 2013, p. 45).

This idea can be traced back to changes in the understanding of the relationship of science and society in the last quarter of the 20th century (Bauer et al. 2007) and the loss of citizens' trust in technocratic decision-making. Consequently, STS scholars proposed to strengthen the role of non-scientific actors in decision-making in order to anticipate this development. This is reflected in concepts like "Mode 2 Knowledge Production" (Gibbons et al. 1994) or "Post-Normal Science" (Funtowicz and Ravetz 1993). These concepts are reactions to the increasing demand for more integrative, inter- and transdisciplinary methods regarding pressing technology-related questions (Nowotny et al. 2001). They also take into account that new technological developments are not linear[19] but shaped by society (Kiran 2012).

Consequently, Hellström (2003) calls for TA to "include elements of speculative foresight through public involvement" (ibid., p. 380) in order to understand modern risks as systemic and as "threats which seem inevitable yet impossible to predict" (ibid., p. 380). The difficulties of risk assessment make it necessary to think of "'pre-emptive' knowledge management or knowledge improvement tools for integrating and utilizing tacit understandings" (ibid., p. 382). Thus, public involvement would "help shed light on the consequences of new proposed technologies, resolve problems of risk assessment and management practice, provide new cognitive frameworks for complexity reduction and suggest new ways of managerially drawing the boundaries of emerging technological systems" (ibid., p. 382). Moreover it was proposed, that public engagement in the governance of NEST should be strengthen by the formation of new organizations that span the boundary between knowledge production and public action (Barben et al. 2008, p. 779) which was also emphasised by Harald König and coauthors:

> Ideally, collaboration between governmental organizations, the academic and industrial synthetic genomics/biology communities, civil society organizations and the public would give rise to a safe yet dynamic web of mutual accountability and responsibility as the basis for a flexible and integrative governance approach to these integrative science fields. (König et al. 2013, p. 21)

In line with a general development towards more public participation also TA followed the "participatory turn" called out by Jasanoff (2003). She argues that more accountability of science is required by society and that this is most prominent in the demand for greater transparency and participation (ibid., p. 236). With concepts such as "Citizen Science" (Hand 2010; Irwin 1995) or "Responsible Research and Innovation" (RRI; von Schomberg 2013) the demand for public participation enters political programmes. This means, for example, that participation is seen as an integral part of research funding on the European Union level (e.g. in Horizon 2020[20]). In von Schomberg's words:

> Responsible Research and Innovation is a transparent, interactive process by which societal actors and innovators become mutually responsive to each other with a view to the (ethical) acceptability, sustainability and societal desirability of the innovation process and its marketable products (in order to allow a proper embedding of scientific and technological advances in our society). (von Schomberg 2013, p. 19)

---

[19]"Linear" referrers here to classic innovation theories which locates the creative potential solely at the side of the researches and developers. The public as potential users is reflected by the market that either accepts or rejects the invention. In more recent theories address the interconnectivity of today's innovation processes.

[20]More information can be found at: http://ec.europa.eu/programmes/horizon2020/ and http://ec.europa.eu/programmes/horizon2020/en/h2020-section/responsible-research-innovation. Accessed 14 Aug 2014.

According to Stilgoe et al. (2013), RRI is characterized by four dimensions of responsibility; the deliberative or inclusion dimension aims at the involvement of stakeholders and citizens within technology development processes for:

> [...] inclusively opening up visions, purposes, questions, and dilemmas to broad, collective deliberation through processes of dialogue, engagement, and debate, inviting and listening to wider perspectives from publics and diverse stakeholders. This allows the introduction of a broad range of perspectives to reframe issues and the identification of areas of potential contestation. (ibid., p. 38)

In the past years TA scholars respond to the concept and employed RRI to answer problems that come with the governance of NEST like synthetic biology. Nevertheless, the concept itself seems is far from being well defined (Heil et al. 2014). Public participation in RRI is presented as vehicle to meet societal needs and ensure successful innovations—and eventually even prevents (technology) conflicts (cf. Grunwald 2012). With respect the motivations describe at the beginning of this section, RRI merge all three motivations. In any case, since RRI becomes more and more dominant in the TA community it is most likely that public engagement will be promoted further.

6 Critics on the Use of Public Engagement in Synthetic Biology

Given the assumption that public engagement in the governance of NESTs can improve decision-making—by making it socially robust, less conflictual, and, thus, more responsive—societal actors are needed to assess the pros and cons of the technological and scientific developments under debate (Radstake et al. 2009, p. 313). Involving societal actors raises a number of normative questions that requires answer with far reaching consequences: Which format will be chosen? Who will be invited? How should the relevant issues and the legitimate arguments be determined? Considering the effects of *successful* public engagement—which possibly mean an exercise that stimulate the public debate in society—Torgersen and Schmidt assume:

> They not only will influence the course and outcome of the engagement exercise but might also propose, if successful, what issues and arguments might become relevant in an ensuing broader public debate. (Torgersen and Schmidt 2013, p. 46)

Thus, advocates of public participation should be much more aware of and reflective about this. Torgersen and Schmidt (2013) further argue that a "public debate over SB 'out there' might develop very differently subject to issues and arguments determined to be relevant and the choice what SB might be compared to early-on" if influence by upstream public engagement (ibid., p. 46). And thus, the choice made for the public engagement exercise "might heavily influence the future public image of the technology and impinge on its commercial success" (ibid., p. 46).

Beside, a fundamental problem seems to be that the wider public (however defined) is apparently insufficiently interested in science and science governance to participate spontaneously and take an active part in the public debate. This seems to be difficult even for stakeholders if there is uncertainty about the eventual nature of new products, processes, benefits, and risks of a technology (Torgersen and Schmidt 2013). Lehmkuhl, too, comes to the conclusion that the difficulty to initiate a public debate on synthetic biology—unless it is catalysed by suitable research results—is the current dilemma: It seems to be desirable to hold a public debate on the pros and cons of synthetic biology now, when it is to some extent still possible to decide which research progress is socially desired. But at the same time, it is difficult to stimulate the debate now, because there is this lack of concrete results which can only be produced through the course of research efforts that would be the object of this debate (Lehmkuhl 2011, p. 50)—again the recurrent theme of the TA in the NEST field: when is the right time to start it.

In the academic debate on early public engagement, the main weakness of initiated public debate is seen in the fact that the deliberative exercises tend to be staged rather than spontaneous "self-arising" public debates. This is due to the fact that the latter would be completely uncoordinated regarding the involved actors and participants, communication channels and formats as well as the choice of topics. In contrast, initiated public debate very much depends on the design made by its organizers (e.g. STS social scientists or specialists in technology assessment) as well as—to a considerable extent—on the expectations of the sponsors of the exercise (Bogner 2012; Torgersen and Schmidt 2013). This is the main restriction for a further use of the results of these exercises in order to predict future public dialogues and draw conclusions on the public perceptions and expectations on new technologies like synthetic biology.

Facing this restriction of early, invited public engagement, the scientific community has tried to find methodical solutions, among others innovative formats, sophisticated selection and recruitment of participants (Sturgis 2014, pp. 38–39). Scholars that promote public engagement with a rather normative motivation complained that thereby "the *how* still trumps the *why*" (Stilgoe et al. 2014, p. 5, emphasis original): beyond the participatory production of knowledge there is only insufficient emphasis on the question, how participatory formats and their results can be reasonably included in the development of governance frameworks or in governance procedures. For them, involving the public becomes meaningless if the results of the exercise cannot produce impact on governance. But in Germany and the most European countries there are no procedures regulated by law—with a few exceptions in planning regulations—that assure that the results of public engagement enter the political decision-making process. And as long as this is left to chance, participative exercise will fail to live up to its high-level expectations (e.g. gain of acceptance, avoidance of conflicts or elicitation of a broad public debate) of those who follow the normative approach of public engagement.

In addition, it is also important to reconsider the image of "the public" and its construction, because this is what is leading or misleading in finding answers to the question how participatory formats can be included in the development of governance frameworks for and in the governance of NEST in a meaningful way (elaborated e.g. by Stilgoe et al. 2014, p. 7).[21] The "knowledge deficit model" (Bauer et al. 2007) is still very predominant and thus public engagement is often used to obtain acceptance. Anyway it is questioned whether familiarity with synthetic biology is related to the technology's evaluation and finally leads to more support from a broader public. Although the empirical data (e.g. Eurobarometer 2010) indicate this, it is also possible that it is "a technophile avant-garde that—because of its affinity to and support of technologies—has heard about synthetic biology in the first place" (Gaskell et al. 2010, p. 35).

7 Conclusions

There are numerous good reasons for the engagement of citizens and stakeholders (in general "the public") in synthetic biology technology assessment. So what might be the best medium for public engagement in synthetic biology? First of all, public engagement exercises—even if they are organized by scientists—create rooms for mutual learning: Scientists learn about the rationales of non-scientific actors and vice versa. Of course especially interested people will join these exercises, but these can act as ambassadors for scientific arguments in the public. This should not be confused with promoting acceptance but rather promoting science (even though the two are close to each other). For example, it has been argued that the public was over-addressed in the field of nanotechnology (cf. Torgersen and Schmidt 2013)—but notably, the grand controversy on this technology has yet not arisen—maybe due to a "normalization" of the debate (Grunwald 2012). The bustling activities of social scientists and philosophers around nanotechnology influenced its trajectories of development remarkably. Even today's versatile promotion of RRI is strongly rooted in this history (Grunwald 2014). Thus, nanotechnology may serve as a role model for public engagement in synthetic biology where early public engagement and debate lead to an objective but also balanced and inclusive debate and eventually also to (better) future applications meeting social needs in the sense of RRI. The early stage provides at least the opportunity to contribute to shaping applications according to social needs—or maybe even decide jointly where the limits for the technology should be drawn (see also Hagen et al. in this volume).

However, diverse and sometimes conflicting motivations and expectations are brought up by the advocates of public engagement and debate. Moreover, sometimes the results of engagement exercises fail to live up to these expectations, especially if they are on a high level (e.g. gain of acceptance, avoidance of conflicts, elicitation of a broad public debate). In the case of synthetic biology, the limitations

---

[21]For a persuasive example see Molyneux-Hodgson and Balmer (2014) who analyse the performance of a synthetic biology research programme that sought to address issues of innovation in the water industry. They found that the conceptualisation of public actors as consumers who are ignorant of the complexities of water and its true value became an innovation barrier.

of early public engagement and debate become very clear. The few applications—mainly in the medical sector—have not elicited any self-propelling debate (cf. Pardo et al. 2009). Little is known about synthetic biology in the public, and the media primarily depicts the positions of scientists.

In a nutshell, there is currently hardly anything that stirs up emotions and so it seems that there is little to discuss (see also Kaiser 2012). This might change quickly, but it is hard to say whether this will occur and what it will be like. However, once such a self-propelling debate has started, it is difficult to influence its direction, as the example of (green) biotechnology has shown—and this is probably the main motivation for scientists in the field of synthetic biology to be supportive of public engagement: They simply fear the hindrance of their research if public controversies lead to restrictive governance. Nevertheless, organisers of such exercises still carry responsibility: By framing their exercise they may also frame the possibly elicit public debate afterwards and thus, should be transparent about their goals and motivations.

References

Albrecht S (2014) Synergene: responsible research implementieren—Zur Umsetzung von RRI am Beispiel eines Projekts zur Synthetischen Biologie. Presentation at the NTA6/TA14 Conference, Vienna (June 6, 2014). http://www.oeaw.ac.at/ita/fileadmin/redaktion/Veranstaltungen/konferenzen/ta14/ta14-albrecht.pdf. Accessed 14 Aug 2014

Barben D, Fisher E, Selin C, Guston D (2008) Anticipatory governance of nanotechnology: foresight, engagement, and integration. In: Hackett EJ et al (eds) The handbook of science and technology studies. MIT Press, Cambridge, pp 979–1000

Bauer MW, Allum N, Miller S (2007) What can we learn from 25 years of PUS survey research? Liberating and expanding the agenda. Public Underst Sci 16(1):79–95

Benner SA, Sismour AM (2005) Synthetic biology. Nat Rev Genet 6:533–543

Bernauer T, Meins E (2003) Technological revolution meets policy and the market: Explaining cross-national differences in agricultural biotechnology regulation. Eur J Polit Res 42(5):643–683

Bogner A (2012) The paradox of participation experiments. Sci Technol Human Values 37(5):506–527

Boldt J, Müller O, Maio G (2009) Synthetische Biologie: Eine ethisch-philosophische Analyse. BBL, Bern

Bonfadelli H, Friemel TN (2011) Medienwirkungsforschung: Grundlagen und theoretische Perspektiven. UVK, Konstanz

Böschen S (2005) Vom Technology zum Science Assessment: (Nicht-)Wissenskonflikte als konzeptionelle Herausforderung. Technologiefolgenabschätzung—Theorie und. Praxis 14(3):122–127

Brüggemann M (2014) Between frame setting and frame sending: how journalists contribute to news frames. Commun Theory 24(1):61–82

Cameron DE, Bashor CJ, Collins JJ (2014) A brief history of synthetic biology. Nat Rev Microbiol 12:381–390

Collingridge D (1980) The social control of technology. Open University Press, London

GEC: German Ethic Council (2011) Werkstatt Leben. Bedeutung der Synthetischen Biologie für Wissenschaft und Gesellschaft. Documentation of a public conference at 23 Nov 2011 in Mannheim/Germany. http://www.ethikrat.org/veranstaltungen/weitere-veranstaltungen/werkstatt-leben. Accessed 14 Aug 2014

Dana GV, Kuiken T, Rejeski D, Snow AA (2012) Synthetic biology: four steps to avoid a synthetic-biology disaster. Nature 483(7387):29

EPTA: European Parliamentary Technology Assessment (2011) Synthetic Biology. EPTA Briefing Note No. 1 (November 2011). http://www.tab-beim-bundestag.de/en/research/u9800/EPTA_briefingnote_A4.pdf. Accessed 14 Aug 2014

ETC Group: Action Group on Erosion, Technology and Concentration (2007): Extreme genetic engineering. An introduction to synthetic biology. January 2007. http://www.etcgroup.org/sites/www.etcgroup.org/files/publication/602/01/synbioreportweb.pdf. Accessed 14 Aug 2014

Eurobarometer (2010) Special Eurobarometer 341/Wave 73.1—Biotechnology. Report. Fieldwork: January–February 2010. European Commission, Brussels. http://ec.europa.eu/public_opinion/archives/ebs/ebs_341_en.pdf. Accessed 14 Aug 2014

Eurobarometer (2013) Special Eurobarometer 401—Responsible research and innovation (RRI), Science and Technology. Report. Fieldwork: April–May 2013. European Commission, Brussels. http://ec.europa.eu/public_opinion/archives/ebs/ebs_401_en.pdf. Accessed 14 Aug 2014

European Commission (2005) Synthetic biology: applying engineering to biology—Report of a NEST high-level expert group. EUR 21796. European Commission/DG Research, Brussels

Funtowicz SO, Ravetz JR (1993) Science for the post-normal age. Futures 25(7):739–755

Fussenegger M (2014) Kunst oder künstlich? Chancen, Risiken und Perspektiven der Synthetischen Biologie. Neue Zürcher Zeitung, 2 Jul 2014. http://www.nzz.ch/wissenschaft/biologie/chancen-risiken-und-perspektiven-der-synthetischen-biologie-1.18334868. Accessed 14 Aug 2014

Gaskell G, Stares S, Allansdottir A (2010) Europeans and biotechnology in 2010: Winds of Change? European Commission/Directorate General for Research, Brussels. http://ec.europa.eu/research/science-society/document_library/pdf_06/europeans-biotechnology-in-2010_en.pdf. Accessed 14 Aug 2014

Gerhards J, Neidhardt F, Rucht D (1998) Zwischen Palaver und Diskurs. Strukturen öffentlicher Meinungsbildung am Beispiel der deutschen Diskussion zur Abtreibung. Westdeutscher Verlag, Opladen

Gibbons M, Limoges C, Nowotny H, Schwartzman S, Scott P, Trow M (1994) The new production of knowledge. The Dynamics of Science and Research in Contemporary Societies. Sage, London

Gill B (1994) Die Vorverlegung der Folgenerkenntnis. Science Assessment als Selbstreflexion der Wissenschaft. Soziale Welt 45(4):430–454

Grunwald A (2010) Technikfolgenabschätzung—eine Einführung. 2. Auflage. Edition Sigma (Gesellschaft—Technik—Umwelt, Neue Folge 1), Berlin

Grunwald A (2012) Synthetische Biologie als Naturwissenschaft mit technischer Ausrichtung. Plädoyer für eine „Hermeneutische Technikfolgenabschätzung". Technikfolgenabschätzung—Theorie und. Praxis 21(2):10–15

Grunwald A (2014) Responsible research and innovation: an emerging issue in research policy rooted in the debate on nanotechnology. In: Arnaldi S, Ferrari A, Magaudda P, Marin F (eds) Responsibility in nanotechnology development, pp 191–205

Guston DH (2001) Boundary organizations in environmental policy and science: an introduction. Sci Technol Human Values 26:399–408

Guston DH, Sarewitz D (2002) Real-time technology assessment. Technol Soc 24(1–2):93–109

Habermas J (1992) Faktizität und Geltung. Beiträge zur Diskurstheorie des Rechts und des demokratischen Rechtsstaats. Suhrkamp, Frankfurt a.M

Hahn J, Seitz SB, Weinberger N (2014) What can TA learn from 'the people'? A case study of the German citizens' dialogues on future technologies. In: Michalek TC, Hebakova L, Hennen L, Scherz C, Nierling L, Hahn J (eds) Technology assessment and policy areas of great transitions, Prague, pp 165–170

Hand E (2010) Citizen science: people power. Nature 466(7307):685–687

Haslinger J, Hocke P, Hauser C (2014) Ausgewogene Wissenschaftsberichterstattung der Qualitätspresse? Eine Inhaltsanalyse zur Nanoberichterstattung in repräsentativen Medien Österreichs, Deutschlands und der Schweiz. In: Gazsó A, Haslinger J (eds) Nano Risiko Governance. Der gesellschaftliche Umgang mit Nanotechnologien. Springer, Vienna, pp 283–310

Heil R, Dewald U, Fleischer T, Hahn J, Jahnel J, Seitz SB (2014) Konzeptionelle Überlegungen für ein Forschungsvorhaben zur Klärung des Verhältnisses von Responsible (Research and) Innovation und TA. Talk at the international conference NTA6-TA14 "Responsible Innovation. Neue Impulse für die Technikfolgenabschätzung? (02–04.06.2014) Vienna, Austria

Hellström T (2003) Systemic innovation and risk: technology assessment and the challenge of responsible innovation. Technol Soc 25(3):369–384

Hennen L (2012) Why do we still need participatory technology assessment? Poiesis Prax 9(1–2):27–41

Irwin A (1995) Citizen science: a study of people, expertise and and sustainable development. Routledge, London

Jasanoff S (2003) Technologies of humility: citizen participation in governing science. Minerva 41(3):223–244

Kaiser M (2012) Commentary: looking for conflict and finding none? Public Underst Sci 21(2):188–194

Kiran AH (2012) Does responsible innovation presuppose design instrumentalism? Examining the case of telecare at home in the Netherlands. Technol Soc 34(3):216–226

Kollek R, Döring M (2012) TA-Implikationen der komplexen Beziehung zwischen Wissenschaft und Technik. Technikfolgenabschätzung—Theorie und Praxis 21(2):4–9

König H, Frank D, Heil R, Coenen C (2013) Synthetic genomics and synthetic biology applications between hopes and concerns. Curr Genomics 14:11–24

Leduc S (1912) La biologie synthétique. In: Poinat A (ed) Étude de biophysique. Peiresc, Paris

Lehmkuhl M (2011) Die Repräsentation der synthetischen Biologie in der deutschen Presse. Abschlussbericht einer Inhaltsanalyse von 23 deutschen Pressetiteln. Deutscher Ethikrat, Berlin. http://www.ethikrat.org/dateien/pdf/lehmkuhl-studie-synthetische-biologie.pdf. Accessed 14 Aug 2014

Mitchelstein E, Boczkowski PJ (2010) Online news consumption research: an assessment of past work and an agenda for the future. New Media Soc 12:1085–1102

Molyneux-Hodgson S, Balmer AS (2014) Synthetic biology, water industry and the performance of an innovation barrier. Sci Public Policy 41(4):507–519

Nowotny H, Scott P, Gibbons MT (2001) Rethinking science: knowledge in an age of uncertainty. Polity, Cambridge

Pardo R, Engelhard M, Hagen K, Jørgensen RB, Rehbinder E, Schnieke A, Szmulewicz M, Thiele F (2009) The role of means and goals in technology acceptance. EMBO Rep 10(10):1069–1075

Parens E, Johnston J, Moses J (2009) Ethical issues in synthetic biology: an overview of the debates. 24 Jun 2009. http://www.synbioproject.org/process/assets/files/6334/synbio3.pdf. Accessed 14 Aug 2014

Pei L, Gaisser S, Schmidt M (2012) Synthetic biology in the view of European public funding organisations. Public Underst Sci 21(2):149–162

Philp JC, Ritchie RJ, Allan JEM (2013) Synthetic biology, the bioeconomy, and a societal quandary. Trends Biotechnol 31(5):269–272

Rabinow P, Bennett G (2009) Synthetic biology: ethical ramifications 2009. Syst Synth Biol 3(1–4):99–108

Radstake M, van den Heuvel-Vromans E, Jeucken N, Dortmans K, Nelis A (2009) Societal dialogue needs more than public engagement. EMBO Rep 10(4):313–317

Rip A, Voß J-P (2013) Umbrella terms as a conduit in the governance of emerging science and technology. Sci Technol Innov Stud 9:39–59

Sauter A (2011) Synthetische Biologie: Finale Technisierung des Lebens—oder Etikettenschwindel? TAB-Brief 39:16–23

Scheufele DA, Tewksbury D (2007) Framing, agenda setting, and priming: the evolution of three media effects models. J Commun 57(1):9–20

Schot J, Rip A (1996) The past and future of constructive technology assessment. Technol Forecast Soc Chang 54:251–268

Star SL, Griesemer JR (1989) Institutional ecology, 'Translations' and boundary objects: amateurs and professionals in Berkeley's Museum of vertebrate zoology, 1907–39. Soc Stud Sci 19 (3):387–420

Stilgoe J, Owen R, Macnaghten P (2013) Developing a framework for responsible innovation. Res Policy 42(9):1568–1580

Stilgoe J, Lock SJ, Wilsdon J (2014) Why should we promote public engagement with science? Public Underst Sci 23(1):4–15

Stirling A (2008) "Opening Up" and "Closing Down": power, participation, and pluralism in the social appraisal of technology. Sci Technol Human Values 33(2):262–294

Sturgis P (2014) On the limits of public engagement for the governance of emerging technologies. Public Underst Sci 23(1):38–42

Szybalski W, Skalka A (1978) Nobel prizes and restriction enzymes. Gene 4:181–182

TA-SWISS (2012): Synthetische Biologie in der Gesellschaft. Eine neue Technologie in der öffentlichen Diskussion. https://www.ta-swiss.ch/projekte/biotechnologie-und-medizin/synthetische-biologie. Accessed 14 Aug 2014

ter Meulen V (2014) Time to settle the synthetic controversy. Nature 509:135

Torgersen H (2009) Synthetic biology in society: learning from past experience? Syst Synth Biol 3 (1–4):9–17

Torgersen H, Schmidt M (2013) Frames and comparators: how might a debate on synthetic biology evolve? Futures 48:44–54

van Doren D, Heyen NB (2014) Synthetic biology: too early for assessments? A review of synthetic biology assessments in Germany. Sci Public Policy 41(3):272–282

von Schomberg R (2013) A Vision of Responsible Research and Innovation. In: Owen R, Bessant J, Heintz M (eds) Responsible innovation: managing the responsible emergence of science and innovation in society. Wiley, Hoboken, pp 51–74

Walk H (2013) Herausforderungen für eine integrative Perspektive in der sozialwissenschaftlichen Klimaforschung. In: Knierim A, Baasch S, Gottschick M (eds) Partizipation und Klimawandel—Ansprüche, Konzepte und Umsetzung. Oekom Verlag, Munich, pp 21–35

Way JC, Collins JJ, Keasling JD, Silver PA (2014) Integrating biological redesign: where synthetic biology came from and where it needs to go. Cell 157(1):151–161

Wilsdon J, Willis R (2004) See-through science: why public engagement needs to move upstream. Demos, London

Public Engagement in Synthetic Biology: "Experts", "Diplomats" and the Creativity of "Idiots"

Britt Wray

1 Introduction

One common narrative that has arisen in policy and media concerning synthetic biology describes the field as aiming to make biology easy to engineer (Jefferson et al. 2014). The hope is that by making biology easy to engineer, scientists will be able to harness the sustainable production capacities of natural systems and direct them to make valuable products, such as biofuels, drugs or other valuable chemicals. Synthetic biologists want to rid biological systems of their complexity in order to make the engineering of biology a set of routine and standard practices (Endy 2005).

Synthetic biologists are often described as working predominantly (but not exclusively) with simple organisms such as yeast and bacteria in order to engineer them to produce outputs that are helpful in the clinic, the biotechnology industry and in basic molecular research (Church et al. 2014). However, the diverse field is quickly moving towards re-engineering more complex organisms as well. Examples of this can be seen acutely in the emerging mission to re-create certain species that we've lost, known as de-extinction. In de-extinction, scientists aim to create facsimiles of extinct species. One way that they aim to do this using the tools of synthetic biology is by editing genes that code for the unique phenotypic traits of extinct animals into their closest living relatives' genomes. The scientists involved then plan to bring modified embryos that contain those genes to term inside of surrogate mothers from the closely-related living species (Sherkow and Greely 2013). The interventions that synthetic biology and related genome-editing

---

B. Wray (✉)
University of Copenhagen, Karen Blixens Vej 4, 2300 København S, Building: 14.4.09, Copenhagen 2300, Denmark
e-mail: hello@brittwray.com; vml231@hum.ku.dk

© Springer International Publishing Switzerland 2016
K. Hagen et al. (eds.), *Ambivalences of Creating Life*, Ethics of Science and Technology Assessment 45, DOI 10.1007/978-3-319-21088-9_9

177

techniques afford are therefore not (at least theoretically) restricted to using "simpler" organisms. As humans re-design nature according to our own ambitions, "such changes may entail a fundamental rethinking of the identity of the human self and its place in larger natural, social and political orders" (Pauwels 2011, p. 116).

The idea that synthetic biologists can make biology easy to engineer supposes that they have enabled a "de-skilling" of science, therefore creating greater access to genetic engineering technologies for non-trained professionals, which in turn increases its susceptibility for potential "dual use" (Schmidt 2008; Kelle 2009).[1] Dual use technologies produce desired and undesired outcomes. Dual-use topics in synthetic biology range from biofuel and cheap drug production on the one hand, to potential novel weaponry for bioterrorism and microbes that could escape the lab to invade surrounding ecosystems on the other.

The societal implications of the potential futures that synthetic biology may bring forth are myriad, and there are more individuals who want to weigh in on that discussion than only the scientists and engineers involved. Over the last several years, synthetic biology has witnessed a wide range of creative communicators take part in its debate. At a time when the (at least ideological if not always practical) "de-skilling" of biological engineering is being brought to bear in synthetic biology, diverse individuals are claiming more participation and openness in communities and systems that comment on, and shape, the field.

Synthetic biology has been noticeably proactive, as an emerging technoscience, in inviting interdisciplinary creative talent to contribute to its discourse (Calvert 2010; Marris 2015). This ethos can be traced in synthetic biology through the activities of a cultural movement like DIYbio (Penders 2011), the increasing participation of artist and other non-scientist researchers at events like the International Genetically Engineered Machines Competition (Agapakis 2014),[2] and the growth of bioart projects that explore synthetic biology (Kerbe and Schmidt 2013). Specific instances can be found in numerous projects of experimental engagement that concern the field. These can range from events like the Kopenlab biohacking festival in Copenhagen,[3] to bioart workshops in Genspace (a community lab in

---

[1]There are a variety of questions raised by synthetic biology that circulate in policy and media that extend far beyond its emergence as a "de-skilling" science. These include social justice for the global south, responsible innovation, ontological implications, modes of regulation and governance, its impact on the bio-economy, and more. In this chapter I am using the example of "de-skilling" and the related term of "dual-use" to make a general point about societal concerns, but do not mean to regard these concerns as the only topics that have garnered attention and debate.

[2]2014 marked the first year that iGEM had a competitive track for teams whose projects explicitly mobilized art and design in synthetic biology. http://2014.igem.org/Tracks/Art_Design. Accessed 01 Sept 2014.

[3]The Kopenlab festival was comprised of "a collaborative space for citizen science, DIYbio, contemporary art and maker culture." It took place as part of Science in the City during the Euro Science Open Forum, Copenhagen, 2014. http://kopenlab.dk. Accessed 14 June 2014.

Brooklyn),[4] to institutionally-funded interdisciplinary residency programs like Synthetic Aesthetics.[5]

In each their own way, individuals who participate in these activities have entered the discourse of synthetic biology as interdisciplinary actors, who often (but not always) lack traditional scientific expertise. In the case of DIYbio, professional scientists have often offered their expertise to DIYbiologists, and can sometimes be found actively participating in DIYbio spaces (Grushkin et al. 2013).[6] However, they take a non-institutional approach to practicing science by doing so, which offers an alternative to traditional assumptions about how science should be carried out. Similarly, when artists and designers get involved in the field, they inject their own disciplinary modes of thinking and making into the discourse, and expose professional scientists to other ways of looking at, and questioning, the subject of their work. Together, these diverse practices establish a polyphony. This polyphony helps to diversify and complicate the public understanding of synthetic biology in ways that could not be afforded through an exclusively institutional and scientific framing.[7]

This increasingly polyphonous set of practices in synthetic biology intersects with the imaginaries of what the field means that bubble over into public consciousness through their representational circulation at science festivals and genetic engineering competitions, on blogs, in galleries, on the airwaves, in social media and more. In effect, it is not only professional science communication that recognizes itself as such that is crafting and telling public stories about synthetic biology to diverse non-expert audiences.

---

[4]Genspace has been serving New York City as an outreach centre promoting citizen science since 2009. In 2010 it became the first-ever community biotechnology laboratory with a Biosafety Level One facility. http://genspace.org/. Accessed 05 Sept 2014.

[5]Synthetic Aesthetics is an international research project investigating the crossover between art, design, social science and synthetic biology. http://www.syntheticaesthetics.org. Accessed 19 Nov 2014.

[6]Biohacking, which includes the DIYbio community, is a heterogenous "scene" that does not adhere to any one movement, mission or aim. At the 2015 PACITA Technology Assessment Conference in Berlin, German biohacker Rüdiger Trojok gave an elucidating talk about the complexities of the biohacker identity. He explained that although biohackers around the world seem to celebrate the non-institutional practice of biotechnology, they are far from a unified community. For example, some groups in North America (more closely aligned with DIYbio) affiliate their work with the possibility for commercialization, while others in Europe and Asia align more closely to an anti-capitalist and activist ethic.

[7]Literary critic Mikhail Bakhtin developed a theory of polyphony to describe Dosteovsky's tendency to write novels wherein his characters would each speak for themselves and not act in service of any other character's will (Bakhtin and Caryl 1984). This creates the condition that a discussion can never be finalized according to an individual's views, meaning that a true polyphony constitutes a collection of many voices with each their own distinctiveness. Polyphonous discussions may be aspired to for democratic purposes, so that discourses evolve according to inputs from many voices without any one voice acting in the service of another. Although the processual "de-skilling" of synthetic biology may be one important aspect for the diversification of the field, it is not the sole factor. Long traditions of synthesizing life in fiction, or experimentation to make life from non-living components in chemistry for example have also played in a role in stratifying its polyphony over time.

2 Science Communication: Seductive Science and Scary Science

For my purposes here, I am referring to science communication as communication about science "that facilitates conversations with the public that recognize, respect, and incorporate differences in knowledge, values, perspectives, and goals." (Nisbet and Scheufele 2009, p. 1767). With its emphasis on tolerance for difference in the public realm, I consider science communication in this vein to be closely related to discourses on public engagement that seek to foster two-way—dialogic as well as open-ended—conversations between various experts and publics. Too often, science communication is considered as the delivery of information to uninformed or under-informed publics, usually through methodically crafted discursive texts: lectures, scripts, chatter, or some other instance of language. However, thanks to various practitioners and scholars (e.g., Elam and Bertilsson 2003; Irwin and Michael 2003), we know that public engagement with science need not be reduced to a one-way street where experts deliver informational goods to ignorant audiences (the "deficit model" of science communication). Even in instances where more dialogic forms of communication take place, such as when two-way discussion is fostered between experts and laypublics, or communication experiences are more participatory than prescriptive, science communication scholars have been criticized for overlooking the affective dimensions of these sites. As Davies (2014) demonstrates, science communication scholars focus far too narrowly on the discursive elements (or language in use) of science communication when trying to make sense of it. Davies challenges the idea that the potential impacts on the participants in a science engagement experience depend largely on the discursive elements in that activity, media or event. Instead she shifts our gaze to the importance of site, embodiment, materiality, and emotion in understanding how science experiences make people feel in the moment and afterwards—including how they feel about science itself.

But what makes science worth engaging with for lay publics? What do we—diverse and dynamic publics—get out of science events, media, and other forms of engagement, especially when our day-to-day lives don't depend on interacting with scientific ideas? Davies offers some ideas on what might make it worthwhile:

> People (whether scientists or laypeople) generally participate in public engagement because they want to—because they find some satisfaction or enjoyment in talking about nanotechnology at a museum forum event, experiencing the spectacle of the Body Worlds exhibitions, or participating in a policy-oriented discussion. There is, we might say, a hedonism of science as leisure and pleasure, and it is this latent and largely unacknowledged reservoir of emotion that powers many of the encounters between scientific knowledges and publics. (Davies 2014, p. 101)

But a danger looms where pleasure is the only lens through which we evaluate the worth of a science communication experience. This can reinforce the old idea that knowing more about science leads to loving it: that knowledge precedes and causes interest and connection rather than the other way around. Contemporary science engagement is much more layered and complex than that simple tenet suggests. The idea that science engagement aims to make people love science smacks of getting people to love broccoli. Certainly it can be good for you, but not everyone's a fan. Some want it every day, some never, while others ebb and flow depending on the setting, occasion, and social context. In our everyday lives, we regularly feel emotional about ideas that we don't necessarily "love" or identify with. It can bring us a lot of personal satisfaction to engage that way, without leading to some kind of further behaviour or change of heart. Similarly, lay publics can feel all sorts of things about science, appreciating their experiences of contact with it beyond binary feelings of good or bad. Davies argues that the gulf between exposure to science and loving science is worth exploring.[8]

> Many scientists do think that interested publics will like science better, and become a more accepting market for its products (or perhaps be recruited into it). But is this dynamic the only one structuring expressions of interest, pleasure and delight? Can we understand them in any other terms? I would suggest that pleasure in public engagement is indeed a more complex phenomenon—one that requires further attention in order to account for and understand its role and meaning. (Davies 2014, p. 101)

Scholars of science communication and public engagement should start looking at the everyday experiences of individuals that breed identification with and interest in science, and learn what (if anything) flows from those experiences over time. As Davies says, "It is precisely when science is not taken too seriously that engagement with it becomes powerful" (p. 102). Which makes me wonder: Must we become "science lovers" for the impacts of engaging with science to be powerful in our lives? Or might we be swimming in a whole other ocean of affects, yet to be fully explored?

These questions are relevant to the discussion about synthetic biology and its public facing narratives on two levels. Firstly, the debate about synthetic biology often gets linked to the controversy around genetically modified crops that came before it (Kronberger et al. 2011). Proponents of the field strongly aspire to avoid similar controversy in the public discussion of synthetic biology. This amounts to what Claire Marris has called "synbiophobia-phobia," or the fear of a synthetic biology-fearing public. In this case, proponents of the field see "public attitudes" as a major obstacle to the contributions synthetic biology can make to the "public good" (2015). This connects to Davies' argument that science engagement often—misguidedly—aims to make publics "love" science. There is much to be explored in the space between love and fear that occupies communications about synthetic biology, and doing so might strengthen the degree of reflective engagement that is really taking place.

---

[8]I recently wrote about this idea for The Evolving Culture of Science Engagement, a collaborative research blog between MIT and Culture Kettle. http://www.cultureofscienceengagement.net/blog/2015/2/2/guest-post-embodying-engagement-with-science. Accessed 5 May 2015. Part of this paragraph is excerpted from the post. I am grateful to Peter Linett for the opportunity.

However—and this is my second point—"the public" is often imagined to react in certain ways before their voice is even made present in the debate. Marris argues that those imaginaries get inserted into public discussions in the place of their public's actual convicted views because, "They are omnipresent as disembodied, imagined, publics but absent as actual persons or organisations" (2015, p. 90). Marris gives accounts of leadership councils and national research forums on synthetic biology which have excluded real lay publics from attending, but demonstrates that when a place is made for publics, the very identity of who "the public" is can sometimes be misplaced. While installed as a social scientist in synthetic biology research networks, Marris and her colleagues have been labelled as "members of the public" at engagement fora. Their roles as social scientists involved in the field are to assess the ethical, legal, and social implications of synthetic biology, which is altogether different from standing in as a representatives for generally interested and/or concerned citizens. This red flag should signal us to slow down and consider whose voices are being included in the debate, and consider what it is that they are truly being asked.

At other times, the communications are not so ambiguous. Character sketches in film, TV and other popular media influence how publics come to think about the identities of different professionals, and in the case of synthetic biology, its practitioners have not escaped this sort of stereotyping. In order to better understand their stereotype, Meyer et al. analyzed 48 big budget films that relate to themes of engineering life and depicted the scientists doing it (2013). Their analysis traced films that portrayed the Frankenstein-like genius of old sci-fi thrillers to more contemporary storytelling. They found that science storytelling experienced a shift somewhere in its filmic history whereby films now couple the image of a scientist to industry and entrepreneurial spirit, in place of the old "mad scientist" image that Frankenstein brings to mind.

Referring to the contemporary image of the biotechnological scientist as "Frankenstein 2.0", the researchers note that Frankenstein 2.0 refutes the old stereotype of a bad or mad scientist and sketches a prominent modern character whose academic excellence is challenged by his or her own, or his or her superior's entrepreneurial drive. The most prominent recurring character found in the contemporary films they examined was a scientist whose research is driven by commercial gain. Sometimes, the scientists depicted can only reach their goal of commercial gain by knowingly crossing ethical boundaries. This develops a stereotype of the life-engineering scientist as a "worker" fulfilling business goals, who will possibly sacrifice their own scientific ambitions and morality to get there. The researchers' concluded that the way scientists engaged in genetic engineering are depicted in film has to also be seen in light of the actual societal debate about genetic technologies. When scientists are shown to be primarily profit-driven, such illustrations inform the debate about the ethical and social implications of biotechnology, and ideas about who will eventually take responsibility for them.

Although many real world synthetic biologists might not identify with their filmic stereotypes as discovered by these researchers, they cannot escape its circulation in the public imagination. It is important that public engagement practitioners and science communicators working on synthetic biology understand this image, the message it conveys, and take responsibility for how their own projects work to challenge or reinforce it. Stereotypes of emerging technologies that are little known to general publics, such as synthetic biology, influence how the technology will be perceived, debated and ultimately, accepted or disputed. But stereotypes—like any other social phenomena—are malleable, and creative communicators bear both the privileges and responsibilities of influencing public opinion about the matters of concern that could be attached to them.

3 Matters of Concern

Bruno Latour has said that "a matter of concern is what happens to a matter of fact when you add to it its whole scenography, much like you would do by shifting your attention from the stage to the whole machinery of a theatre." (2008, p. 39). Latour argues that this is the general action of "science studies" upon the sciences, which introduced a new gaze upon scientific fields that were once considered indisputable and "simply there" on their own. Science studies turned the sciences into moving phenomena that are available for dispute once the observer zooms out to focus on how they are always tethered to other social and technological apparatuses. When it comes to matters of concern, everything looks different than how we were first told they did.

In synthetic biology, the most directly obvious matters of concern are often talked about in two distinct ways that balance great hope with great fear. On the one hand, synthetic biology is said to one day possibly be able to "heal us, heat us and feed us" (Osbourne 2012). This has become common rhetoric for talking about synthetic biology and is a narrative commonly put forth in introductory writings about the field (Marris 2015). On the other hand, upon further scrutiny, we are told that synthetic biology could lead to a host of unintended social and environmental problems. Among those often listed are new forms of bioterrorism, biosafety issues, patenting concerns, as well as a growing cultural mindset that relies on "techno-fixes" instead of tackling root problems (Jefferson et al. 2014).

The hopes and perils that get communicated along with the principle that synthetic biology aims to make biology easy to engineer bolster the speculative promises of the field. This can strengthen the circulation of myths about its potential dangers in a fashion that drowns out more accurate and nuanced discussion of the field (Marris and Rose 2012). For example, an interdisciplinary consortium gathered at Kings College in London to investigate and discuss the degrees to which threat, risk and safety concerns in the discourse on synthetic biology are valid, or conversely, blown out of proportion. They concluded largely that there is too much concern over hot air. Specifically, they reported that the common narrative found in numerous media stories about synthetic biology frames its advances as "de-skilling" science, making it of greater access to non-trained professionals, and therefore susceptible to potential "dual use." This idea then asserts a strong element of technological determinism: the belief that science progresses in a linear forward-marching fashion, acquiring new successes and problems along the way in a predictive path (Jefferson et al. 2014). However, where there is little evidence to support such predictive and inescapable advancement in practice, specialists and publics are left to grapple with "speculative ethics" (Nordmann 2007).

Technological determinism paints a picture whereby synthetic biology continues to advance and accrue successes in de-skilling biology towards an unstoppable future. However, the researchers discuss that this type of thinking fails to consider the myriad real scientific challenges that synthetic biologists face while trying to make biology easy to engineer in the lab (Jefferson et al. 2014). The implication is that this failure overlooks the importance of tacit knowledge, which only expert experience can cultivate over countless hours of intimate toiling with their research, study organisms, and lab methods. This therefore significantly limits the degree to which *just anyone* can do it well. Highly skilled professionals are still very much integral to the field's successes. However, this has generally been dropped from public-facing narratives that promote synthetic biology's revolutionary potential to open up access to wider communities of practitioners. I am writing from the perspective of a practicing science communicator, and am interested here in how we might slow down the ways we talk about synthetic biology to make sure we're "getting it right" to the best of our abilities. There is room here to do a better job.

4 Slowing Down with Interdisciplinarity

Within spaces of academic expertise, several strategic interdisciplinary working teams have been formed to develop and assess synthetic biology practices "up stream" from their widespread deployment in culture and science at large. These working groups often comprise ethicists, philosophers, anthropologists, sociologists, and STS scholars that work in collaboration with scientists to address problems that may arise as the technology develops. They also work to establish deeper understandings of what is actually happening in the present state of affairs.

Rabinow and Bennet have said that these interdisciplinary models of synthetic biology and society interaction require significant refashioning when compared to similar initiatives in the biosciences that have come before (2012). They argue that earlier iterations—as seen in The Human Genome Project for example, the largest ELSI (ethical, legal, and social implications research) project to date—are not satisfactory models to follow for interdisciplinary collaboration in synthetic biology. They argue that because synthetic biology marks a new "post-genomic era" where nature is re-invented rather than merely studied, we must similarly, therefore, acquire new tools to deal with it responsibly (2012, p. 267). But it is also because ELSI research frameworks have often had the effect of bringing critical interdisciplinarity to a field *after the fact*—as an assessment that gets "tacked on" and can be "checked off"—once the humanities and social science scholars have arrived to examine the science and engineering that's already in motion.

This "down stream" approach has been criticized for missing a rich opportunity that's afforded when divergent perspectives are brought together to consider futures before they materialize, and must then be responded to (Balmer and Bulpin 2013). As a result, the shift to post-ELSI frameworks that deal with "upstream" interdisciplinarity and Responsible Research and Innovation (RRI) have become increasingly common, bringing interdisciplinary thinkers together at early stages in research networks.[9] This shift has been tied in with the fact that "social scientists are becoming a required component of synthetic biology research programmes in Europe, the US and beyond" (Calvert 2013, p. 176). It has become common that grants may not be awarded to synthetic biology researchers without the presence of social researchers on their team. Their involvement on paper is perceived as accounting for "cheques and balances" of the research in a responsible manner, especially when the social scientists arrive at an earlier rather than later stage in a synthetic biology project's development.

In theory, interdisciplinary collaborations between social and natural scientists raise critical questions on all sides of the disciplinary divides before scientific research evolves too quickly and unintended effects must then be undone. Despite the noble aims of interdisciplinarity in the field, some social scientists have reported that they suffer from constrained power in their relationships with synthetic biologists. Rabinow and Bennet gave a concerned account of their experiences as anthropologists employed at Synberc—a large American multi-university synthetic biology research centre—where they felt that their humanistic perspectives were met with dismissal, disinterest, and at times, hostility, from their scientific collaborators (2012). Jane Calvert, a social scientist who often works with synthetic biologists has said that she has at times felt that what she has to offer has been snubbed by the scientific community. She advised that non-scientists attached to the field should embrace an "ethics of discomfort" (Calvert 2013, p. 189). A manifesto was even written by a group of social scientists actively working in European synthetic biology centres, calling for new experimental forms of collaboration with scientists so that more fruitful outcomes might be discovered for all parties involved (Balmer et al. 2012).

Conflict in such collaborations is of interest to me here because of the spaces it renders visible for experimenting with new forms of communication in communities that work on synthetic biology. However, as a colleague of mine once mentioned, the pseudo re-invention of "experimental collaboration" with good etiquette may be nothing more than the end of independent funding for critical scholars.[10] This line of thinking is also echoed by the aforementioned researchers who question the validity of a post-ELSI shift. Though quite possible, I am not ready to accept such a cynical stance here. As I will show below, Isabelle Stengers

---

[9]Some have questioned the validity of this shift (Myskja et al. 2014).

[10]I am thankful to Dr. Kristin Hagen for this comment.

has some ideas to offer about slowing down thinking that can relate to conflict in communication. Her ideas might be of interest to those working across disciplinary divides, especially if they want to revitalize interdisciplinary communication through experimentation.

5 Experts, Diplomats and Idiots

Stengers has prompted expert groups to think about how it is possible to put forth proposals in their fields that do not claim "what ought to be" but provoke a "slowing down" of common reasoning. The aim of this is to create opportunities for alternative ways of being and acting around matters of concern that are relevant for their work (Stengers 2005). For Stengers, slowing down thinking in this way is about finding room to become ever so slightly more aware of the problems and situations that mobilize a group of experts to work on a problem in the first place. She is explicit in saying that her ideas are intended for practicing experts, rather than generalists. They are therefore of relevance to discussions about interdisciplinary collaborations that involve pluralist view-points from multiple disciplines in synthetic biology networks.

Although Stengers focuses her thinking on expert communities, for the purpose of her analysis she also discusses two other figures that I will refer to: the "diplomat" and the "idiot" (Stengers 2013). For Stengers, experts are the ones whose practices are not threatened by the debates that concern them, because experts are expected to present what they know and are not expected to foresee the ways that their knowledge might be taken up by others in different situations. Synthetic biologists themselves are the most obvious examples of experts in their field that operate in this way.

Contrastingly, diplomats are communication vessels for those whose identities are threatened by their practices. They serve to destabilize any homogenous nods towards unified beliefs about how science can advance "the general interest." Instead, diplomats provoke experts to think about the unexpected, dark, or skimmed-over possibilities of their beloved practices, the very existence of which make them experts. They call expert assumptions into question and show how things could have been otherwise. I therefore argue that the active diplomatic roles in synthetic biology are multiple, involving a wide variety of social researchers, artists, designers, and creative communicators who engage publics about the science in process. They also interfere with scientific thinking by posing critical questions to expert practitioners about their approaches to their research, as can be seen in the aims of post-ELSI research. The conflicts found in uncomfortable interdisciplinary research collaborations are small evidence of this function of threat that the "diplomatic" perspectives of social researchers can bring into synthetic biology. Additionally, the diverse community of practitioners who produce communications and engagement experiences with synthetic biology (for example, artists, designers, and DIYbiologists) can produce ideas that go against the grain of technological progress that expert narratives hold dear, even if they are working outside of collaborations that carry an explicitly "post-ELSI" spirit.

Furthermore, Stengers mobilizes the ancient Greek character of "the idiot" to make an argument about the function of slowing down thought in expert areas of scientific progress. The original term "idiot" was meant to describe someone who did not speak in ancient Greece and was therefore cut off from community relations. However, with the help of Deleuze's philosophical remediations of the figure, Stengers uses a refreshed version of the idiot to represent one who resists the way in which a situation is presented. The idiot does not feel the urge to fall into consensus with any situation at hand, and importantly, by not doing so, he or she galvanizes others around them to call their own situations into question as well. The idiot does not interact with science as such, but with the politics and economies of it. The idiot is therefore a catalyst of political thought and action through their refutations, moving others to ask "what is more important that we should be doing?" But the idiot is not righteous and does not know the answer. For Stengers, the idiot is key, not because he or she creates unending perplexity for a situation at hand, but because the idiot makes us slow down and question our own assumptions. The idiot's actions make us look for what ever made us feel we were authorized "to believe we possess the meaning of what we know" in the first place (2005, p. 995).

And so, diplomats and idiots are not the same, but they both "jam the system" of expert knowledge, and shift the automatic associations it carries that bind speed to concepts like efficiency, innovation, and growth. Returning now to Jefferson et al.'s argument that much of the public discourse of synthetic biology is based on technological determinism, something is needed to "jam the system" in how we think and talk about synthetic biology *in public* and *with publics*. Where is the idiot who can ask those who make deterministic statements about synthetic biology how they came to believe they possess the meaning of what they know? Diplomats (who are found in critical scholarship that questions the assumptions and beliefs in synthetic biology) and idiots (who 'misbehave' and disregard the way things are designed to function (see Michael 2012) can help us slow the pace to see better what we think we know. The actions of both figures—albeit in different ways—create space for pause, reflection, and the careful choosing of how to direct concerns and questions towards the science itself.

The question then is, how can science communicators and engagement practitioners create opportunities for a slowing down of thinking that publics can partake in when pondering a particular story about an emerging technoscience? What can be done to facilitate "diplomatic" and "idiotic" behaviour in a communication landscape that all too often, still defers to the deficit model of science communication, and relies squarely on the expert voice as the default of reason? It's all fine and well to have esoteric theoretical notions as inspirations for approaches to creative communications, but figuring out how to employ such theories in practice is important to try and resolve. For the remainder of this article, I will argue that there are rich sites of opportunity for science communication in the case of synthetic biology to create narratives more "diplomatically" and "idiotically" than it often does, pointing to examples that have already achieved this.

6 Concern in Science Communication and Public Engagement

The gesture of *slowing down* that's offered here is an idea intended to guide towards avenues of thinking and doing that might enable general audiences to develop a personal analytic ethics towards synthetic biology; one that encourages listening, caring and empathy exchange between different polyphonous points of view. Returning to Sarah Davies' line of thought, we should be paying attention to "the non-discursive—to the role of, for instance, the emotional, material or creative within public engagement" (2014, p. 94). For scholars of public engagement and science communication, slowing down might help us notice these often overlooked elements, and identify new opportunities for creative intervention.

Davies argues that when science communication focuses too much on the content and discursive elements of public engagement work it misses an opportunity to dive into the affective ability of the non-rational and chaotic elements of that content to make people literally *feel some way* about science. Science dialogue events for example can be "dramatic and emotional" when people come face to face in public to discuss a scientific topic (2011, p. 94). However science communication researchers may boil down the effectiveness of engagement events to a series of evaluation criteria that overlook any of its sensory elements. Science communication and engagement projects ranging from lectures to workshops, gallery exhibits, films, radio broadcasts, interactive installations and more have inherent sensory dimensions that prop up our experience of them. Admittedly, some of these formats enable more participation with publics than others, but even those that communicate more unidirectionally than not have non-discursive elements to them. The smell of an art installation, the colours on a screen, the tone of voice being used, the speed at which one travels through an exhibit, and all other embodied aspects of engagement work are perceived on levels that often remain invisible to scholarship of them. But as you may recognize from personal experience, these material aspects can significantly affect how you come to think and feel about the experiences and ideas caught up in such works.

Creative communicators themselves should not forget the importance of how materials, sites, and temporalities in their own work inform and shape the public's experience—and eventual understanding—of a field like synthetic biology. Although artists and designers may possess special knowledge from their education about the affects such elements produce (aesthetically, phenomenologically, emotionally), other creative communicators of science (e.g., writers, journalists, panelists, speakers) might do well to explore their potential as well.

The way in which creative public engagement work about synthetic biology is materially presented has real possibility to influence how audiences think about the field. For example, at a bioart exhibition in Vienna called Synth Ethic that featured works commenting on synthetic biology, researchers found that the majority of visitors to the exhibition sensed no ethical problem with the use of bacteria and other "lower" organisms in any works of living art. However, when animals of higher biological complexity or humans were posed as part of the artwork, real ethical discomfort ensued. The gallery visitors expressed a general need for boundaries to be implemented around how far certain biotechnologies and their practical attachments should be allowed to be developed, along the lines of organismal complexity (Kerbe and Schmidt 2013).

The registers of representation in the works at the exhibition differed (and were perceived differently) most obviously in terms of the complexity of the living organisms involved. Therefore the organisms used had the power to sway public opinion in terms of the degree of "wrongness" of the artwork, reverberating into opinions about the more traditional practice of the science itself. In this sense, the artist's choice of organism becomes a decisive keel in the way that people think about the technology employed. It becomes a material choice that the artist must then take some responsibility for. In this sense, the choices of the artist become tools that can, in part, influence the debate.

This leaves me with two questions that I find help me in my own science engagement productions: (1) How might creative communicators—from a diversity of disciplines—benefit from looking at the ripple effects that artistic and material choices have in spaces of public imagination, discussion and dissent? More evaluative research is needed in order to carry this out properly. And (2), How does that knowledge stay the same, or change, across different media platforms, art/science collaborations, and beyond? I'm not suggesting that consideration of this line of questioning will serve all science engagement practitioners well, or in any particular way. I am suggesting that these questions are just one potential direction, among many, that could be helpful to think about for building reflexivity into science engagement experiences. We need to commence public engagement projects at narrative design stages that we deliberately locate far away from assuming what publics might be thinking (Marris 2015). Similarly, we should distance our work from any aims to make publics feel affectionate for science, instead creating space for emergent feelings to flow forth from engagement experiences (Davies 2014). In this sense, I am arguing for more deliberate experimentation with how creative communication can deepen and slow down the ways we tell and share stories about synthetic biology in "diplomatic" and "idiotic" ways.

7 Art and Design in the Agora of Synthetic Biology

In ancient Greece, the agora was the city's public meeting place: the heart of artistic, political, spiritual and athletic life. As one can imagine, it was where topics that carried significance for debate were openly discussed among different publics. As Eleanor Pauwels points out, in synthetic biology, "matters of concern create an 'agora'; they create political conditions for dissenting imagination" (2010, p. 1447). Today, synthetic biology is a matter of concern to heterogenous groups in heterogenous ways, and creates a modern metaphorical agora at their sites of intersection (Pauwels 2010).

Scientists, policy makers, anthropologists, philosophers, science hobbyists, artists, designers, watch-dog organizers and other concerned citizens can be found opening up, closing down, and straddling space for deliberation in the synthetic agora. The polyphony found there, in different ways with each voice, adds complexity to the matters of concern at hand. This arrangement of expert and non-expert scientists in biotechnological debates echoes what researchers have shown to be trending in European science policy and communication: there has been a shift towards the language and involvement of nonscientists, such as laypeople and citizen scientists in public engagement, participation and dialogue strategies on science (Lengwiler 2007). The work being done here amounts to an unstable moving target that comprises diverse worldviews from different disciplines. Critical public engagement practices that flow forth from these sites can explore topics much broader than the "de-skilling" and "dual use" narratives that have become so familiar. At the same time, there is a danger is cloaking such polyphonies as solutions that should be embraced when combatting communication break-down or disciplinary tensions in evolving fields like synthetic biology. There are, nevertheless, real threats bound up in new technologies that specialists must address with the skills that they've honed.

But when it comes to the value of polyphonic thinking, Andrew Pickering incites hope in the function of art as something that can allow us to slow to a hover, think critically, and resist "high modernist adventures" as James Scott would put it (Scott in Pickering 2010). High modernist adventures "aim at the rational reconstruction of large swathes of the material and social worlds and remind us of their often catastrophic consequences" (2010, p. 392). Centuries of industrialization on one hand and climate change on the other would be a stock example. In synthetic biology, interdisciplinarity across the sciences, social sciences and humanities that includes art and design can help us slow down to acknowledge both sides of the supposed opposition at play. That opposition being the fruits of the shift to a 21st century model of bio-fabrication versus the plethora of unintended consequences it has been argued synthetic biology may present. But interdisciplinarity across the arts and sciences, on account of its sheer existence, does not point out simple solutions. Each collaboration or intervention across the disciplinary divides is its own story, and should be treated accordingly.

For example, artists Oron Catts, Ionat Zurr and Corrie van Sice exhibited a work called "The Mechanism of Life—After Stéphane Leduc" at the Science Gallery in Dublin as part of the "Grow Your Own" exhibition. The artwork reflected on the origin story of how we have come to use the term synthetic biology (2013). It is often written that the first person to use this term was a French scientist named Stéphane Leduc. He believed that he had found evidence in the lab that life was merely a chemical mixture, devoid of any metaphysical forces (Keller 2003). In the early 20th century, Leduc created organic-looking life-like entities in the laboratory by mixing metals, oils and inks. His creations had filamentous growth patterns, which are now understood to be caused by osmotic pressure, but at the time were used to speculate about the ability to engineer of life from scratch. For Leduc, these life-like blobs showed that living things could be created according to a chemical recipe, giving rise to what he called "La Biologie Synthétique" (1912).

Catts, Zurr and van Sice's work commented on today's synthetic biology, which although paradigmatically different from Leduc's early version of it, is still underpinned by a belief that that life can be reduced to its interlocking chemical parts, and consequently, engineered. This mindset allows synthetic biologists to design and build living systems as though they are guided by simple mechanistic processes that do not amount to something greater than the sum of their parts. To explore this critique, the artists re-created Leduc's early chemical experiments in the gallery using a rapid prototyper, or 3D printer. Rapid prototyping has been highly anticipated as a 21st century tool that will revolutionize the sustainable fabrication of goods through distributed manufacturing. Synthetic biology has been the subject of this same narrative in its marketing, where it is said that we will one day be able to make everything from medicines to perfumes at home with desktop DNA synthesizers.

In the artwork, the rapid prototyper mixed chemicals that Leduc had worked with more than a century earlier. But where Leduc claimed that their interlocking reactions constituted synthetic life, the artists exposed them to the public in a gallery as nothing more than what they are: droplets of mixed chemicals. By doing this experiment in an art setting, visitors could see that these droplets of an "engineered life force" only lasted a few seconds before dissipating into an entropic, murky, chemical soup. If Leduc saw the same soup come to life after his synthetic form of "life" had died, its memory doesn't hold up well in our contemporary imagination. The artwork pries open the closed nature of "performative statements" about synthetic biology that step in at any point, from an expert position, to not only describe the world it is commenting on but do something in it. In synthetic biology for example, performative statements that declare its ability to make biology easy to engineer, end up unifying its nearly irreconcilably diverse practitioners in real life through this stated vision (Bensaude Vincent 2013). The artwork opens up rather than closes down questions by bringing the past and present into the same field of consideration. This reminder of now-debunked historic thinking is enlightening, because as Bensaude Vincent has argued, "futuristic visions are so attractive that they blind the past" (2013, p. 28).

When read through Stengers, on the one hand, the artists here become the *diplomats* whose work questions the rhetoric of the *expert* (an historical scientist). On the other hand, the droplets become the *idiots* that force viewers to question the situation in front of them, without telling them what to think. The work comments on a tension between what experts say they know—that life can be designed through mechanistic reactions—what diplomats reveal—the limits of expert knowledge—and what the idiots force us to re-evaluate by breaking the system. "The Mechanism of Life—After Stéphane Leduc" is one small example by which these ideas can be unveiled to publics through "diplomatic" and "idiotic" behaviour.

I do not mean to imply though that from following a legacy of bioart into the present and bringing non-traditional players into scientific spaces, that critical discourse is automatically generated for richer public discussions of synthetic biology. Interdisciplinary experimentation does not inherently allow for better descriptions of the world than that provided by science alone in synthetic biology. It is not enough to say that bringing artists into the lab brings critique into the lab; they're not the same thing. Similarly, making hacker spaces for DIYbio does not necessarily attain the ideal of the dawn of a democratic biotechnological era. In some cases it might not do anything other than fetishize the transformation of biotechnologies into personal technologies for new commercial markets, as has occurred with personal computing (Tochetti 2012). Rather, there are asymmetries at play here that we might like to take note of. Slowing down to listen to the nuanced discussions that are laced throughout the discourse, but are not always prioritized, might help us do that.

Synthetic Aesthetics is a useful case to explore as an experimental example of public engagement in synthetic biology. Synthetic Aesthetics is one of the most visible and ambitious projects of interdisciplinary collaboration to date that brought artists, designers and social researchers into the matrix of synthetic biology in order to work towards open-ended, inquiry driven outcomes that can stimulate public discussion. It was a research and creation residency program funded by the US National Science Foundation and UK Engineering and Physical Sciences Research Council that brought synthetic biologists together with artists, designers, a social scientist and a philosopher to produce works that question how people might like the technology to be used.

Since synthetic biologists and designers share an interest in creating new solutions for found problems, it has been argued that as biology becomes a product of design choices over evolutionary pressures, collaborations between synthetic biologists and designers may be able to shed light on "how to design life well" (Ginsberg et al. 2014). The Synthetic Aesthetics residents considered this question at a time when international efforts were (and still are) being made to find solutions for new sustainable manufacturing techniques, carbon neutral fuels, cheap drugs and rare high-value materials. Through six paired teams consisting of one synthetic biologist and one artist or designer each (including Oron Catts, mentioned earlier), their collaborations sought to explore and/or question how and why we might now work to attain these techniques and commodities through biological design.

The project launched in 2010, and engaged the artist, designer and scientist residents to produce interdisciplinary works that emerged partly in the artist or designer's studio and partly in the scientist's lab over several months. Their creations and collaborative processes were studied by the resident philosopher, social scientist and overseeing design and science fellows who helped facilitate the initiative. To conclude the residency experiment, Synthetic Aesthetics became a book that uses the project as a frame through which to explore the role of *design* in a world where scientists are "designing nature." It fostered an interdisciplinary space for discussion and included writings from the social sciences, humanities, science, engineering and design researchers attached to the project (Ginsberg et al. 2014). The residents also made several public appearances throughout the duration of the program, sharing their projects and processes of collaboration with audiences at galleries, festivals, conferences and in on-air interviews. The project outcomes ranged from speculative packaging that grows itself, to cheeses produced with bacterial strains found on the human body (although the bacteria didn't even require genetic engineering for their creation), to philosophical reflections on the cultural motivations for the field. Though not an exhaustive list, these project outcomes show how varied the residents' approaches to experimentally interrogating synthetic biology were. Each team made artifacts or conceptual works that engaged their publics through varying levels of provocation, humour and critique.

In some instances, the creative outcomes of Synthetic Aesthetics are tied into a larger creative movement known as speculative design. Speculative design uses design as a conceptual tool in the vein of art to produce objects and experiences that explore and present possible futures to audiences. Speculative designs work at the nexus of where the technological and the social intertwine, prodding people to interpret what they think such possible future technologies suggest for their own lives, through narratives that the designs themselves propose (Dunne and Raby 2013).

One widely circulated example of speculative design in synthetic biology, which opens up the question of "how to design life well" comes from designers Alexandra Daisy Ginsberg (who was design fellow on the Synthetic Aesthetics project) and James King. The designers created a probiotic yogurt drink containing synthetic microbes that can detect and report health problems in the gut by speckling one's faeces with a specific colour that correlates to a particular issue that the bacteria sensed. The yogurt drinkers would therefore know they need to visit the doctor while they otherwise might have remained oblivious until other, perhaps more unfortunate symptoms made themselves apparent. Their project—$E.\ chromi$—was developed with the 2009 iGEM competition-winning team from Cambridge University and is based on real research in synthetic biology where synthetic bacteria are able to produce the pigments they exhibit in their "Scatalog.[11]" However, the yogurt drink does not (yet) exist as a market product. The Scatalog has travelled with the designers to several exhibitions, conferences and festivals where it serves as a catalyst for conversation with publics. Peering at the multi-coloured excrement, viewers can weigh in on if they think this would be a good social use of the technology, opening a window to further discussion about the implications of synthetic biology itself.

Based on their potential for future synthetic biology-based product formation, speculative designs have been accused of being disguised advertisements for the field that sell synthetic futures to audiences in the present. The well-crafted, sleek and seductive designs show how an engineered biology can function not only on the genetic, but social scale. But because they tell stories about how emerging technologies might affect our lives, speculative designs have been subject to much casual debate at synthetic biology conferences, festivals and workshops I've attended about how they risk being instrumentalized as marketing tools. Despite these off-record debates, many speculative designers, such as those involved in Synthetic Aesthetics, have stated clearly that their work opens up speculative

---

[11]http://www.echromi.com/. Accessed 04 March 2015.

futures that invite emergent interpretations to be formed in the minds of their audiences, rather than prescribe judgment about how the technology should be supported.

If we are to take that argument to heart, speculative design has more in common with scientific media about synthetic biology (for example, as produced by the press) than may be obvious at first glance. As Todd Kuikens and Eleanor Pauwels show, "the press (concerning synthetic biology) may not tell the public what to think, but by covering topics it often tells them what to think about" (Kuiken and Pauwels 2012, p. 2). Similarly, speculative design becomes another platform for "thinking about" the field. However, its propensity towards marketing synthetic biology (by supporting expert narratives) or critiquing it (diplomatically, idiotically) cannot be understood at face value and must be considered on a case-by-case basis.

Interdisciplinary constellations across the arts and sciences seem to have caught on for purposes of experimentation, idea incubation and public engagement concerning synthetic biology. For example, the European Commission funded Studiolab, a 3-year research initiative that merges the art studio with the research lab, where participants focus heavily on bridging the divide between science, art and design in synthetic biology. They too have made a publication from their findings about bioart, biodesign, DIY biohacking, ecological art, and A-life art.[12] Similarly, the European Commission funded the larger 4-year umbrella project Synenergene, which mobilizes a wide variety of art, design, and hacker approaches to fostering public dialogue about synthetic biology. Although the involvement of artists and designers in an emerging technoscientific field like synthetic biology may sound highly esoteric, there are several research programs, residencies, granting agencies and individuals supporting this type of interdisciplinary experimentation. Similar to what can be found in the biohacking scene, these artistic activities introduce individuals to the field through non-traditional means, either as creators, participants or witnesses to interdisciplinary works that are displayed in diverse cultural settings (ex: galleries, blogs, documentaries). From having followed the field's growing activities in this sense since 2010, it seems reasonable to believe that these types of interdisciplinary projects will continue to evolve for the time being, attracting wide-ranging publics to take part in their unfolding.

8 Making Room for Science Communication to Get More Diplomatic and Idiotic

Professional scientists and their publics are accustomed to being invited to interact in formalized engagement events (for example: the museum lecture with time for Q and A). But what room is there to allow individuals' feelings from both sides of the

---

[12]http://studiolabproject.eu/synthetic-biology. Accessed 04 March 2015.

professional/public divide to become a bit more subjective, personal and playful? Would the affects of communication change if the fora of communication were different? What power could more sensory-based science communication (that takes inspiration from the works of artists, designers and other creative communicators) have in diversifying the widely circulated discourse, like "dual use," that dominates public discussion (Marris et al. 2014)? How can creative approaches to public engagement with synthetic biology help reduce the effects of "synbiophobia-phobia" (Marris 2015), bringing expert practitioners into more sober relationships with their publics?

By slowing down the claims in public-facing science stories and allowing for personal inquiries and feelings to shape their surface through experimental productions, can we foster productively sensitive (although always subject to failure) dynamics between communicators and publics? Can this process be helped by lessons that are being learned in interdisciplinary synthetic biology research networks that include social scientists, about what it means to collaborate and communicate *well* across the disciplines? This line of questioning is in no way meant to lead to a path that can "smooth things over" or homogenize discussion into some form of consensus between various publics on matters of concern in synthetic biology. It is meant to enlighten and entertain and perhaps inject some surprise into how we discuss our biotechnological futures in the "post-genomic era," through experimental public engagement.

When compared to discourses on other matters of concern involving technoscience (such as geoengineering for example), the societal discussion on synthetic biology is still young. This helps make it particularly well positioned to benefit from experimental and innovative approaches to communication that raise it in the public imagination, and can then circulate back into spaces of expert work. The philosophical push to slow things down in order to experiment with multiple ways of narrativizing the field might help to involve more individuals in the field's unfolding. This inquiry takes 'post-ELSI' research seriously, which declares that we need new experiments in knowledge production between scientists, social researchers and their publics that are "pluralist, reflexive, and promote mutual learning" (Balmer et al. 2012; Fitzgerald et al. 2014; Rabinow and Bennet 2012). Perhaps by focusing on how public engagement and science communication can make one *feel*—and not simply learn—in order to dialogically respond and pose further open-ended questions, practitioners can find routes towards Stengers' sense of *slowing down* the processes we use to think about expert technoscience. By focusing sincerely on the affective qualities of the engagement experience, communicators might be able to embody sites and experiences that challenge our biased responses like the diplomat or idiot would, creating space for multiple potentials to come forth in the place of over-confident, pinned-down narratives. This could help us move beyond deferral to expertise without also considering the other voices that polyphonies show are always present, but often enjoy less privilege of exposure.

References

Agapakis CM (2014) Designing synthetic biology. ACS Synth Biol 3:121–128. doi:10.1021/sb4001068

Balmer AS, Bulpin K, Calvert J, Kearnes M, Mackenzie A, Marris C, Martin P, Molyneux-Hodgson M, Schyfter P (2012) Towards a manifesto for experimental collaborations between social and natural scientists. https://experimentalcollaborations.wordpress.com/. Accessed 05 Dec 2014

Balmer AS, Bulpin KJ (2013) Left to their own devices: post-ELSI, ethical equipment and the International Genetically Engineered Machine (iGEM) Competition. BioSocieties 8:311–335. doi:10.1057/biosoc.2013.13

Bakhtin M, Caryl E (1984) Problems of Dostoevsky's poetics. University of Minnesota Press, Minneapolis

Bensaude Vincent B (2013) Between the possible and the actual: philosophical perspectives on the design of synthetic organisms. Futures 48:23–31. doi:10.1016/j.futures.2013.02.006

Calvert J (2010) New forms of collaboration: synthetic biology, social science, art and design. Presentation at Synbio in society: toward new forms of collaboration? Woodrow Wilson Center, Washington DC, USA, 12 May 2010

Calvert J (2013) Collaboration as a research method? Navigating social scientific involvement in synthetic biology. In: Doorn N, Schuurbiers D, van de Poel I, Gorman ME (eds) Early Engagem. New Technol. Open. Lab. Springer, Dordrecht, pp 175–194

Catts O, Zurr I, Van Sice C (2013) The mechanism of life—After Stéphane Leduc. In: Science Gallery, https://dublin.sciencegallery.com/growyourown/mechanismlife%E2%80%94afterst%C3%A9phaneleduc. Accessed 16 Dec 2014

Church GM, Elowitz MB, Smolke CD et al (2014) Realizing the potential of synthetic biology. Nat Rev Mol Cell Biol 15:289–294. doi:10.1038/nrm3767

Davies SR (2014) Knowing and loving: public engagement beyond discourse. Sci Technol Stud 27 (3):90–110

Dunne A, Raby F (2013) Speculative everything: design, fiction, and social dreaming. MIT Press, Cambridge

Elam M, Bertilsson M (2003) Consuming, engaging and confronting science the emerging dimensions of scientific citizenship. Eur J Soc Theory 6:233–251. doi:10.1177/1368431003006002005

Endy D (2005) Foundations for engineering biology. Nature 438:449–453. doi:10.1038/nature04342

Fitzgerald D, Littlefield MM, Knudsen KJ et al (2014) Ambivalence, equivocation and the politics of experimental knowledge: A transdisciplinary neuroscience encounter. Soc Stud Sci 44 (5):701–721. doi:10.1177/0306312714531473

Ginsberg AD, Calvert J, Schyfter P et al (2014) Synthetic aesthetics: investigating synthetic biology's designs on nature. The MIT Press, Cambridge, MA

Grushkin D, Kuiken T and Millet P (2013) Seven myths and realities about do-it-yourself biology. SYNBIO 5. Woodrow Wilson Center

Irwin A, Michael M (2003) Science, social theory and public knowledge. Open University Press, Maidenhead, Philadelphia

Jefferson C, Lentzos F, Marris C (2014) Synthetic biology and biosecurity: how scared should we be? Kings College London

Kelle A (2009) Synthetic biology and biosecurity. EMBO Rep 10(1S):S23–S27. doi:10.1038/embor.2009.119

Keller EF (2003) Making sense of life: explaining biological development with models, metaphors, and machines, 1st edn. Harvard University Press, Cambridge, MA

Kerbe W, Schmidt M (2013) Splicing boundaries: the experiences of bioart exhibition visitors. Leonardo. doi:10.1162/LEON_a_00701

Kronberger N, Holtz P, Wagner W (2011) Consequences of media information uptake and deliberation: focus groups' symbolic coping with synthetic biology. Public Underst Sci 0963662511400331. doi:10.1177/0963662511400331

Kuiken T, Pauwels E (2012) Beyond the laboratory and far away. Policy brief. Woodrow Wilson Center for International Scholars

Latour B (2008) What is the style of matters of concern? Uitgeverij Van Gorcum

Leduc S (1912) La biologie synthétique. In: Poinat A (ed) Étude de biophysique. Peiresc, Paris

Lengwiler M (2007) Participatory approaches in science and technology: historical origins and current practices in critical perspective. Sci Technol Human Values 33:186–200. doi:10.1177/0162243907311262

Marris C (2015) The construction of imaginaries of the public as a threat to synthetic biology. Sci Cult 24(1):83–98. doi:10.1080/09505431.2014.986320

Marris C, Rose N (2012) Let's get real on synthetic biology—opinion—11 June 2012—New Scientist. http://www.newscientist.com/article/mg21428684.800-lets-get-real-on-synthetic-biology.html#.VI399aaGNMw. Accessed 14 Dec 2014

Marris C, Jefferson C, Lentzos F (2014) Negotiating the dynamics of uncomfortable knowledge: the case of dual use and synthetic biology. BioSocieties 9:393–420. doi:10.1057/biosoc.2014.32

Meyer A, Cserer A, Schmidt M (2013) Frankenstein 2.0.: identifying and characterising synthetic biology engineers in science fiction films. Life Sci Soc Policy 9:9. doi:10.1186/2195-7819-9-9

Michael M (2012) "What are we busy doing?" Engaging the Idiot. Sci Technol Human Values 37:528–554. doi:10.1177/0162243911428624

Myskja BK, Nydal R, Myhr AI (2014) We have never been ELSI researchers—there is no need for a post-ELSI shift. Life Sci Soc Policy 10:1–17. doi:10.1186/s40504-014-0009-4

Nisbet MC, Scheufele DA (2009) What's next for science communication? Promising directions and lingering distractions. Am J Bot 96:1767–1778. doi:10.3732/ajb.0900041

Nordmann A (2007) If and then: a critique of speculative nanoethics. NanoEthics 1:31–46. doi:10.1007/s11569-007-0007-6

Osbourne G (2012) Speech by the chancellor of the exchequer, Right Honourable George Osborne MP, to the Royal Society, November 9, London, Royal Society

Pauwels E (2011) The value of science and technology studies (STS) to sustainability research: a critical approach toward synthetic biology promises. In: Jaeger CC, Tàbara JD, Jaeger J (eds) European research on sustainable development. Springer, Berlin, pp 111–135

Pauwels E (2010) Who let the humanists into the lab. Val UL Rev 45:1447

Penders B (2011) DIY biology. Nature 472:167

Pickering A (2010) The cybernetic brain: sketches of another future. University of Chicago Press

Rabinow P, Bennett G (2012) Designing human practices: an experiment with synthetic biology. University of Chicago Press

Schmidt M (2008) Diffusion of synthetic biology: a challenge to biosafety. Syst Synth Biol 2:1–6. doi:10.1007/s11693-008-9018-z

Sherkow JS, Greely HT (2013) What if extinction is not forever? Science 340:32–33. doi:10.1126/science.1236965

Stengers I (2005) The cosmopolitical proposal. In: Latour B, Weibel P (eds) Making things public: atmospheres of democracy. MIT Press, pp 994–1003

Stengers I (2013) Introductory notes on an ecology of practices. Cult Stud Rev 11:183–196

Tochetti S (2012) DIYbiologists as "makers" of personal biologies: how MAKE Magazine and Maker Faires contribute in constituting biology as a personal technology. Journal of Peer Production. http://peerproduction.net/issues/issue-2/peer-reviewed-papers/diybiologists-as-makers/. Accessed 16 Dec 2014

Early Engagement with Synthetic Biology in the Netherlands—Initiatives by the Rathenau Instituut

Virgil Rerimassie

1 Introduction

In the past, many developments in biotechnology have stumbled on (various) societal concerns, certainly when looking at the European Union. Consider, for instance, the fierce—and still unsettled—public controversy on agricultural biotechnology in the European Union (Levidow and Carr 2010). In the wake of such polarized discussions, the need to align technological developments more explicitly with societal values has been growing stronger. Against this backdrop, several organizations have initiated early assessments of potential ethical, legal and societal implications (ELSI) of emerging technologies and stimulated early public engagement thereon. This particularly applies to synthetic biology. Synthetic biology stands for the latest phase in the development of biotechnology, in which scientists are gaining increasing control over the fundamental biological building blocks, allowing the design of biological systems which display functions that do not exist in nature (NEST High-Level Expert Group 2005). Synthetic biology is developing very quickly and may help in finding solutions for important societal challenges, such as providing sustainable energy and realizing a biobased economy. At the same time, synthetic biology is not without risks and moreover raises challenging ethical questions (cf. Stemerding and Rerimassie 2013). Given these tensions and previous experiences with biotechnologies, early engagement activities seem justified, even though synthetic biology is still (predominantly) confined to the laboratory.

The Rathenau Instituut, the Dutch office for technology assessment (TA) and science system assessment, is one of the organizations that addressed synthetic biology early on in its development. This chapter aims to describe and reflect on

V. Rerimassie (✉)
Rathenau Instituut, PO Box 95366, 2509 CJ The Hague, The Netherlands
e-mail: v.rerimassie@rathenau.nl

© Springer International Publishing Switzerland 2016
K. Hagen et al. (eds.), *Ambivalences of Creating Life*, Ethics of Science and Technology Assessment 45, DOI 10.1007/978-3-319-21088-9_10

how the Rathenau Instituut has facilitated public engagement with synthetic biology. To illustrate the context in which such public engagement activities have been undertaken, the chapter starts with a sketch of the state-of-the-art of synthetic biology and the institutional position of the Rathenau Instituut. The timeframe during which the described activities have taken place ranges from about 2006 to (early) 2015. However, rather than listing a chronological description of the activities of the Rathenau Instituut with regard to synthetic biology, the activities are examined within a framework developed by Van Est et al. (2012) to analyze the governance of nanotechnology in the Netherlands. This framework distinguishes public engagement activities relating to three different spheres of the science and technology governance landscape: the *political* sphere, the *science and technology* sphere and the *societal* sphere. Furthermore, it distinguishes between *informing* and *engaging* activities.

2 Context of Engagement Activities

In order to understand the engagement activities of the Rathenau Instituut it is useful to draw a picture of the context in which such activities have taken place. Two factors that enable and constrain the activities will be discussed: the current state-of-the-art of synthetic biology and the institutional position of the Rathenau Instituut.

2.1 The State-of-the-Art of Synthetic Biology

The state-of-the-art of the technology that is made subject to public engagement activities is a major factor in how such activities can be set up. Public engagement regarding a well-established technology comes with different challenges than public engagement regarding a technology that is still predominantly confined to the laboratory. For instance, a public engagement activity concerning nuclear power may be enabled by the presence of a broad range of active stakeholders, who can easily be mobilized. On the other hand, its effectiveness with regard to achieving a constructive dialogue may be constrained by the vested interests. In contrast, the effectiveness of public engagement activities regarding technologies that are still in an experimental phase may be constrained by the lack of active stakeholders. At the same time, the lack of vested interests and tensions might enable a meaningful open discussion.

Synthetic biology falls in the latter category. The field is developing rapidly, but is so far predominantly confined to the laboratory. Correspondingly, the number of stakeholders engaged with synthetic biology is still limited. The public debate and even awareness about this emerging technology is limited both in the Netherlands

(Stemerding and Van Est 2013) and internationally (European Commission 2010; Pauwels 2013; Rerimassie et al. 2015). This does not detract from the fact that meanwhile an international debate on synthetic biology is taking shape, although mainly in academic circles (Rerimassie et al. 2015).

2.2 The Institutional Position of the Rathenau Instituut

The history of the Rathenau Instituut can be traced back to the demand of the Dutch government and Parliament to set up a bureau that would signal and study both the potential positive and negative societal aspects of science and technology. Moreover, it should stimulate societal opinion making and bring these insights and opinions into the political decision making process (Ministerie van Onderwijs en Wetenschap 1984). The Dutch Parliament did not plea for an organization within or near the Parliament itself, but opposed the idea to set such an organization up within a ministry. Rather, it had to be placed more at arm's length of the government, as to guarantee its independence (Van Est 2013). The formal description of the technology assessment task of the Rathenau Instituut reads as follows:

> [t]he role of the institute is to contribute to societal debate and the formation of political opinion on issues that relate to or are the consequence of scientific and technological developments. This specifically includes the ethical, social, cultural and legal aspects of such developments. In particular, the institute facilitates the formation of political opinion in both chambers of the Parliament of the Netherlands and in the European Parliament (OCW 2009, p. 1, derived from Van Est 2013).

According to Van Est[1] (2013) the institute's position towards the realms of Parliament, government, science and society has a dual nature. On the one hand it is positioned in the 'heart' of the scientific community, namely the Royal Netherlands Academy of Arts and Sciences (KNAW). However, the institute is not doing research for scientific reasons primarily, but aimed at contributing to the societal and political debate. In addition, it has a rather autonomous position within the KNAW, such as having an independent board. The relationship with the political realm is dual: on the one hand, it is positioned at some distance from the political process. This is in contrast to e.g. the French TA bureau, the 'Office Parlementaire d'Evaluation des Choix Scientifiques et Technologiques', where members of parliament conduct assessments themselves, or the Office of Technology Assessment at the German Bundestag (TAB), where activities are closely monitored by members of Parliament (Ganzevles et al. 2014). On the other hand, Parliament and the ministries are the main clients of the institute. Unsurprisingly, it is also physically located in The Hague, the political center of the Netherlands. Finally, the institute has no formal bonds with any societal organization, but at the same time is

---

[1]One of the coordinators of the Rathenau Instituut.

dedicated to stimulating public debate and henceforth actively searches for connections with relevant societal actors.

Evidently, this institutional position influences the organization's public engagement strategies regarding synthetic biology. Its independent position vis-à-vis government and Parliament allows freedom to determine its strategies but does not guarantee an audience.

3 Informing and Engaging in Different Social Spheres

In order to highlight and better understand the Rathenau Instituut's public engagement activities in synthetic biology, they will be structured along a framework developed by Van Est et al. (2012). In line with this framework, public engagement is broadly understood in the sense that it encompasses all kinds of activities aimed at bringing in a 'public perspective' into the development of an emerging technology. Thus:

> '[p]ublic perspective' signifies all sorts of ethical, social and regulatory issues, which go beyond 'narrow' innovation and economic aspects of S&T development (Van Est et al. 2012, p. 7)

Public engagement activities are divided into two categories, namely those aimed at *informing* and those aimed at *engaging*. In brief, activities aimed at informing can be understood as one-way communication, while those aimed at engaging encompass two-way dialogue in which actors interact to identify problems and stimulate the development of desirable solutions.

Furthermore, the activities are structured along three spheres of the science and technology governance landscape: *the political sphere, the science and technology sphere,* and *the societal sphere.* The political sphere primarily encompasses Parliament, but also the government (ministries, agencies and their civil servants). Next, activities may also be aimed at the realm of science and technology, e.g. university or industry researchers and technology developers. Consider for instance the Dutch tradition of constructive technology assessment, aimed at broadening the design of new technologies through the feedback of technology assessment activities into the actual construction of technology (Schot and Rip 1997). Finally, the societal sphere encompasses activities concerning civil society, trade and labor unions and (members of) the general public. An overview of the different types of activities and spheres is provided in Table 1.

In the remainder of the chapter, key public engagement activities of the Rathenau Instituut concerning synthetic biology (primarily in the Dutch context) will be discussed along these distinctions. However, it should be noticed that these activities will rarely fit only one specific category. In fact, looking for synergy and seeking to establish connections between the different spheres turns out to be an important feature of several engagement activities. Nevertheless, the framework provides a useful tool to understand them and discuss their impact.

Table 1 Overview of types of activities aimed at informing or engaging in order to integrate ethical and social aspects into the societal sphere, the science and technology sphere, and the political sphere

	Societal sphere	S&T sphere	Political sphere
Informing	*Aim*: One-way communication to inform lay citizens *Label*: Public understanding of science	*Aim*: ELSI-research to timely signal problems and inform researchers to stimulate development of desirable solutions *Label*: classical ELSI-research, upstream reflection	*Aim*: TA research to timely inform MPs *Label*: Classical parliamentary TA
Engaging	*Aim*: Two-way communication between citizens, experts and policy makers; TA to stimulate the public debate on science and technology *Label*: Participatory TA, public dialogue, upstream public engagement	*Aim*: Engaging scientists in a two-way dialogue with citizens and stakeholders to identify problems, and stimulate the development of desirable solutions *Label*: Constructive TA, real-time TA, upstream public engagement	*Aim*: TA to timely engage MPs in the political debate on science and technology *Label*: Participatory parliamentary TA

This table originally appeared in iJETS (International Journal of Emerging Technologies and Society) 2012, 10, p. 8, as part of Van Est et al. (2012). Reprinted with permission

4 Informing and Engaging in the Political Sphere

4.1 Early Informing Activities

The Rathenau Instituut started examining synthetic biology quite early on in its development. The introduction to the field can be traced back to 2006, when a Rathenau Instituut researcher attended the Synthetic Biology 2.0 conference in Berkeley, California. This experience was the most important source of inspiration for the report 'Constructing Life' (De Vriend 2006), which was one of the first reports that addressed the potential societal impact of synthetic biology. In 2007 the institute published a Dutch version of the report (De Vriend et al. 2007) and a 'Message to the Parliament' (a brief summary of the study and its policy recommendations) based hereon (Van Est et al. 2007). As a result of these efforts, members of the Dutch Labor Party (Partij voor de Arbeid) asked parliamentary questions to draw the attention of the Cabinet to synthetic biology (Parliamentary Papers 2007). In its response, the Dutch cabinet underscored the importance to monitor the developments in the field and requested several advisory bodies to examine the developments, such as the Dutch Health Council and the Commission on Genetic Modification. However, in the subsequent years (and up until now) synthetic biology did not become a real topic of debate in Parliament, which is not really surprising, since synthetic biology is still predominantly confined to the laboratory.

During this period, the Rathenau Instituut has closely monitored the developments in the field and participated in a couple of international projects dedicated to analyzing the potential impact of synthetic biology, such as 'Synthetic Biology for Health, Ethical and Legal Issues' (SYBHEL) from 2009 to 2012. In addition, synthetic biology has played an important part in activities on NBIC-convergence (the synergetic convergence of nanotechnologies, biotechnologies, information and communications technologies and cognitive sciences) (Van Est and Stemerding 2012). In this context, the institute published the book 'Life as a Construction Kit' (Swierstra et al. 2009a, b), which was launched during the Dutch Societal Dialogue on Nanotechnology (Maatschappelijke Dialoog Nanotechnologie), an initiative of the Dutch government to stimulate broad discussion on nanotechnology in which viewpoints and opinions could be expressed by all kind of stakeholders and publics (Van Est et al. 2012). The institute also participated in the European project 'Making Perfect Life', dedicated to informing the European Parliament on NBIC-convergence and related challenges (Van Est and Stemerding 2012). Yet, a societal or political debate on synthetic biology did not emerge.

4.2 Looking for Novel Approaches to Facilitate Political Engagement

Meanwhile, the developments in synthetic biology kept pushing forward; more groups became active in the field (e.g. ERASynBio 2014) and important scientific breakthroughs were realized, such as the creation of a bacterium with a fully synthetic genome by the group of Craig Venter (Gibson et al. 2010). In addition, many TA organizations published reports on the broad range of societal and political questions synthetic biology may give rise to (Rerimassie et al. 2015). How do we, for instance, weigh potential benefits of synthetic biology versus its potential safety and security risks?

In addition, the type of questions raised by synthetic biology cannot always self-evidently be answered from established political ideologies. Consider, for instance, the tension between the potential applications of synthetic biology dedicated to sustainability at the expense of 'naturalness'.[2] This tension is particularly troublesome for green oriented parties that promote both naturalness and sustainability and are used to them going hand in hand, rather than having to choose one at

---

[2]See for example the recent petition "Synthetic is not natural", launched by a number of NGOs, including the ETC Group and Friends of the Earth, urging the company Ecover to 'keep extreme genetic engineering out of "natural" products' (ETC Group et al. 2014). This firm announced plans to shift from palm kernel oil to an algal oil as a basic ingredient for their soap products. To this end, the algae would be modified by means of synthetic biology. In contrast to the aforementioned NGOs, for Ecover the oil represents a 'natural' and sustainable alternative for the unsustainable palm kernel oil (Stemerding and Jochemsen 2014).

the expense of the other. The Rathenau Instituut therefore perceived the need to establish political engagement on such dilemmas, before they become urgent.

So in 2011, almost five years after of the publication of 'Constructing Life', the Rathenau Instituut decided to actively start promoting political debate on synthetic biology. However, due to the stage of development of the field and the lessons learned from recent experiences with the political debate on nanotechnology (Van Est et al. 2012), the institute did not consider the time right to incite a parliamentary debate on synthetic biology (nor would it have been likely to be successful). Members of Parliament have limited time and need to prioritize. Since synthetic biology is still mostly confined to the laboratory, we expected that members of Parliament would be unlikely to prioritize synthetic biology over more urgent issues. Therefore, the Rathenau Instituut started looking for novel approaches to facilitate political awareness and discussion on synthetic biology: it decided to target the world of political parties rather than Parliament.

When looking beyond the elected officials of a political party—which are often the primary addressees of TA—we find a network consisting of, for instance, policy advisors, political think tanks (or scientific bureaus), and political youth organizations (PYOs). Such bodies could fulfil a valuable role in examining emerging technologies from the perspective of the political party they are connected to, in a timely manner. With this in mind, the Rathenau Instituut reached out to Dutch political youth organizations and the international Genetically Engineered Machines competition (iGEM).

4.2.1 Future Synthetic Biologists

iGEM is the global student competition for teams in the field of synthetic biology. In this competition, students use standardized and interchangeable genetic building blocks (BioBricks™) to design microorganisms with new properties. iGEM began in 2003 as a summer course for students at the Massachusetts Institute of Technology. In 2004 the course was transformed into a competition in which five different teams participated. In 2011 the competition had grown into a full blown international competition, in which no less than 160 teams participated from 30 countries (iGEM 2011). In spite of limited means and the short timeframe, the projects are nonetheless often very impressive. Therefore, iGEM is often considered a poster child for the potential of synthetic biology.

Due to the explosive growth of iGEM, the organization decided in 2011 to regionalize the competition into three preliminaries (or 'jamborees' in iGEM jargon). The European-African jamboree was to be held in Amsterdam. In order to facilitate political engagement, the Rathenau Instituut seized this opportunity to organize a youth debate on synthetic biology: a 'Meeting of Young Minds' between future politicians and future synthetic biologists, in which the iGEM participants were seen as future synthetic biologists.

An important part of the work of iGEM teams is the so-called 'policy and practices' (previously called 'human practices') element. This implies that the iGEM participants do not only work on their project inside the laboratory, but also need to pay close attention to the societal aspects of their research. The idea of a Meeting of Young Minds therefore resonated well with the culture of iGEM (Rerimassie and Stemerding 2014).

4.2.2 Future Politicians

The Rathenau Instituut sought the future politicians in the circles of Dutch PYOs. Seven of the nine Dutch PYOs were willing to formulate a tentative political view on synthetic biology and enter into debate with each other and representatives from iGEM teams. The PYOs however, had little to no knowledge about synthetic biology. Therefore, the Rathenau Instituut undertook several support actions. First, relevant studies were made available on the website of the institute. In addition, an expert meeting was organized together with the iGEM team from the Technical University Delft. Furthermore, the Rathenau Instituut developed future scenarios on synthetic biology in the form of techno-moral vignettes: brief 'snapshots' of a future situation in which synthetic biology is applied, but at the same time raises moral questions (Lucivero 2012).[3] Since this was the first time the PYOs would learn about synthetic biology, it was important that the Rathenau Instituut provided multiple perspectives of synthetic biology and avoided giving a biased view. Correspondingly, some of the experts that took part in the expert meeting stemmed from the field of synthetic biology itself, but others came in from the perspectives of risk assessment, intellectual property and philosophy.

Prior to the debate, the PYOs were asked to draft a political pamphlet, in which they outlined their views on synthetic biology. This provided valuable input on how to organize the debate and served as an important preparation for the PYOs. In addition, these pamphlets provided the institute with extra material for analysis, since the debate was not likely to allow all of the viewpoints of the PYOs to be discussed.

4.2.3 Analyzing the Meeting of Young Minds

The well-attended event generated interesting results, in particular by shedding light on how Dutch political parties actually might think about synthetic biology. An analysis of the debate was featured in the Rathenau Instituut report 'Politiek over leven' (Rerimassie and Stemerding 2013). In 2014, an updated English version of the report, called 'SynBio Politics' was published (Rerimassie and Stemerding 2014).

---

[3]For more information and examples see: www.rathenau.nl/SynBio. Accessed 14 Dec 2014.

4.3 Informing Policy Makers

Next to informing Parliament and reaching out to PYOs, the Rathenau Instituut undertook several actions to inform Dutch policy makers. In 2011, the institute organized a workshop aimed at examining issues surrounding the risk assessment of synthetic biology in collaboration with the Netherlands Commission on Genetic Modification (COGEM). During this workshop, experts in synthetic biology and risk assessment discussed whether in the short term synthetic biology applications can still be adequately assessed using the current assessment framework for GMOs, and what kind of problems may arise in the future (COGEM 2013).

More recently, the Rathenau Instituut has joined hands with the Dutch Institute for Public Health and the Environment (RIVM), which in 2014 labeled synthetic biology explicitly as a focal point (RIVM 2014a). Together, the Rathenau Instituut and the RIVM organized an event to inform civil servants from various ministries on the developments in synthetic biology. During this meeting two prominent Dutch synthetic biologists presented their research, and Dutch iGEM teams presented their projects (RIVM 2014b). Both the Rathenau Instituut and RIVM aspire to make this a yearly event, in order to keep policy makers up-to-date.

5 Informing and Engaging in the Societal Sphere

5.1 Trying to Inform the General Public

Public awareness of synthetic biology in Europe and in the Netherlands is still quite low. The most recent Eurobarometer on biotechnology (European Commission 2010) showed that 83 % of Europeans had never heard of synthetic biology. In order to increase public awareness, the Rathenau Instituut participates in science communication activities. On a small scale, the institute collaborated in 2010 with the iGEM team of the Technical University Delft to organize an educational workshop for children and their parents (iGEM TU Delft 2010). This event, in fact, turned out to be the first step towards further intensive collaboration with the iGEM community. On a larger scale, the institute recently contributed to the synthetic biology edition of an educational quarterly magazine dedicated to the life sciences (BWM 2014).

However, it is important to note that science communication as such is no formal task of the institute. It rather aims to broaden the knowledge base on emerging technologies by providing information on the societal dimensions of science and technology. Ideally, organizations that are primarily concerned with science communication then draw from this information. For instance, the popular Dutch science communication website 'Kennislink' dedicated a theme page to synthetic biology. The page draws heavily from several recourses of the Rathenau Instituut to educate on the developments of the field, such as the aforementioned techno-moral vignettes (Kennislink 2014).

5.2 Mobilizing Civil Society

In addition to informing activities directed towards the general public, the Rathenau Instituut has made an effort to mobilize civil society organizations. In 2013, the institute published two reports. One was the aforementioned report 'Politiek over leven' (Rerimassie and Stemerding 2013), which—next to an analysis of the 'Meeting of Young Minds debate'—gave an overview of recent developments in synthetic biology and associated ethical, legal and societal issues. The purpose of the report was to provide a knowledge base for a broad array of stakeholders. The other report—'Geen debat zonder publiek' (Stemerding and Van Est 2013, 'No debate without public', translation VR)—provided an analysis of the emerging societal debate on synthetic biology in the Netherlands, United Kingdom, Germany and the United States. Based on these two reports, the Rathenau Instituut wrote the two-page brief 'Synthetische biologie vereist samenspraak' (Stemerding et al. 2013, 'Synthetic biology requires deliberation', translation VR). In this brief, the Rathenau Instituut called for Dutch civil society to actively engage with synthetic biology, in particular because the technology is still in a phase in which it can be more easily steered.

5.2.1 Political Cafe on Synthetic Biology

In order to mobilize civil society the institute also organized a 'Political cafe' in January 2013, during which the reports were presented and a public debate was held. As the name of the event indicates, this engagement activity was also intended to address the political sphere. In fact, the event was intended to bring together several (active and potential) stakeholders in the domain of synthetic biology, such as synthetic biologists, policy makers, philosophers and STS-scholars, in addition to civil society organizations and political organizations, which were the main target group. The event was a success in the sense that it was well-attended, notably by policy makers and other civil servants that started sharing the institute's sense of urgency. However, it did not succeed in mobilizing most of the civil society organizations and political think tanks that were specifically targeted. Yet, other civil society organizations such as Stichting Natuur & Milieu (Foundation for Nature and the Environment) and Stichting Christelijke Filosofie (The Dutch Foundation for Christian Philosophy) actively participated in the event. In doing so, they contributed to understanding how societal organizations might think of synthetic biology.

5.2.2 Lorentz Workshop on Synthetic Biology and the Symbolic Order

In fact, the engagement of the Dutch Foundation for Christian Philosophy led to further collaboration with the Rathenau Instituut, which culminated in the organization of a fruitful stakeholder workshop. During this 'Lorentz workshop' philosophers, STS-scholars, synthetic biologists and other stakeholders examined how synthetic biology challenges the 'symbolic order', the stock of twin concepts we use to categorize our reality, such as the distinction of 'natural' and 'artificial' (Stemerding and Jochemsen 2014; Swierstra et al. 2009a, b).

6 Informing and Engaging in the Science and Technology Sphere

6.1 Informing Synthetic Biologists

One interesting feature in the previously described public engagement activities is the active involvement of synthetic biologists. Many synthetic biologists indeed seem quite reflective towards the societal aspects of their research. In any case, technology assessment practitioners, STS-scholars etc. are often represented at synthetic biology conferences, in order to inform synthetic biologists from their perspective. Correspondingly, the Rathenau Instituut has been represented at various national and international synthetic biology conferences. In 2007 the institute presented and discussed its report 'Constructing Life' (De Vriend 2006) at the Synthetic Biology 3.0 conference in Zürich (SB 3.0 2007). In the Netherlands, the Rathenau Instituut presented, for example, at conferences organized by the Dutch Biotechnology Society NBV (2012, 2015). Most activities regarding synthetic biologists are, however, more intensive than informing alone.

6.2 iGEM as a Responsible Research and Innovation Laboratory

As discussed earlier, the Rathenau Instituut interacted with the iGEM community on several occasions. The success of iGEM and the important role of 'human practices' in the competition makes the iGEM community an important ally in public engagement.

6.2.1 Meeting of Young Minds 2012

In 2011 the institute organized the 'Meeting of Young Minds' (described above) between the iGEM participants and Dutch political youth organizations. In 2012 the institute also organized such an event, but this time two iGEM teams were selected to stage a debate about a topic of their choice. The iGEM team of University College London simulated an ecological synthetic biology crisis and tested whether effective containment of the crisis was possible. The iGEM team of the TU Delft organized a stakeholder discussion on dilemmas of dual-use research, inspired by the controversy concerning publications of research on the H5N1 bird flu virus (Rathenau Instituut 2014).

6.2.2 Collaboration in SYNENERGENE

More recently, the Rathenau Instituut started participating in SYNENERGENE, an international project dedicated to stimulating responsible research and innovation (RRI) and dialogue with regard to synthetic biology, funded by the European Commission under the 7th Framework Programme (SYNENERGENE 2014). In collaboration with other SYNENERGENE partners, the Rathenau Instituut will conduct a series of 'real-time technology assessments' to explore possible futures for the development of synthetic biology. In order to do so, a number of iGEM teams working on particular creative and significant ideas for innovation were asked to develop future scenarios. The teams developed two kinds of scenarios, based on their own iGEM project: application scenarios and techno-moral vignettes, which respectively relate to the plausibility and desirability of the envisaged synthetic biology application (Stemerding 2014). The Rathenau Instituut and partners are assisting the teams by providing guidelines and regular advice. The future scenarios will be used by the institute and other SYNENERGENE partners in stakeholder workshops. Additionally, the development of future scenarios intervened in the iGEM project itself. By thoroughly scrutinizing the envisaged application on its plausibility and desirability, the iGEM teams were enabled to identify different innovation pathways. This allowed them to work out the pathway that seemed technically and societally the most robust and thus would be most likely to achieve the 'right impacts' (cf. Von Schomberg 2013). A quote of the 2014 iGEM team from the Technical University of Eindhoven, one of the selected teams, can help to illustrate this:

> (…) we wrote an Application Scenario and described some of the possible outcomes in three Techno-Moral Vignettes. While working on these pieces our view on our own project dramatically changed. New possible applications came to mind as well as new ethical issues and questions (iGEM TU/e 2014).

Perhaps more importantly, this collaboration thus allowed SYNENERGENE partners to educate future synthetic biologists on potential ethical, legal and societal issues related to synthetic biology and enhanced their reflexivity on such issues (Betten and Rerimassie 2015).

7 Conclusion and Outlook

The aim of this chapter was to discuss and examine public engagement activities that the Rathenau Instituut has initiated on synthetic biology. The institute has undertaken several informing and engaging activities in different spheres of the science and technology governance landscape, and these activities have often aided in establishing connections between these different spheres.

Synthetic biology may contribute to addressing various societal challenges, but at the same time is not without risks and raises tough societal and ethical questions. Society faces the challenge of collectively determining how synthetic biology should develop and what conditions should be taken into account. The Rathenau Instituut has aimed to contribute to this deliberation process since its early years, by shedding light on synthetic biology from various societal perspectives, thus broadening the knowledge base on the emerging field. The institute also aimed to mobilize and build bridges between stakeholders that may play an important role in the future governance of synthetic biology in a timely manner.

One important observation about the various public engagement activities is that synthetic biologists have shown great willingness to participate. Synthetic biologists seem to be quite reflexive towards societal values, which may benefit the facilitation of meaningful public engagement. Hopefully, it may also inspire scientists in other domains.

How the future (Dutch and international) debate on synthetic biology will evolve remains yet to be seen. Will it mirror the intense (and occasionally hostile) experiences with earlier biotechnologies, characterized by great distrust among stakeholders, or will we be able to find a renewed 'tone of voice'? In this regard, a quote from DWARS, the political youth organization connected to the Dutch Green Party (GroenLinks) may be inspirational. In their political pamphlet for the 'Meeting of Young Minds' they state:

> Within GroenLinks, genetic modification evokes the same sense of resistance as nuclear energy does. The commission however, does not intend to dismiss synthetic biology in advance. It proposes to explore where genetic modification fell short, and synthetic biology may contribute to the common good. Dismissing synthetic biology beforehand would constitute a missed opportunity and does not match the progressive nature of GroenLinks. Rather, a critical, resolute and sober approach is much more appropriate. In particular, because synthetic biology offers opportunities to save human lives. (DWARS 2011, translation VR)

"Critical, sober and resolute". Indeed, words that are quite at odds with descriptions of earlier biotechnology debates. Hopefully, words like these will be used in the future to characterize the deliberation on synthetic biology.

Acknowledgments I would like to thank Dirk Stemerding, Rinie van Est, Marieke Ruitenburg, Pol Maclaine Pont and the organizers and participants of the TA summer school 'Analyzing the Societal Dimensions of Synthetic Biology' for informing and engaging with me on the content and structure of this chapter.

References

Betten AW, Rerimassie V (2015) "Eindelijk iets dat wel betekenis heeft". Over morele reflectie bij toekomstige synthetische biologen. Podium voor Bio-ethiek, 22 (1):18–20

BWM (2014) Synthetische biologie. Creatief met cellen. Cahier Biowetenschappen en Maatschappij 33(4):59–67

COGEM (2013) Synthetic biology update 2013. Anticipating developments in synthetic biology. COGEM, Bilthoven

De Vriend H (2006) Constructing life. Early social reflections on the emerging field of synthetic biology. Rathenau Instituut, The Hague

De Vriend H, Van Est R, Walhout B (2007) Leven maken. Maatschappelijke reflectie op de opkomst van synthetische biologie. Rathenau Instituut, The Hague

DWARS (2011) Visiestuk Synthetische Biologie. Een DWARSe visie op de toekomst van synthetische biologie, DWARS, Utrecht

ERASynBio (2014) Next steps for European synthetic biology: a strategic vision from ERASynBio. European research area network for the development and coordination of synthetic biology in Europe. European Commission, Brussels

ETC Group, Friends of the Earth, Organic Consumers Association et al (2014) Synthetic is not natural. Keep extreme genetic engineering out of "natural" products. http://www.syntheticisnotnatural.com/. Accessed 28 May 2015

European Commission (2010) Special Eurobarometer 340: Science and technology. European Commission, Brussels

Ganzevles J, van Est R, Nentwich M (2014) Embracing variety: introducing the inclusive modelling of (Parliamentary) technology assessment. J Responsible Innovation 1(3):292–313

Gibson D, Glass GI, Lartigue C et al (2010) Creation of a bacterial cell controlled by a chemically synthesized genome. Science 329(5987):52–56

iGEM TU Delft (2010) Workshop on synthetic biology. http://2010.igem.org/Team:TU_Delft#page=Education/Workshop_on_synthetic_biology. Accessed 14 Dec 2014

iGEM (2011) Teams registered for iGEM 2011. http://igem.org/Team_List?year=2011. Accessed 14 Dec 2014

iGEM TU/e (2014) Introduction SYNENERGENE. http://2014.igem.org/Team:TU_Eindhoven/Synergene/Introduction. Accessed 14 Dec 2014

Kennislink (2014) Thema: synthetische biologie. http://www.kennislink.nl/thema/synthetische-biologie. Accessed 14 Dec 2014

Levidow L, Carr S (2010) GM food on trial. Testing European democracy. Routledge, London

Lucivero F (2012) Too good to be true? Appraising expectations for ethical technology assessment. Ph.D thesis, University of Twente

Ministerie van Onderwijs en Wetenschap (1984) Beleidsnota integratie van wetenschap en technologie in de samenleving. Ministerie van Onderwijs en Wetenschap, The Hague

NBV (2012) Chemistry of life. http://nbv.kncv.nl/chemistry-of-life-engineering-biology/parallel-session-8.127157.lynkx. Accessed 14 Dec 2014

NBV (2015) Biotech @Work. http://nbv.kncv.nl/biotech-a-work-2015.176259.lynkx. Accessed 13 Mar 2015

NEST High Level Expert Group (2005) Synthetic biology: applying engineering to biology. Report of a NEST High-Level Expert Group. Office for Official Publications of the European Communities, Luxembourg

OCW (2009) Instellingsbesluit Rathenau Instituut. Staatscourant, nr. 11024, 22 juli

Parliamentary Papers (2007) Annex to Papers of the Parliament of the Netherlands, Aanhangsel Handelingen II 2007/2008, nr. 2070800670. http://parlis.nl/kvr29998. Accessed 3 April 2015

Pauwels E (2013) Public Understanding of synthetic biology. Bioscience 63(2):79–89. doi:10.1525/bio.2013.63.2.4

Rathenau Instituut (2014) Meeting of young minds 2012 http://www.rathenau.nl/themas/thema/project/synthetische-biologie/meeting-of-young-minds-2012.html. Accessed 14 Dec 2014

Rerimassie V, Stemerding D (2013) Politiek over leven: in debat over synthetische biologie. Rathenau Instituut, The Hague

Rerimassie V, Stemerding D (2014) SynBio politics: bringing synthetic biology into debate. Rathenau Instituut, The Hague

Rerimassie V, Stemerding D, Zhang W, Srinivas KR (2015) Discourses on synthetic biology in Europe, India and China. In: Ladikas M, Chaturvedi S, Zhao Y, Stemerding D (eds) Science and technology governance and ethics. A global perspective from Europe, India and China. Springer, Dordrecht

RIVM (2014a) Synthetische biologie http://www.rivm.nl/Onderwerpen/S/Synthetische_biologie. Accessed 14 Dec 2014

RIVM (2014b) Verslag debat Verantwoord omgaan met synthetische biologie. RIVM, Bilthoven

SB 3.0 (2007) Synthetic Biology 3.0 Conference Proceedings. ETH Zürich, Zürich

Schot J, Rip A (1997) The past and future of constructive technology assessment. Technol Forecast Soc Chang 54:251–268

Stemerding D (2014) iGEM as laboratory in responsible research and innovation. J Res Res Innovation 2(1):140–142

Stemerding D, Jochemsen H (2014) Niet natuurlijk? Cahier Biowetenschappen en Maatschappij 33(4):56–57

Stemerding D, Rerimassie V (2013) Discourses on synthetic biology in Europe. Rathenau Instituut, The Hague

Stemerding D, Van Est R (2013) Geen debat zonder publiek. Het opkomende debat over synthetische biologie ontleed. Rathenau Instituut, The Hague

Stemerding D, Rerimassie V, Messer P (2013) Het Bericht. Synthetische biologie vereist samenspraak. Rathenau Instituut, The Hague

Swierstra T, Boenink M, Walhout B, Van Est R (2009a) Leven als bouwpakket. Ethisch verkennen van een nieuwe technologische golf. Klement, Zoetermeer

Swierstra T, Van Est R, Boenink M (2009b) Taking care of the symbolic order. How converging technologies challenge our concepts. Nanoethics 3(3):269–280

Synenergene (2014) What is Synenergene? http://synenergene.eu/information/what-synenergene. Accessed 14 Dec 2014

Van Est R (2013) Political TA: opening up the political debate. Stimulating early engagement of parliamentarians and policy makers on emerging technologies—Attempts by the Rathenau Instituut. In: Doorn N, Schuurbiers D, van de Poel I et al (eds) Early engagement and new technologies: opening up the laboratory, Springer, Dordrecht, pp 137–153

Van Est R, Stemerding D (2012) European governance challenges in 21st century bio-engineering. European Parliament, Brussels

Van Est R, De Vriend H, Walhout B (eds) (2007) Bericht aan het Parlement. Synthetische biologie: nieuw leven in het biodebat. Rathenau Instituut, The Hague

Van Est R, Walhout B, Rerimassie V, Stemerding D, Hanssen L (2012) Governance of nanotechnology in the Netherlands: informing and engaging in different social spheres. Int J Emerg Technol Soc (iJETS) 10:6–26

Von Schomberg R (2013) A vision of responsible innovation. In: Owen R, Heintz M, Bessant J (eds) Responsible innovation. Wiley, London, pp 51–74

A Critical Participatory Approach to the Evaluation of Synthetic Biology

Inna Kouper

1 Introduction

Synthetic biology is an ongoing fuse of the previous attempts in chemistry, genetics, and bioengineering to understand and control the evolution and development of living organisms. Sometimes called extreme genetic engineering, it raises profound questions about humans' abilities to control their own environment and about the boundaries between living and non-living. The ideas of design, standardization and re-use applied to the living world create uneasiness and a sense of ethical uncertainty. Who will be making decisions about the boundaries of scientific investigations? Who should be contributing to the discussions about synthetic biology implications and oversight? How can we make sure that policy decisions incorporate concerns of various stakeholders?

In this chapter I address these and other questions by combining methods of historical, linguistic, and sociological research. Drawing on the results of my doctoral dissertation that examined the discourses of synthetic life (Kouper 2011), I examine forms, themes, justifications, and stakeholders that have contributed to the discussions of synthetic biology in the 20th–21st centuries. I argue that the evaluation of synthetic biology and similar emerging technoscientific areas benefits from a discursive perspective that is coupled with critical social theory. Such a perspective is built into the framework proposed in this chapter.

By emphasizing social critique and participation, I hope to further embed synthetic biology into the space of social dilemmas that need to be negotiated as opposed to the space of scientific problems that need to be solved. While societal research within the frameworks of ethical, legal, and social implications (ELSI) is already an ongoing part of synthetic biology, more attention to public perceptions and deliberations as well as to the debates about science visions is needed

I. Kouper (✉)
Indiana University, Bloomington, IN, USA
e-mail: inkouper@indiana.edu

© Springer International Publishing Switzerland 2016
K. Hagen et al. (eds.), *Ambivalences of Creating Life*, Ethics of Science and Technology Assessment 45, DOI 10.1007/978-3-319-21088-9_11

(Shapira et al. 2015). The evaluation of synthetic biology as a deeper ELSI problematization will benefit from being placed into a broader social and cultural context and from going beyond the dichotomies of experts and ordinary public or fears and safeguards.

2   The Framework: Decisions, Contexts, Participation, and Critique

Technoscientific systems are complex systems in which resources and agents interact and generate professional and societal outcomes. Science begins in the lab and then goes beyond it as its achievements become accepted, commercialized, and regulated within the larger society. Some scientific advancements are met with resistance and create tensions. Proper evaluation frameworks of societal impacts of technoscience can help to alleviate tensions, to negotiate or reconcile opposing views, and to ensure the co-development of the scientific and societal goals.

Attention to the development of evaluation frameworks in synthetic biology is rather selective. The most attention has been paid to ethical issues and the issues of risks and safety (Cho et al. 1999; Peter D. Hart Research Associates, Inc. 2008; Deplazes et al. 2009; Bennett et al. 2009; Presidential Commission for the Study of Bioethical Issues 2010). Are these approaches enough to address the complexities of synthetic biology as a technoscientific system? Will they be sufficient to compare its impacts across communities and over time? Do other approaches (e.g., broader calls for social and environmental assessments) provide enough details to facilitate evaluations?

Some authors (e.g., Deplazes et al. 2009) argue that a framework that incorporates multiple participants into the discussions about synthetic biology has already been established, and that it is a matter of the individual interest and participation to advance further interaction. They cite Garfinkel et al. (2007) as an example of how discussions can be structured. A closer look at Garfinkel's report and its approach reveals that it includes twice as many scientists and policy makers than any other stakeholders. Twelve participants that contributed to the discussions that generated the report had a natural science background, mostly biology, chemistry and medicine. Four participants had a social science background and two participants were coming from the law background.

The societal impact of synthetic biology goes beyond technical and scientific feasibility, risks, or ethical questions narrowly construed. Synthetic biology raises larger questions of the autonomy of living beings and priorities in life meanings and boundaries. It also raises questions about the limits of human actions and responsibilities of individuals and social groups that go beyond moral permissibility. Even though many of these questions have been previously raised in the context of biotechnology, the increasing possibilities of synthetic biology to serve human goals and the fast pace of technoscience expand this array of questions and call for mechanisms to facilitate continued discussions.

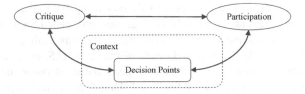

Fig. 1 The critical participatory framework for the evaluation of synthetic biology: an overview

The critical participatory framework for analyzing and evaluating synthetic biology proposed below draws on two key concepts, participation and critique, and places decision points that are part of the evaluative efforts in a historical and cultural context (see Fig. 1).

The framework is connected to citizen empowerment and can be supported with various theories and methodologies. The components of participation and critique concern all stakeholders, including the decision makers, the media, and the publics; they are aimed at turning the debates and analyses into self-reflecting and self-problematizing modes rather than the modes of criticizing or evaluating others. A deep consideration of the concepts of participation and critique opens up our critical and linguistic sensibilities and allows us to question the stable "true" or "natural" meanings of events and decisions vis-à-vis the ideas of rationality, knowledge, discourse, and power. The arrows that go in both ways between the components of the framework emphasize the importance of their interaction, even though the nature of such interactions can vary and be asymmetrical.

The framework is grounded in the discourse analysis perspective, particularly in a critically-oriented set of approaches that were developed on the intersection of history, linguistics, sociology, and anthropology. Critically-oriented discourse analysis goes beyond texts as products and aims to focus on the struggle between social forces and on knowledge that can lead to emancipatory change (see, for example, Fairclough 1989, 1992, 2001; Van Dijk 1993). In what follows, each of the components of this framework is illustrated and discussed in detail.

3 Decision Points

Synthetic biology discourse revolves around the acquisition of knowledge and techniques that allow scientists to create and extensively modify living organisms. Along with great expectations, such knowledge raises fears, so among the first decision points are the questions of whether to allow synthetic biology investigations or not. The ETC group (2007), for example, raises some of the fundamental questions of its permissibility: "… Is synthetic biology socially acceptable or desirable? Who should decide? Who will control the technology, and what are its potential impacts?". Such questions expand the policy debates by inviting more actors into the discussion and by potentially increasing the range of options open for consideration. Several social and citizen groups have called for a moratorium on the release and commercial use of synthetic organisms until a comprehensive oversight framework is developed (Friends of the Earth U.S. et al. 2012).

Once synthetic biology is accepted as permissible, decision points that are concerned with the questions of "What?" and "How?" proliferate. Decisions have to be made with regard to organisms in experiments and their treatment and release, the methods of investigation and their cost and effectiveness, the benefits and risks, and many other aspects of emerging technologies. My analysis of the earlier debates in synthetic biology (Kouper 2011) demonstrates that decisions with regard to synthetic biology were made and justified by emphasizing predominantly one or more of the following: (a) the availability of new products and instruments, (b) the value of scientific knowledge and understanding, (c) risks, consequences, and uncertainties, (d) life improvements, and (e) inevitability and self-evidence of synthetic biology. The list is characteristic of the discussions that took place at a particular time and is somewhat biased toward positive justifications. Over time, it will change as those arguments branch into new or more nuanced arguments or become decision points themselves.

The products and instruments justification argues that the synthesis of organisms and living components helps to create commercially available products and services that can be used by people in their professional and daily lives, including medicine, food, computing, water production, air cleaning, and so on. A public lecture on synthetic biology from the American Association for the Advancement of Science envisioned such products and services as follows:

> Vaccine factories that can rapidly respond to viral threats such as bird flu. Biologically produced photocells to meet energy demands. Plants that act as sensors for explosive materials. Houses of living wood that repair themselves. Self-producing pocket calculators. Bacteria that will pan for gold. These are some of the possibilities from the nascent field of synthetic biology. (Baker 2005, p. 1)

Decisions regarding such products concern their practical and commercial value as well as the balance between risks and benefits. With microorganisms used as factories or "in the wild" there is always a danger of unpredicted safety and environmental consequences or what Snow and Smith (2012) called the "unintended sources of harm". Additionally, the development and sale of products in some economies can disrupt other economies, especially along the "developed—developing" country divide.

The knowledge and understanding justification emphasizes the intrinsic benefit of synthetic biology as a quest for learning about the unknown. Such claims are often used to argue for the importance of synthetic biology as a discipline. Learning about life and understanding it better is one of the most important goals for humans. The claims of knowledge sometimes confuse *understanding*, which is concerned with seeking meaning of various expressions of human actions (cf. *Verstehen* as it was used by W. Dilthey, M. Weber, M. Heidegger and others), with *explanation*, a discovery of causal relationships among the objects of the physical world.

Bourgeois (1976) argued that a significant component of *Verstehen* is empathy. Empathic understanding goes beyond observations of behaviors and explications of their causes, it looks for insights about motivations and intentions, i.e., about internal mental states. Inanimate objects do not have internal mental states; only complex living organisms, particularly humans, can be thought of as having internal mental states.

The discourse of synthetic biology focuses predominantly on microorganisms, therefore, it may be premature to insist on the inclusion of the issues of consciousness and mental states into the discussion. At the same time work in synthetic biology involves human actions and decisions that have a moral orientation. If the understanding of life through the separation and subsequent assembly of parts becomes a dominant approach, there is a danger that the moral component will be largely ignored in future investigations and that understanding of motivations and intentions will not be part of the sought-for knowledge of life. Ultimately, shifting the focus to mechanical understanding may result in the exclusion of empathy from decision-making (see also the Sect. 6 below for further problematization of the dominant approaches in synthetic biology).

Concerns over implications and consequences are often used to argue both for and against synthetic biology. A common element in these concerns is uncertainty about possible outcomes. The negative consequences include damage to the environment, bioterrorism and biotechnological wars, harmful effects on human health as well as on the well-being of other living creatures, and threats to biodiversity. Scientists and activists are probably most vocal in discussing the consequences, but the ways these two groups approach the discussion are different. Activists argue primarily for in-depth assessments of social, ethical, and environmental impacts.[1] Fast commercialization of unregulated technologies, according to these views, promotes private ownership of biological organisms and can have a dramatic negative effect on poorer economies, lowering export prices, depriving people in developing countries of their jobs, and advancing unsustainable, U.S.-centric solutions.

Scientists, on the other hand, insist on the continuation of research and demonstrate their eagerness to be involved in public debates about its consequences. Their statements are grounded in the assumption that further developments in the chemical and biological synthesis are inevitable (hence, the inevitability or self-evidence justification), therefore, it is more productive to focus on the positive and to try to minimize the negative. In addition to explicit statements about further synthesis of biological organisms being unavoidable, claims of inevitability or self-evidence of synthetic biology can be communicated via certain grammatical and rhetorical forms, such as verbs in present and simple future tense, modal verbs and constructions that communicate obligations, such as *should have to*, and the overall assertive form of the statements. As one of the most prominent scientists in this

---

[1]See, for example, a series of case studies developed by the ETC group: http://www.etcgroup.org/tags/synbio-case-studies. Accessed 10 Oct 2014.

research area has argued, the scientific community needs to develop and promote rigorous ethical and safety standards as well as "discuss the benefits of synthetic engineering to balance the necessary, but distracting, focus on risks" (Church 2005).

Claims of better life often co-occur with other justifications, particularly with the products and instruments justification. Such claims usually emphasize the potential to solve problems on a larger scale, such as alleviating hunger, conquering diseases, remediating climate change, satisfying human needs, and offering improved nutrition and longer and healthier life span. This volume's chapter by Verseux and co-authors, for example, describes exciting opportunities of using synthetic biology approaches to create permanent space colonies so that humans can become independent from Earth as our home planet.

Many claims and justifications that surround decisions in the discourse on synthetic biology can be aggregated into two larger themes that characterize problematic fields or clusters of uncertainty. These themes represent overlapping discussions in which participants raise questions and determine what decisions need to be made at each point in time. The two larger themes are the Promethean/Creator theme and the Management and Responsibility theme (see Figs. 2 and 3).

The Promethean/Creator theme is concerned with the uncertainty regarding human ambitions of achieving greater knowledge of life and biology and with the outcomes and ultimate consequences of such ambitions. In Greek mythology, Prometheus was a powerful deity, a titan who helped and protected humanity (Roman and Román 2010). To help humans survive, Prometheus stole fire from the gods, gave it to humans, and subsequently received severe punishment for his actions. There are five conceptual clusters in this theme, with each of the clusters

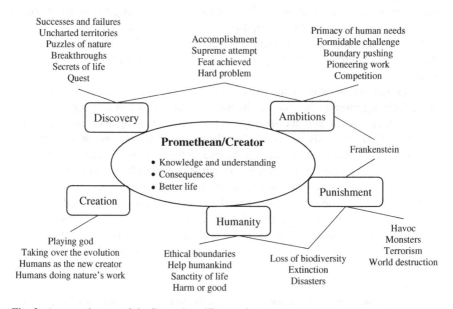

Fig. 2 A semantic map of the Promethean/Creator theme

raising questions that may turn into decision points. The clusters and possible problematic fields or questions are listed below:

1. *Creation.* The creation cluster focuses on the role of humans in the creation of nature, including living creatures, and raises questions about whether human investigations, including the studies in bioengineering and synthetic biology, transgress the ethical, scientific, or theological limits and become unwise or irresponsible (Dabrock 2009).
2. *Discovery.* The discovery cluster refers to the desire to acquire knowledge, use it, and share it with others. Similar to the ambitions cluster, it is based on the heroic terminology of feats, accomplishments, and hardships. It is also grounded in the rhetoric of travel and adventure, e.g., when synthetic biology is described as a quest, a search for solutions to the puzzles of nature, or as a trip into uncharted territories. The discussions in this cluster raise questions of access to and prioritization and acceptance of scientific discoveries, i.e., whether the boundary-pushing research merits more funding, whether its assessments are premature, and how we can ensure equal access to the means of discoveries as well as to their outcomes.
3. *Ambitions.* The ambitions cluster signifies great ambition in advancing synthetic biology and a struggle against more powerful forces, such as gods, nature, and uncertainty. The concerns over great ambitions and discoveries, associated with the creation of life, are often tied to possible punishment. One of the broader questions with regard to synthetic biology and human ambitions is how our cultural and social norms align with our ambitions and what criteria can be used to prioritize one over the other.
4. *Punishment.* The punishment cluster includes fear of negative consequences as well as the generalized fear of monsters and destruction. While most of the clusters in the Promethean/Creator theme combine rational and irrational argumentation, the questions of punishment for human ambitions and proper safeguarding and mitigation techniques require more attention to their emotional component. This, in turn, may prompt longer deliberations and more diversity in legislative and communicative approaches. Similar concerns are relevant to the next cluster, humanity.
5. *Humanity.* The last cluster, the humanity cluster is concerned with well-being of the humankind and insists on actions being justified by the love for and protection of humanity. This cluster, characterized by affective vocabulary and empathic attitudes towards other living forms, still has a limited representation in the synthetic biology discourse, even though some questions, such as the ethical concerns as well as the questions of how sustainability and communal values can be part of synthetic biology, have already been raised in the discussions and research literature (Rabinow and Bennett 2012; Engelhard et al. 2013).

The second larger theme, the Management and Responsibility theme, approaches synthetic techniques and life as scientific subjects and focuses on the management of boundaries between science and society, between feasible and hard to

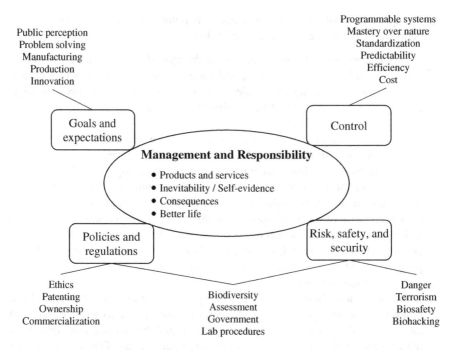

Fig. 3 A semantic map of the Management and Responsibility theme

reach, and between socially and culturally desirable and prohibited. The semantic map of this theme is presented in Fig. 3. The clusters and possible problematic fields for this theme are listed below:

1. *Goals and expectations.* The goals and expectations cluster focuses on specifying how synthetic biology can contribute to science and to the larger society as well as how to manage societal expectations with regard to it. The questions of whether synthetic biology delivers on its promises, how "science" and "the public" interact, and who manages the expectations and how are raised within this cluster.

2. *Control.* The control cluster emphasizes the necessity to direct and manipulate behaviors of living organisms, because this would allow humans to have full power over their environment. Usually, full control over the environment is considered to be beneficial to humans. To achieve mastery of nature, scientists need to find ways to standardize living forms and their components, to reduce cost of their manufacture, and to increase programmability, interchangeability, and predictability of biological systems. The problematic cluster of control can be connected to the cluster of ambitions in the Promethean/Creator theme as it raises the questions of desirability of full control over nature in the context of the existing societal norms and values. The cost and benefit factor is another consideration.

3. *Risk, safety, and security.* The risk, safety, and security cluster is concerned with managing the mostly negative consequences of synthetic biology, such as the escape of modified organisms into the environment or the risks of biohacking and bioterrorism. The concerns over strategies of minimizing the risks and increasing security can turn into the choices between alternative strategies, for example, a *laissez faire* strategy of letting scientists deal with risks, a required researcher-led assessment strategy, and a required independent assessment strategy (Douglas and Savulescu 2010).

4. *Policies and regulations.* The policies and regulations cluster somewhat overlaps with the risk, safety and security cluster as it focuses on such issues as the role of government and other regulating agencies in scientific developments, forms of ownership with regard to living organisms and biological systems, and the distribution of responsibility with regard to the development and oversight of synthetic organisms. Some scientists delegate their own ethical and practical responsibilities to other agencies and argue that since their work had been "green-lighted by government offices, the National Academies and an independent ethics review board" (Weiss 2008), there is no reason for not proceeding with research. Others consider their involvement in public discussions to be a sufficient condition for continuing research in synthetic biotechnologies.[2] Governance of synthetic biology has generated many discussions, and many questions need to be carefully addressed before choosing a governance framework or approach (Hurlbut 2015).

To summarize, a multitude of decision points surrounds synthetic biology. The decisions need to be made with regard to the following:

- Feasibility assessment or whether certain technologies are possible from the scientific and economic standpoints.
- Moral permissibility, or whether synthetic biology is acceptable and desirable as a course of scientific and societal actions and whether it fits with our current ethical sensibilities.
- Outcomes, risks, and benefits, or whether we can identify what will happen with the overall approach and particular technologies in synthetic biology.
- Access and ownership, or who can study and conduct research in synthetic biology, who has the rights to use its results, and who can access and disseminate information about it.
- Measurements of impact, or how to evaluate the role of synthetic biology in our well-being, environmental health, sustainability, and consensus.
- Societal needs and expectations, or how we want synthetic biology to fit within our understandings of science and technology role in society.
- Awareness and education, or how to disseminate information about synthetic biology and how to identify stakeholders and reach out to them.

---

[2]See, for example, a press release by the J. Craig Venter Institute in 2008, http://www.jcvi.org/cms/research/projects/synthetic-bacterial-genome/press-release/. Accessed 21 Apr 2015.

4 Context

Context shapes the setting for events, actions, and statements. Knowing the context means to know how certain arrangements and practices came into being and how certain decisions may affect them. In the case of synthetic biology it is crucial to understand the historical background and lessons from related disciplines as well as recognize the multiplicity of genres, communities, and social structures that affect decision-making.

The ideas of perfecting, conquering and improving nature go back in centuries. As Newman (2004) points out, from the ancient to modern civilizations, humans imitate, compete with, perfect, and re-create the nature. And most, if not all such attempts throughout the centuries, from the Greeks to alchemy and Judaic golem creation to early chemical experiments and modern genetics, generated vigorous debates around the legitimacy of such attempts as well as their moral and onto-logical limits. In the 20th century attempts to improve nature and synthesize new life forms are traced to Hugo de Vries and Stephane Leduc, who experimented with breeding, chemically grown molecules, and genetic manipulations (Campos 2009). Awareness of the immense implications of their work and fears of negative out-comes are evident in the de Vries contemporaries' statements of man as a creator, creating life in a test-tube and "trembles" between the inanimate existence and incipient vitality (Campos 2009, p. 9).

Over the decades, experiments with improving and synthesizing organisms branched into biotechnological approaches to breeding and mutation, attempts to create organic forms from non-organic, standardization of biological parts, and DNA syntheses.[3] The term "synthetic biology" appears and disappears from common usage over time. Thus, according to Campos (2009), it was on the rise in the 1930s, disappeared in the 1970s, and re-appeared in the 2000s. The re-appearances of the term, particularly in the 2000s, indicate renewed attempts to claim autonomy for this area of scientific endeavors; that is why it is often presented as unprecedented, revolutionary, fascinating, and very valuable. The grand vision of synthetic biology promises to help humans with understanding of the origin and history of life, creating new complex life forms, controlling evolution, and improving human condition.

At the same time, there is also an awareness of limitations of what synthetic biology can deliver as well as fear that exaggerated promises will increase skep-ticism and generate fear and resistance. Contrary to the grand vision that can be found is some earlier accounts of synthetic biology of the 2000s, many later accounts avoid references to life and other broader connections and limit the def-initions to biological parts, constructs, and components. Lam et al. (2009), for example, describe synthetic biology as focusing on DNA-based engineering, gen-ome minimization, and protocell creation.

---

[3]See chapter by Carlos G. Acevedo-Rocha in this volume for a more detailed account of synthetic biology and its varieties.

The tensions between the rhetoric of grand vision and attempts to contain the field within the reasonable and achievable scientific claims generate disparate, sometimes contradictory statements within synthetic biology. The field progresses quite quickly, but many theoretical, methodological, and technological challenges have not been resolved or even yet addressed. As some scientists have pointed out earlier (Rasmussen et al. 2003) and continued to acknowledge later (e.g., Gardner and Hawkins 2013), the domain lacks theoretical and experimental grounding, it has no agreement about which scientific problems to address, and it is insufficiently related to reality. Despite the scientific disagreement, products of synthetic biology get commercialized rapidly. Faith in the engineering side of synthetic biology, i.e., the strive to make cells or organisms work in certain ways even if the mechanisms are not fully understood, drives the fast development and funding of this type of research.

In the 20th–21st centuries discussions about synthetic biology spread via many channels and genres. Along with the established genres of newspaper, magazine and journal articles, information disseminates through websites, blogs, online forums, and many other forms. Patents play a crucial role in framing of synthetic biology and affect its development in the future, especially, the rights to the creation and use of synthesized organisms and their products and derivatives. Patenting of living organisms is already authorized by several court decisions and accepted patents (Demott and Thomas 1980; Holtug 1998). The majority of patent applications originates from companies, with universities and research institutes following closely (Van Doren et al. 2013).

Established genres have their own rules, which need to be taken into consideration in outreach and deliberations. Thus, research articles belong to the formal scientific discourse, which is known for its high usage of technical terms as well as for its actual and rhetorical impersonality and objectivity. Magazine and newspaper articles deliver news and commentary and may be more colorful in language. Journalistic writing, or media discourse, relies on non-technical metaphorical language and uses the inverted pyramid as a structuring device with the most important information going first and details following (Vos 2002).

Newer genres and the digital counterparts of the traditional genres blur the boundaries between the various types of writing as they combine news reporting, explanations, and personal evaluations and expand the array of contributors. Thus, the article "Synthetic biology's malaria promises could backfire," published on scidev.net by social scientist Claire Marris, combines a short analysis of the claims of the semi-synthetic artemisinin project with some vivid evaluation statements of the role of synthetic biology products in the advancement of the field. Marris argues that this form of artemisinin "is not cheaper than the existing source, nor alleviating a shortage, it will—at best—play a relatively small part among the multitude of factors that determine whether or not people suffer and die from malaria" (Marris 2013). This example illustrates the changing role of the journalist as a mediator between science and the general public due to the increased visibility of other writers, including researchers, pundits, and informed non-specialists, especially on the Internet. Another excerpt below illustrates individuals' contributions to the discussions expressed publicly via a blog:

> With WHO's blessing Keasling's synthetic artemisinin replaces the natural product: an economical disaster for African families who have invested all their meager resources in Artemisia annua plantations, lured by the promise of big profits. (Lutgen 2014)

Bourdieu's concept of field helps to conceptualize the multiplicity of actors as well as the impact of institutions and material conditions on their actions. According to Bourdieu (1969, 1989), individual agents act in conditions that have their own laws and structures. The system of agents in structured conditions can be described as a social field, which, similar to a magnetic field, involves various forces. Each agent occupies a certain position that is characterized by a specific type of participation and has a certain amount of power or authority within the field. Depending on the amount and type of power, agents are able to influence other agents, the field and the distribution of power.

The development of fields is a historical process. As areas of human activity become more and more differentiated, society organizes into relatively autonomous fields with their own types of legitimacy, i.e., their own rules as to what defines an authority. Modern societies consist of multiple fields, such as the economic field, the cultural field, and the bureaucratic field (Bourdieu 1981, 1993, 1994). These are the larger fields that are loosely associated with the economy, arts and culture, and governance in our society. Synthetic biology discussions also straddle across multiple autonomous fields that are smaller in scale, including the fields of journalism, science, religion, business, personal (private) sphere, politics/policy, and digital media. Each of these fields is characterized by its own logic of what is valuable and important and who makes contributions and how.

The field of digital media, for example, acquired its relative autonomy through the emphasis on the novelty of technologies in use and on the rapid changes in social and communicative practices facilitated by these technologies. It overlaps with other fields, but it has its own unique characteristics that distinguish it from others and justify its conceptualization as an autonomous field that strives for recognition and establishment of its own system of power, rewards, and distinctions. Contrary to other fields that already have stable structures and actors in certain social positions, this field's system of actors, structures, and positions is under development. Consequently, digital media may be more open to changes and disruptions. Innovative forms of communication can reach younger people or professional groups that are outside of the traditional forms of communication.

Many actors participate in digital media, including scientists, journalists, entrepreneurs, politicians, and others. In the digital media field they combine the power and authority they have in other fields, e.g., science or journalism, with the new authority acquired from being a blogger, a quoted expert, a moderator, or an online community member. Some participants do not use capital from other fields. For example, a discussion from 2008 of the news about a synthesized DNA on Slashdot.com, a community-oriented news site, was initiated by a person who provided only a nickname and no other information. This example illustrates that the digital media have different dynamics with regard to who accesses and disseminates information as well as to who provides the content and determines the credibility of messages.

The scientific field consists of agents involved in the systematic observation-based studies and rules that facilitate consistency and accumulation of knowledge. Primary agents in this field are scientists, who engage in knowledge-producing activities and maintain significant autonomy from other fields. In the analysis of science as an autonomous field an important question is whether the social sciences and humanities constitute their own field or fields of production. The social sciences and humanities have structures of authority and capital that are similar to other academic disciplines within a university setting. Sometimes, social scientists and humanity scholars are pressed to conform to scientific epistemologies and produce "empirical" and "objective" research. At the same time, the rules of what constitutes knowledge in those domains as well as the channels that are used to communicate this knowledge are different. Therefore, these domains should have autonomy in expressing their position on synthetic biology and its implications.

The issues of life and its creation are germane to the field of religion, which can be defined as a social institution that relies on faith as the foundation of understanding, knowledge, and behavior and is characterized by a system of beliefs, rules, and practices that refers to God or other supranatural entities in its guidance and understanding of the world and the role of humans in it. Nevertheless, we rarely find materials that address the topics of synthetic biology from a religious perspective by agents who belong to the religious field. Even though the concern of "playing God" is quite common in the discussions of synthetic biology (Engelhard et al. 2013), the discussion from the religious perspective is separated from the rest of the discussion.[4] The ordinary practitioners of faith and religion may identify themselves as agents of other fields, e.g., education, politics, or law, and therefore advance their views in indirect ways, via an implied agenda. Lack of authoritative religious sources in public discussions indicates the importance of studying the structure and boundaries of the public sphere as well as examining how certain voices are represented in it.

Business and entrepreneurship, a field that is connected to the concepts of goods production and consumption, markets, and capitalism, plays an important role in the development of synthetic biology. Companies that commercialize the results of synthetic biology research often involve members of other fields. *Synthetic Genomics, Inc.*, for example, was founded by scientists and researchers to commercialize methods and approaches from genetics and other bioengineering domains. PATH, an international, nonprofit health organization based in the U.S., describes itself as blending "the entrepreneurial spirit of business, the scientific expertise of a research institution, and the passion and on-the-ground experience of

---

[4]The Roman Catholic Church, for example, publishes its own guidance on bioethical questions on its website, see, for example, http://www.vatican.va/roman_curia/congregations/cfaith/documents/rc_con_cfaith_doc_20081208_dignitas-personae_en.html. A project on synthetic biology and religion led by G. Bennett (https://labs.fhcrc.org/cbf/Project_Areas/religion/SynBioReligion/Index.html) aims to map how U.S. religious organizations view developments in synthetic biology, but it has not published any results yet. Both accessed 21 Apr 2015.

an international NGO".[5] Commercial products based on the scientific advancements of synthetic biology are entering the market. In addition to antimalarial treatments manufactured with a semisynthetic artemisinin by *Sanofi S.A.*, a multinational pharmaceutical company, there are other products such as vanillin and saffron, bioisoprene (rubber), or surfactants, all generated from synthesized microorganisms.[6] The product range follows the previous biotechnological developments, and even though most of the products promise to improve the quality of life or save the environment, the outcomes will likely depend on which markets they enter and what forms of production they eventually challenge or replace.

Governance of synthetic biology is within the purview of another social field—the field of policy/politics, which is concerned with general governance, decision-making, and relationships between citizens, the state, and other bureaucratic agencies. As Sheila Jasanoff observed, the government regulation of biotechnology depends on the national and cultural context; the decisions can be legitimated by references to scientific or technical expertise or to institutional rationality (Jasanoff 2005). While Europeans, according to Jasanoff, acknowledge and appreciate the interactions between science and politics, the Americans tend to separate them, thereby creating a stronger sense of uncertainty about how the new biotechnological advancements should be governed in a democratic society. If synthetic biology poses a threat to the very distinctions between nature and culture and moral and immoral, it is not clear how the government can remain neutral or "science-based" and not permit moral arguments in the debate (Kaebnick et al. 2014). Approaches to policy-making need to consider the contributions of a broader constituency and the possibility of ongoing deliberations and decision-making as well as mechanisms for communities to govern themselves.

To summarize, context as a component of the evaluation framework of synthetic biology includes a range of historical, cultural, economic, and political forces that need to be taken into account. Attention needs to be paid to the following:

- Purposes of evaluating synthetic biology and its implications
- History of the development of synthetic biology and similar or adjacent fields
- Interdisciplinary connections among disciplines and areas of research
- Genres and sources of discourse that are part of the synthetic biology discussions
- Channels used to disseminate information and their organization
- Structures of labor and rewards used in science, industry, and mass media
- Economic and demographic trends that affect societal needs and outcomes
- Differences in governance approaches and structures across communities and countries.

---

[5]PATH—our approach, http://www.path.org/about/index.php. Accessed 21 Apr 2015.

[6]See https://www.bio.org/articles/current-uses-synthetic-biology for examples of products of synthetic biology. Accessed 21 Apr 2015.

5 Participation

Participation refers to the analysis of who participates in the evaluation and discussions and in what capacity. Participation as a form of social practice is connected to democracy and citizen empowerment. The underlying goal of increasing participation is the shift from hierarchical, top-down and authoritative forms of decision-making to more equalized, inclusive and distributed forms.

Participation in science is complicated by the changing pace and character of technoscientific innovations, flows of knowledge and expertise and globalization of citizen networks and governance (Leach et al. 2005). Historically, participation in science was described via the canonical account of public communication of science, which stated that science is too specialized and complicated to be understood by the public, therefore it needs to be translated into simpler words by mediators (Bucchi 1998). A constructivist approach challenges this account and argues that the public has the right to reject or resist the attempts of the dominant elites to maintain "social control via public assimilation of "the natural order" as revealed by science" (Wynne 1995).

Scientists are among the most prominent voices in the discussions of synthetic biology.[7] They speak not only through their published research, but also via reviews, commentaries, and expert opinions in the news. Journalists write about science as part of their profession. They write for traditional publications, such as newspapers, but also they publish essays and books, and post online. The voice of the public represents members of the public as a speaking subject. It is the most heterogeneous and amorphous category as it refers to people who act as members of a larger population or community and not as members of professional organizations or the state. The public voice is usually represented in digital genres, predominantly in short messages in forums and comment sections in blogs and websites. The public also speaks through surveys and opinion polls.[8]

Scientists, journalists, and the publics are the major voices that contribute to and are often recognized within the discussions and evaluations of synthetic biology. However, other voices contribute to the discussions too. The scholar voice can be defined as represented by social scientists and individuals occupying positions at the departments of ethics, philosophy, or history of science. It has already been mentioned that social sciences and humanities constitute their own social field with its own structures of knowledge and capital. Scholars are concerned with wider societal implications of technoscience and with the study of fundamental questions of existence, knowledge, and matter. In the societal debates about synthetic life, the

---

[7]Here I define the scientist as a professional doing research in natural and applied sciences to allow other scholarly voices to be identified separately.

[8]See, for example, the polls of the Synthetic Biology project: from 2008 http://www.wilsoncenter.org/sites/default/files/finalhart_final_re8706b.pdf, 2009 http://www.synbioproject.org/library/publications/archive/6410/ and 2010 http://www.synbioproject.org/library/publications/archive/6655/. All accessed 06 Jun 2015.

scholar voice can offer logical and conceptual contributions as to what it means to be able to create life and what arguments can be advanced for or against it.

Scholars investigate impacts of synthetic biology on individuals and communities. They possess what Collins and Evans (2002) identified as interactive expertise, a type of expertise that is acquired when one studies a domain from a social perspective, using social sciences and humanities methods and theories. Even though the social researcher cannot do what a practitioner from that domain can do, he or she understands enough to be able to talk about it intelligently. The implication of emphasizing interactional expertise as a legitimate type of expertise is that it allows for commentary and opinion from individuals outside of the group of contributory, i.e., scientific expertise. Social scientists, philosophers, and ethicists become a legitimate group of experts who can weigh in the discussions of societal impacts of synthetic biology and, in addition to helping scientists engage with ethical, legal, social, and other issues (Edwards and Kelle 2012), they can challenge some of the assumptions that come from the scientific field, in other words, be advocates, intermediaries, translators, connoisseurs, critics, or reformers (Calvert and Martin 2009). Calvert and Martin (2009) identify a number of initiatives, such as SYNBIOSAFE[9] in the European Union or SynBERC[10] in the United States, where social researchers both contribute to the discussions and study the consequences of synthetic biology, but also collaborate and potentially influence the production of scientific knowledge.

Another two distinct voices are the policy maker and the activist voices. A policy maker is an individual or an agency associated with funding, governing or decision-making bodies. For example, the Alfred P. Sloan Foundation, a non-profit grant-giving institution in the United States, represents policy makers. Another policy-making institution is the Synthetic Biology Project[11]—an initiative of the Foresight and Governance Program of the Woodrow Wilson International Center for Scholars. Representatives from these and other institutions contribute to the debates and ultimate outcomes through their research and evaluations and by reviewing others' contributions and distributing funds.

An activist is an individual or an agency that actively advances a cause or protests against the establishment. The ETC group, an international organization with offices in the United States and Canada, is an example of a vocal activist group that contributes to the debates on synthetic biology and other emerging technologies that potentially impact the environment and human life. On its "About" page[12] the ETC group describes its goals as "the conservation and sustainable advancement of cultural and ecological diversity and human rights". This organization has consistently argued against the expanding application of the engineering approach to the

---

[9]http://www.synbiosafe.eu/, Accessed 06 Jun 2015.

[10]http://www.synberc.org/, Accessed 06 Jun 2015.

[11]http://www.synbioproject.org/ Accessed 06 Jun 2015.

[12]http://www.etcgroup.org/content/mission-etc.-group Accessed 06 Jun 2015.

matters of life and nature without proper safeguards and limitations. In a case study on artemisinin, the ETC group argued that the production of semi-synthetic variety favors pharmaceutical companies and disempowers local producers whose income depends on growing the *Artemisia annua* plant, a natural source of artemisinin (ETC group 2014).

The democratic processes of decision-making call for many voices to be included in the debates on synthetic biology. However, even if all possible stakeholders are identified, what does it mean to facilitate participation of all of them? Is it even possible? Cornwall (2008) argues that while deep and wide participation may seem ideal, in practice it is virtually impossible to achieve. Choices need to be made to achieve the balance between broad participation and practical limitations based on the purpose at hand. The discourses of synthetic biology demonstrate patterns of unequal participation from various stakeholders and shows limited ways of influencing and possibly changing the outcomes. Despite many calls for more public debates, for participatory science, and for equal opportunities in decision-making, it is counterintuitive to argue that actors in such different positions as scientists, policy makers, and members of the public and with such varying amounts of economic, symbolic, and social capital can enjoy equality in participation.

Participation is also restricted by orientation toward intellectual production and expertise (Bourdieu 1993). Producers of intellectual goods, such as scientists or artists, define their own criteria of what constitutes a legitimate act of production and how it should be evaluated. Non-producers, i.e., the public at large, consume products created by others and have less control over the rules of consumption and legitimacy of products. Considering these differences and limitations that are imposed by the fact that actors belong to different social fields, the questions of equal participation should be transformed into the questions of meaningful participation according to one's field and position in it or, more importantly, into the questions of changing the fields and the distribution of power within and among them.

It is crucial to the effectiveness of the debates in synthetic biology to consider the barriers to wider participation. Many barriers exist that prevent members of the public from engaging in the production and dissemination of science, including educational, economic, and cultural barriers. The power of the public remains to be the subversive power of the masses, which establish their voice in discourse by responding and voting collectively rather than by relying on intellectual or cultural authority. To promote public engagement, the rhetoric of participation has to incorporate the values of intellectual and cultural equality and rational debates, i.e., the values promoted by the Habermas' notion of the public sphere. The venues for public participation would have to offer equal opportunities of authorship and other contributions for all participants, acknowledge the existing barriers, and accommodate points of views of marginalized actors.

To summarize, participation invokes the concepts of identity, citizenship, voice, authority, democracy, and consensus. In considering participation as part of the evaluation framework, it is important to address the following questions:

- Who contributes to synthetic biology discussions, decision-making, information dissemination, and the production of discourse and goods?
- Who is excluded and who excludes themselves from participation?
- Who makes decisions and what structures are in place to enforce and challenge these decisions?
- What are the barriers to and outcomes of participation?
- What forms of participation are considered and what are their intentions and outcomes (e.g., access to information, planning, decision-making, consultations, representation, assessments)?

6 Critique

Critique is a systematic analysis that examines the validity and limits of the existing claims. In its contemporary form it has been exemplified in the critical theory school of thought, particularly, in the works of the Frankfurt school earlier in the 1930s and later in the works of Jürgen Habermas and Michael Foucault. Critical theorists are often concerned with the analysis and critique of the modern Western society and its crises and with the examination of rationality as an indispensable instrument of human emancipation and freedom. The analysis provided below applies arguments from Habermas, Foucault and other critical theorists to the problematics and discourses in synthetic biology.

From a critical perspective, the discussions of synthetic biology are characterized, among other things, by the following:

- A variety of goals, genres, stakeholders, and ideological approaches, some of which inevitably compete with each other
- Historical (dis)continuity in the discourse, when vigorous debates happen at some points in time and "quiet", implied, or hidden discussions happen in between
- More active engagement of some actors (e.g., scientists) compared to others (e.g., scholars and intellectuals, members of the publics) due to the personal and professional preferences as well as due to the distributions of economic, intellectual, and social capital
- Focus on the more immediate and practical issues and on the regulation of particular products or methods due to the economic, political, and regulatory pressures
- Dominance of the technological and engineering approaches to life, i.e., the approaches that consider life as a device or a mechanism that can be understood by pulling it apart and looking at how various parts work together. Once such a mechanism is understood through deconstruction, it can be re-created, improved, and manufactured in desired quantities:

Before a mechanic can assemble a steam engine, he must know the how and why of boilers, cylinders, and pistons and so the chemist who would create a protein must lay the groundwork by finding out all that he can about the amino acids of which it is made. That is why the Los Angeles experimenters are testing their solubility in many liquids; studying their behavior in acids and alkalies; bombarding them with heat, light, X rays, and other radiations; studying their colors; testing their electrical properties with sensitive meters; and inspecting their crystal forms with microscopes. When the properties of each one are completely known, and all can be made to order, chemists will be prepared for the supreme attempt to put them together into products like those of nature. (Martin 1936, p. 18)[13]

Why are the discourses of synthetic biology dominated by the engineering approaches to life, by the scientists' voices, and by the creation and management explanatory themes? The development and dominance of those aspects fits with Foucault's framework of the orders of discourse. Foucault pointed out that discourse exists within the established order of things and that institutions give ordered discourses their power over the social reality and individuals in it (Foucault 1971, 1972, 1973, 1980a, 2002). Among the rules that govern the discourse production are the rules of exclusion that concern what is prohibited. These rules include rituals that determine what can be said and how, and grant exclusive rights to speak about a particular subject.

In synthetic biology discussions, scientists are granted an almost exclusive right to speak about the living and nature, because it is the subject of their professional activities and deep knowledge. The scientists' dominating position in the discourse is legitimized by their status of knowledge producers as well as by the larger condition defined by Foucault as the regime of truth centered on science (Foucault 1980b). What is considered true and false is a historically conditioned distinction, and science has been part of this distinction as a form of the will to truth characterized by the focus on observable, measurable, classifiable, and, we might add, constructible objects and by certain functions imposed on the knowing subject (e.g., look and verify rather than read and interpret). The status of biotechnologies as technoscience, i.e., as an area that combines the development of theoretical knowledge with techniques of its application, adds even more weight to the legitimacy of synthetic, genetic, and molecular biologists. They not only explain how things work, but also change how the world works on an unprecedented scale.

The engineering approach to biology is supported by innovation and technology-driven institutions of business and experimental laboratories. It is driven by the pressing needs for solutions in agriculture, energy, medicine, and warfare. The desire to commercialize knowledge and provide practical solutions determines the drive to manipulate the living matter and use it as a material for economic production. The will to scientific truth transforms into the will to

---

[13]The dominance of the technological and engineering approaches to life in synthetic biology has been characteristic of the 20th and early 21st century discussions. If other approaches, such as systems biology or computational approaches further develop their methods and practical applications, perceptions on life can become more complex and diverse.

engineering truth and becomes one of the most efficient ways to define and apply knowledge as solutions with measurable expenses and outcomes.

It is difficult to challenge actors and positions that are granted the right to seek truth and solve hard problems. If the synthetic biology discourse remains within the paradigm of commercialized science, we may end up with skewed evaluations of synthetic biology that privilege the engineering point of view while excluding the needs of other communities and ignoring the larger issues of public and environmental protection. Viewing the discourse as institutionally ordered and historically conditioned helps to introduce change and promote the visibility of new actors and social fields, which challenge existing configurations of authority and discursive approaches. Non-university models of scientific knowledge production and review, citizen participation and involvement,[14] and digital modes of information dissemination are among the potential points of change.

The decision points in synthetic biology are concerned with the more practical and shorter-term evaluations, such as risks and benefits or ethical aspects of particular techniques. However, an equally important if not the more important component is long-term and it is concerned with possible and ongoing redefinitions of life in mechanistic terms. What does it mean that life is being approached as a device? What are the implications of the dominance of this view for humanity and society? Are our ethical guidelines for lower-order and higher-order organisms enough to make decisions with regard to with whom to experiment and how?

A view that biological components of life can be controlled and manipulated in a manner similar to manipulations of devices and mechanisms can be interpreted within the frameworks of feminist science studies and critical social theory. Haraway (1997), for example, re-introduced the concept of cyborgs to ask how we engage with forms of life when the lines between human, machine, and organic nature are permeable and revisable, and examined whether biology and biotechnology as technoscientific discourses contribute to the strategies of accumulation of wealth and power.

Haraway used the OncoMouse, a type of mouse that is genetically modified to make it more susceptible to cancer and thereby better for cancer research, as a representative example of technoscience. She argued that this mouse represents a shift in the practices of knowledge production, a shift that was described as "condensation, fusion and implosion" of subjects and objects, technical and organic, as well as the implosion of informatics, economics, and politics (Haraway 1997). The implosion produces creatures of nature with no nature that become examples of extreme objectivity, i.e., of the knowledge spaces that are grounded in the engineered, culture-free, and fully operationalized spaces. Haraway asked whether such "empty" space could serve as a ground for moral discourse, for

---

[14]See, for example, the project *Biocurious* (http://biocurious.org/)—an attempt to create a community biology lab for amateurs and promote open and affordable innovations in biotechnology. Accessed 06 Jun 2015.

decisions regarding healthy, good, right, and so on. She argued that through its very artifice the engineering discourse serves as the moral discourse:

> Precisely as fully artifactual, the nature of no nature gives back the certainty and legitimacy of the engineered, of design, of strategy, and intervention. The nature of no nature is the resource for *naturalizing* technoscience with its vast apparatuses for representing and intervening, or better, representing *as* intervening. (p. 103)

Such legitimation of the engineered and artificial imposes the view that there is no unlimited individuality in life and that the full control of variability and individuality is the goal of a rational and progressive society. Such a view obscures nonproprietary and nontechnical meanings of life and puts forward the relationships of labor, ownership, and commodity exchange. Scholars of the feminist and post-structuralist schools of thought offer elaborate accounts of the shifts toward technological approaches in science in general and in biology in particular. To a lesser extent, though, they address the questions of why it is important to foreground nonproprietary and nontechnical meanings. Their concerns and considerations frequently revolve around certain groups or populations, such as women or ethnic minorities, which become disadvantaged due to existing technoscientific developments and excluded from the distribution of wealth and other resources. To expand these considerations, it is useful to incorporate Habermas' critique of modernity grounded in his theory of communicative action, which offers rather universal reasons of why overemphasis on technology and control over the means of production is problematic and potentially disastrous for modernity.

According to Habermas, the evolution of society is the continuous differentiation and rationalization of various spheres conceptualized through the terms of system and lifeworld (Habermas 1984a, b). Systems are fully rationalized spheres of society. They are formal organizations and institutions, such as businesses or governments, which are grounded in a specific form of rationality, in instrumental rationality, which approaches objects and subjects as means and not as ends. Lifeworld (*Lebenswelt*) is a concept used in philosophy and the social sciences to describe the world as it is lived or experienced by humans together. Husserl (1970) used the concept of the lifeworld in his phenomenological analysis of consciousness to describe how each of us and all of us together can actively co-exist because we have the common horizon of all our experiences. Synthesizing previous philosophical and sociological approaches to examining the collective human experience, Habermas (1984b) defines lifeworld as a repertoire of experiences, practices, and understandings, or a "totality of interpretations" from which participants draw as they communicate, reach understanding, establish norms, and reproduce social order. Communicative claims within a society become valid when not only they are acknowledged as true or false, but also are accepted as fitting with commonly shared experiences of the lifeworld.

As society becomes differentiated and its cultural, social, and personal spheres undergo diversification, social cohesion and coordination become more fluid and unpredictable. Habermas saw two ways of addressing the issues of complexity and unpredictability, or contingency in his words: via communicative action, i.e., via discussions aimed at reaching understanding and agreement, or via the steering media of money and power, which circumvent understanding and agreement and substitute them with the goals of reaching ends and achieving efficiency and control. Mediation through money and power rather than language is necessary to achieve efficiency in complex systems, but it alienates individuals, who do not feel responsible. When economic, political and other systems are uncoupled from the lifeworld, i.e., they no longer require common understanding and consensus, the society experiences crises of identity and legitimation.

Placed in the context of Habermas' lifeworld—system distinction, the system takeover of the concepts that technosciences deal with, including the concepts of life, nature, creation, or harm, may lead to crises of identity and legitimation. These are complex concepts that combine elements of the objective, subjective, and intersubjective worlds. The engineering approaches along with the emphasis on products and instruments advance the goals of control, efficiency, and production, and remove other goals, such as co-existence or empathy, from consideration.

According to Habermas, in order for society to develop in a stable and progressive manner, actors must be able to connect newly arising situations with meanings, groups, and generalized competences that already exist in the lifeworld. Such connections facilitate cultural reproduction, social integration, and socialization. A crisis occurs when individuals cannot interpret or justify information and decisions by using existing interpretative repertoires; they do not know how to act and what to do because the existing interpretative repertoires do not provide enough guidance; or they feel that groups with which one could identify are not legitimate or valid anymore.

The reluctance of the public to rely on scientists as experts and the polarization of opinions with regard to genetic engineering are examples of the legitimation crisis. As scientists and the state might be interested in the advancements of synthetic biology as a way to achieve particular system-oriented ends, such as boosting economy or solving the energy crisis, they increasingly rely on a particular interpretive repertoire, i.e., on the engineering tropes that are best suitable for such goals. But the accumulating scientific knowledge does not fit within the existing lifeworld structures and cannot be accepted as legitimate until the scientific and non-scientific meanings get reconciled through extensive communication or forced via systemic measures, e.g., via ratified laws and policies.

At the same time, scientific controversies are also a crisis in solidarity as the conflicting attitudes towards scientific issues erode the identity of certain collectives, such as scientific experts, activists, citizens, or local communities. As these groups include members who hold different opinions and can be for or against certain approaches to an issue at question, the membership in those groups becomes problematic. For a researcher, for example, raising questions about the role of values and politics in scientific research or about personal responsibility for ones actions can be interpreted as questioning their group solidarity.

Another factor that contributes to tensions surrounding scientific progress is the so called expert-layperson divide. On one hand, the expert-layperson distinction can be taken at its face value and viewed as an inevitable outcome of the division of labor and professional specializations. It is inevitable that those who contribute to the production of knowledge know more about their object of studies; therefore, they have more authority and weight in discourse. On the other hand, such distinctions can be viewed as means through which power is exercised and certain types of knowledge become more valued than others (Foucault 1982). The movement towards more equal participation of various members of society in science communication would have to (a) counteract professional expertise with other kinds of expertise, i.e., economic, linguistic, interactional, or cultural, and (b) elevate the status of other competences and convincingly argue that the knowledge-producing status of scientists is not sufficient for them to maintain their privileged position in the evaluations of synthetic biology that have wide societal implications.

Foucault described several objectives pursued by those who exercise power, such as the maintenance of privileges, the accumulation of profits, and the exercise of a function. While such objectives can be presumed to exist and influence current activities in knowledge production, some if not most of the justifications that are employed in the synthetic biology discourse mask those objectives and promote disinterestedness and universalism as the foundational norms of science. The objectives focus on the production of goods, advancement of knowledge, and improvement of human life. It is difficult to object to those goals. Further research of power relations specified by Foucault can reveal how particular forms of institutionalization, e.g., the autonomy of scientific institutions and the development of non-university labs, and degrees of rationalization, e.g., the technologization of scientific production and communicative resources, contribute to the exercises of power in relation to science in the public sphere.

To summarize, critique offers a particular perspective on the approaches to the evaluations of synthetic biology. As a form of reflective practice that involves both understanding and explanation, it helps to show the character and limits of existing issues and claims in synthetic biology with the ultimate goals of self-transformation or societal change. It invokes the concepts of reason and rationality, justice, and power and addresses the following questions:

- What underlying values and ideological positions contribute to decision-making?
- Who benefits from the decisions that are being made and from the outcomes?
- Who has the power to make things happen?
- What views and values dominate the discourse and which ones are invisible or marginalized?
- What are the foci and reasons of resistance and opposition?
- What cultural and social contingencies need to be taken into account when making decisions about science?
- Are there any opportunities for change and is such a change desirable?

7 Conclusion

The critical participatory framework proposed in this chapter places decisions that need to be made with regard to the analysis and evaluation of societal impacts of synthetic biology in the historical, social, cultural, political, and economic contexts. It argues that the decisions in context need to be examined from the perspectives of participation and critique to make sure that the complexities and interactions in practices and discussions of synthetic biology are addressed on various levels. Figure 4 condenses the framework into a single diagram that can guide specific evaluation designs and implementations. The analysis provided in this chapter is a snapshot that illustrates the use of the framework and the trends in themes, meanings, and decisions within a certain historical period. Critical awareness of innovations in synthetic biology would necessitate more analyses over time to monitor developments and decision-making and to offer adequate solutions.

The factors and questions suggested as crucial in each component of the model are by no means comprehensive. They are the initial relevant points that will allow the evaluators to become sensitized to particular issues in order to develop a focused approach suitable for addressing a more specific problem or a question. The analytical procedure suggested by this framework involves multiple iterations of going back and forth between decision points, context, participation, and critique in an attempt to evaluate societal impacts and justify decision-making. The framework can also be used to create mappings that are comparable over time and across systems, for example, to compare nanotechnology and synthetic biology.

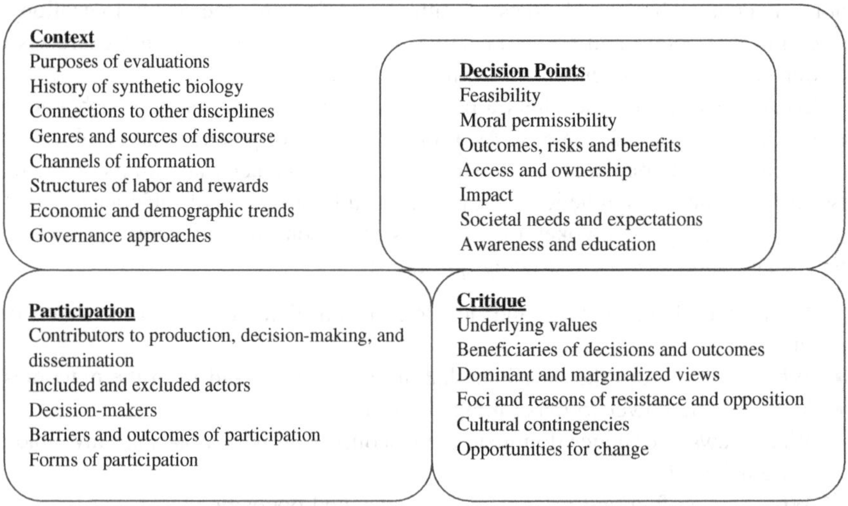

Context
Purposes of evaluations
History of synthetic biology
Connections to other disciplines
Genres and sources of discourse
Channels of information
Structures of labor and rewards
Economic and demographic trends
Governance approaches

Decision Points
Feasibility
Moral permissibility
Outcomes, risks and benefits
Access and ownership
Impact
Societal needs and expectations
Awareness and education

Participation
Contributors to production, decision-making, and dissemination
Included and excluded actors
Decision-makers
Barriers and outcomes of participation
Forms of participation

Critique
Underlying values
Beneficiaries of decisions and outcomes
Dominant and marginalized views
Foci and reasons of resistance and opposition
Cultural contingencies
Opportunities for change

Fig. 4 The critical participatory framework for the evaluation of synthetic biology: components details

References

Baker C (2005) Synthetic biology: hardware, software, and wetware (Summary). American Association for the advancement of science, DoSER public lecture series.

Bennett G, Gilman N, Stavrianakis A, Rabinow P (2009) From synthetic biology to biohacking: are we prepared? Nat Biotechnol 27:1109–1111. doi:10.1038/nbt1209-1109

Bourdieu P (1969) Intellectual field and creative project. Soc Sci Inf 8:89–119. doi:10.1177/053901846900800205

Bourdieu P (1981) The specificity of the scientific field and the social conditions of the progress of reason. In: Lemert CC (ed) Columbia University Press, New York, pp 257–292

Bourdieu P (1989) Social space and symbolic power. Soc Theory 7:14–25

Bourdieu P (1993) The field of cultural production: essays on art and literature. Columbia University Press, New York

Bourdieu P (1994) Rethinking the state: genesis and structure of the bureaucratic field. Soc Theory 12:1–18

Bourgeois W (1976) Verstehen in the social sciences. J Gen Philos Sci 7:26–38

Bucchi M (1998) Science and the media: alternative routes in scientific communication. Routledge, New York

Calvert J, Martin P (2009) The role of social scientists in synthetic biology. EMBO Rep 10(3):201–204. doi:10.1038/embor.2009.15

Campos L (2009) That was the synthetic biology that was. In: Schmidt M, Kelle A, Ganguli-Mitra A, de Vriend H (eds) Synthetic biology: the technoscience and its societal consequences. Springer, Dordrecht, pp 5–21

Cho MK, Magnus D, Caplan AL et al (1999) Ethical considerations in synthesizing a minimal genome. Science 286(80):2087–2090. doi:10.1126/science.286.5447.2087

Church G (2005) Let us go forth and safely multiply. Nature 438:423. doi:10.1038/438423a

Collins HM, Evans R (2002) The third wave of science studies: studies of expertise and experience. Soc Stud Sci 32:235–296

Cornwall A (2008) Unpacking "participation": models, meanings and practices. Community Dev J 43:269–283. doi:10.1093/cdj/bsn010

Dabrock P (2009) Playing god? Synthetic biology as a theological and ethical challenge. Syst Synth Biol 3(1–4):47–54

Demott JS, Thomas E (1980) Test-tube life: Reg. U.S. Pat. Off. Time. http://www.time.com/time/magazine/article/0,9171,924274,00.html. Accessed 5 May 2009

Deplazes A, Ganguli-Mitra A, Biller-Andorno N (2009) The ethics of synthetic biology: outlining the agenda. In: Schmidt M, Kelle A, Ganguli-Mitra A, de Vriend H (eds) Synthetic biology: the technoscience and its societal consequences. Springer, Dordrecht, pp 65–79

Douglas T, Savulescu J (2010) Synthetic biology and the ethics of knowledge. J Med Ethics 36:687–693. doi:10.1136/jme.2010.038232

Edwards B, Kelle A (2012) A life scientist, an engineer and a social scientist walk into a lab: challenges of dual-use engagement and education in synthetic biology. Med Confl Surviv 28:5–18

Engelhard M, Coles D, Weckert J (2013) Case studies—overview of ethical acceptability and sustainability (5.1). http://www.progressproject.eu/wp-content/uploads/2013/05/Progress-Deliverable-5-1-final.pdf. Accessed 2 Mar 2015

Fairclough N (1989) Language and power. Longman, London

Fairclough N (1992) Discourse and social change. Polity Press, Cambridge

Fairclough N (2001) The discourse of new labour: critical discourse analysis. In: Wetherell M, Taylor S, Yates S (eds) SAGE, London, pp 229–379

Foucault M (1971) Orders of discourse. Soc Sci Inf 10:7–30

Foucault M (1972) The archaeology of knowledge and the discourse on language. Pantheon Books, New York

Foucault M (1973) The order of things; an archaeology of the human sciences. Vintage Books, New York

Foucault M (1980a) Power/knowledge: selected interviews and other writings, 1972–1977. Pantheon Books, New York

Foucault M (1980b) Truth and power. Pantheon Books, New York

Foucault M (1982) The subject and power. In: Dreyfus H, Rabinow P (eds) Michel Foucault beyond structural hermeneutics. Chicago University Press, Chicago, pp 208–226

Foucault M (2002) Archaeology of knowledge. Routledge, New York

Friends of the Earth U.S., International Center for Technology Assessment, ETC Group (2012) The principles for the oversight of synthetic biology. http://www.biosafety-info.net/file_dir/15148916274f6071c0e12ea.pdf. Accessed 3 Mar 2015

Gardner TS, Hawkins K (2013) Synthetic biology: evolution or revolution? A co-founder's perspective. Curr Opin Chem Biol 17:871–877. doi:10.1016/j.cbpa.2013.09.013

Garfinkel MS, Drew A, Epstein GE, Friedman RM (2007) Synthetic genomics: options for governance. Rockville, MA

ETC group (2007) Syns of omission: civil society organizations respond to report on synthetic biology governance. http://www.etcgroup.org/sites/www.etcgroup.org/files/publication/654/01/etcnrsloanresponse17oct07.pdf. Accessed 14 Oct 2014

ETC group (2014) Case study: artemisinin and synthetic biology. http://www.etcgroup.org/sites/www.etcgroup.org/files/ETC-artemisinin-synbio-casestudy2014.pdf. Accessed 21 Feb 2015

Habermas J (1984a) The theory of communicative action. Reason and the rationalization of society. Beacon Press, Boston

Habermas J (1984b) The theory of communicative action. Lifeworld and system: a CRITIQUE of functionalist reason. Beacon Press, Boston

Haraway DJ (1997) Modest_Witness@Second_Millennium.FemaleMan_Meets_OncoMouse: Feminism and technoscience. Routledge, New York

Holtug N (1998) Creating and patenting new life forms. In: Kuhse H, Singer P (eds) A companion to bioethics. Blackwell Publishing, Malden, pp 206–214

Hurlbut JB (2015) Reimagining responsibility in synthetic biology. J Resp Innov 2:113–116. http://dx.doi.org/10.1080/23299460.2015.1010770. Accessed 20 Feb 2015

Husserl E (1970) The crisis of European sciences and transcendental phenomenology. Northwestern University Press, Evanston

Jasanoff S (2005) Designs on nature: science and democracy in Europe and the United States. Princeton University Press, Princeton

Kaebnick GE, Gusmano MK, Murray TH (2014) The ethics of synthetic biology: next steps and prior questions. Hastings Cent Rep 44(Suppl 5):S4–S26. doi:10.1002/hast.392

Kouper I (2011) The meanings of (synthetic) life: a study of science information as discourse. Dissertation, Indiana University

Lam CMC, Godinho M, Martins dos Santos VAP (2009) An introduction to synthetic biology. In: Schmidt M, Kelle A, Ganguli-Mitra A, Vriend H (eds) Synthetic biology: the technoscience and its societal consequences. Springer, Netherlands, pp 23–43

Leach M, Scoones I, Wynne B (2005) Introduction: science, citizenship and globalization. In: Leach M, Scoones I, Wynne B (eds) Science and citizens. Zed Books, New York, pp 3–14

Lutgen P (2014) Are artemisia plantations killing fields? Pierre\'s Weblog. https://plutgen.wordpress.com/2014/11/08/are-artemisia-plantations-killing-fields. Accessed 26 Jan 2015

Marris C (2013) Synthetic biology's malaria promises could backfire. SciDevNet 29/10/13. http://www.scidev.net/global/biotechnology/opinion/synthetic-biology-s-malaria-promises-could-backfire.html. Accessed 26 Jan 2015

Martin RE (1936) Life from the test tube promised by new feats of modern alchemists. Pop Sci 128(6):14–19

Newman WR (2004) Promethean ambitions: alchemy and the quest to perfect nature. University of Chicago Press, Chicago

Peter D, Hart Research Associates, Inc. (2008) Risks and benefits of nanotechnology and synthetic biology. http://www.nanotechproject.org/process/assets/files/7040/final-synbioreport.pdf. Accessed 10 Feb 2009

Presidential Commission for the Study of Bioethical Issues (2010) New directions: the ethics of synthetic biology and emerging technologies. Washington, D.C.

Rabinow P, Bennett G (2012) Designing human practices: an experiment with synthetic biology. University of Chicago Press, Chicago

Rasmussen S, Raven MJ, Keating GN, Bedau MA (2003) Collective intelligence of the artificial life community on its own successes, failures, and future. Artif Life 9:207–235

Roman L, Román M (2010) Encyclopedia of Greek and Roman mythology. Infobase Publishing, New York

Shapira P, Youtie J, Li Y (2015) Social science contributions compared in synthetic biology and nanotechnology. J Resp Innov 2(1):143–148. doi:10.1080/23299460.2014.1002123

Snow AA, Smith VH (2012) Genetically engineered algae for biofuels: a key role for ecologists. Bioscience 62:765–768. doi:10.1525/bio.2012.62.8.9

Van Dijk TA (1993) Principles of critical discourse analysis. Discourse Soc 4:249–283

Van Doren D, Koenigstein S, Reiss T (2013) The development of synthetic biology: a patent analysis. Syst Synth Biol 7:209–220. doi:10.1007/s11693-013-9121-7

Vos TP (2002) News writing structure and style. In: Sloan WD, Parcell LM (eds) American journalism: history, principles, practices. McFarland & Co, Jefferson, pp 296–305

Weiss R (2008) Md. scientists build bacterial chromosome. Washington Post A04. http://www.washingtonpost.com/wp-dyn/content/article/2008/01/24/AR2008012402203.html. Accessed 21 Apr 2015

Wynne B (1995) Public understanding of science. In: Jasanoff S, Markle GE, Petersen JC (eds) Handbook of science and technology studies. Sage Publications, Thousand Oaks, pp 361–388

Synthetic Biology—Playing Games?

Leona Litterst

1 Play in Synthetic Biology

Professional synthetic biology aspires to an ideal of planning and controlling. Mostly predefined purposes are pursued in long-term projects in a result-oriented way, and synthetic biology is predominantly grasped as bioengineering (cf. Billerbeck and Panke 2012, pp. 19–40; Köchy 2012b, pp. 140–143). In engineering a purpose- and application-oriented approach is dominant. Contrary to this, terms that explicitly underline the playful component are often used in descriptions of synthetic biology and contribute to the generation of a picture of a new and different scientific field that sets itself apart from the traditional scientific community.[1] According to Engelhard (2011, pp. 52–53), too, there is a change in the research culture of synthetic biology in comparison to gene technology, and the playful component, she holds, is a characteristic that is specific to synthetic biology.

When Schrauwers and Poolman (2013, p. 142) state that DNA is suitable as a matter for playing, they pick up a common stance in synthetic biology: to count genetic as playful elements. The most explicit allusion to modularization and standardization as well as play in synthetic biology is made in connection with

---

[1] In the media, the new quality of synthetic biology seems indisputable (Köchy 2012a, pp. 33–49). However, Potthast (2009, p. 43) deemed it not appreciable whether synthetic biology actually represented a paradigmatic shift. Due to the transitional period in which synthetic biology was at that time and still is at present, it is possible that the alleged paradigmatic shift is just a pretense to awake the public interest and the interest of solvent sponsors, without practicing anything fundamentally new.

L. Litterst (✉)
International Centre for Ethics in the Sciences and Humanities (IZEW), Wilhelmstraße 19, 72074 Tübingen, Germany
e-mail: leona.litterst@izew.uni-tuebingen.de

© Springer International Publishing Switzerland 2016 243
K. Hagen et al. (eds.), *Ambivalences of Creating Life*, Ethics of Science and Technology Assessment 45, DOI 10.1007/978-3-319-21088-9_12

functional standardized DNA sequences, so-called BioBricks. Due to their modular and standardized design, BioBricks can be removed and combined in any order and are therefore frequently compared to Lego Bricks (cf. Benner 2012; iGEM-Team UNIK Copenhagen 2014). BioBricks that are characterized and categorized are collected in an open access internet data base, the so-called "Registry of Standard Biological Parts",[2] which is central in the yearly iGEM (International Genetically Engineered Machine) competition. Teams that participate in the iGEM competition aim at equipping organisms with new features by using available BioBricks or at creating entirely new BioBricks. Bacteria smelling like banana or microbial photobase paper are only two examples that have arisen from this "playground of visions" (Fritsche 2013, p. 16)[3] and illustrate the playful character of scientific research in iGEM. The creativity of the teams and the fun-factor are often pointed out as characteristic for the competition and awake the interest of young scientists. While a playful spirit is often found in institutional synthetic biology laboratories, the competition was originally exclusively offered to undergraduates without scientific degrees, and now even has a pupils' section.[4] This seems to underline the ease with which the techniques can be grasped and practiced.

The ease with which synthetic biology can allegedly be carried out is related to another arena for synthetic biology play: "garage biology" or "do-it-yourself (DIY)" biology, where amateur scientists or artists apply techniques of synthetic biology at home or in non-commercial laboratories (Nature Editorials 2010), or scientists try out some ideas outside the official science institutions. Geneticist Ellen Joergensen, president of Genspace, a non-commercial laboratory in Brooklyn, has pointed out that the reason for people coming to these laboratories is their passion for science and not to earn a living from it (Charisius et al. 2012, p. 59). Lay researchers are free in choosing their subjects and projects, and a lot of creativity is released. Furthermore, they are typically interested in inspiring other people to carry out research in non-commercial laboratories, too. Thus on the Genspace (2014) homepage the programmatic question is posed: "Remember when science was fun?"

The attitudes towards garage biology of some established scientists are sometimes quite more cautious. Some considerate DIY biology as a security risk (cf. Bennett et al. 2009; Kuiken 2013). However, the spirit of play does beside the DIY biology and the iGEM competition also exist in some established labs.

Before I return to the role of playful components in synthetic biology from a critical perspective, I will introduce some aspects of the theory of play and point to the distinction between "playing with ideas" and the so-called "bricolage" as two different modes of creative scientific findings in the sciences.

---

[2]http://parts.igem.org/Main_Page. Accessed 12 Nov 2014.

[3]Orig. "Spielwiese der Visionen".

[4]Additionally, every iGEM-team is accompanied by a supervisor and a parallel overgraduate section developed in recent years.

2 Play in a Broader Perspective

2.1 Theory of Play

There are various kinds of play: play of children, adults and animals, war games, betting and gambling as well as transitional and hybrid forms. There are group games and single-player games, like solitaire, in which the coincidence with which cards are shuffled is the antagonist. Even the juggler plays against centrifugal force, the acrobat against gravity and the hiker against the challenges of nature (Staudinger 1984, pp. 30–31). The "play of light" and playing with thoughts, force, risk and destiny are commonly used phrases. Someone can play a role, a foul or an evil game (Grupe 2001, pp. 466–467).

Staudinger (1984, p. 30) considers playing not as something minor or merely childish, but as an opportunity for the realization of creative humans in freedom. Thus play is meaningful as an expression of freedom, but is not necessary and has no benefit beside itself. It happens in freedom and by choice out of lust, pleasure and abandonment. Human play as a specific form of playing arises, according to Künsting (1990, p. 57), from a process of fluctuations of certain complementary natural and cultural powers. It differs from natural play in quality, as humans disclose in an act of awakening a new world with two complementary modes of being: nature and culture. One widely referred description of the term "play" was given by Huizinga (1949):

> Summing up the formal characteristics of play we might call it a free activity standing quite consciously outside 'ordinary' life as being 'not serious', but at the same time absorbing the player intensely and utterly. It is an activity connected with no material interest, and no profit can be gained by it. It proceeds within its own proper boundaries of time and space according to fixed rules in an orderly manner. It promotes the formation of social groupings which tend to surround themselves with secrecy and to stress their difference from the common world by disguise or other means. (p. 13)

In addition, Staudinger (1984, p. 32) points out that a specific space is essential for playing. Through the specific space, the game can be differentiated from the separate non-play-world. According to Staudinger (ibid., p. 38), a game is always innocent and does not long for "good" or "bad". A moral dimension occurs only in the non-play-world. Contrary to this, partially an immanent morality in the game is assumed (cf. Montada 1988, p. 26). This shows, that morality in play is a contentious issue. Nevertheless play remains embedded in a moral scope of the non-play-world and therefore it takes place in an area of moral conditions of the non-play-world.

Despite the ambiguity of the term, Grupe (1982, p. 122) identified six features of play that apply to most of its usages: no purpose, not ordinary, not necessary, freely, instantaneously and a form of self-realization. Notably the two features "innocence" and "purposeless" of play are relevant regarding play in the sciences.

2.2 The Science Game

According to Staudinger (1984, pp. 32–34) the ideal of "pure" sciences conforms to several criteria of play: Science is voluntary as no one is forced to do research; research in "pure" science is done for mere pleasure and joy in it, without any benefit but success; science is subject to a high extent of regularity, and these rules constitute scientific work as such; forming hypotheses and experimenting are the options to play the game in sciences and thereby research reminds of a challenge or competition in which the combatant is the unknown piece of reality. To recognize and verify this piece of reality, to acquire it corresponding to the rules, may finally lead to the pleasure of success in the "big game in research".

On the other hand, according to Huizinga, the fact that science seeks validation with respect to reality implies that it cannot entirely be counted as a game, but may indulge in play "[...] within the closed precincts of its own method" (Huizinga 1949, p. 203).

One mode of playful scientific finding is relevant: the spontaneous, improvised and almost free scientific "playing with ideas" (cf. Lorenz 1983, pp. 83–84). The physics Nobel laureate Richard Feynman wrote in his autobiography:

> Then I had another thought: Physics disgusts me a little bit now, but I used to *enjoy* doing physics. Why did I enjoy it? I used to *play* with it. I used to do whatever I felt like doing – it didn't have to do with whether it was important for the development of nuclear physics, but whether it was interesting and amusing for me to play with. [...] It was effortless. It was easy to play with these things. It was like uncorking a bottle: Everything flowed out effortlessly. I almost tried to resist it! There was no importance to what I was doing, but ultimately there was. The diagrams and the whole business that I got the Nobel Prize for came from that piddling around [...]. (Feynman et al. 1985, pp. 173–174)

Thus, although sciences cannot entirely be counted as play, playful scientific findings limited in time and space are possible and are definitely practiced: in process-oriented work, perception of form and intuitive thinking. The researcher intuitively follows certain directions at the beginning, but is prepared to desist from his aims at any moment and to follow up other aims that appear during the working process (Künsting 1990, pp. 31–35). This kind of playful approach is spontaneous, hardly controlled, and developed to a good deal out of improvisation. It is mostly applied without awareness of its playful character (ibid., p. 32). Occasionally it is considered as "unscientific", and it is not usually highlighted as a mode of gaining knowledge. It is not liable to tight regimentation, and it is uncommitted, thus the result always remains open. From this point of view, the external purpose of gaining knowledge is only relevant in the non-play-world. Within the actual playful approach, it is not dominant. The internal purpose of play is exclusively the playing itself. However, while "playing with ideas" is "innocent", it remains embedded in the ordinary world of non-play, an area of moral conditions which entails a dimension of responsibility. The scientist remains in the non-play area of moral conditions while temporarily resorting to a sphere of playing which might even be at odds with the responsibility dimension.

However, a second mode of creative scientific finding is also relevant: the so-called "bricolage". It is an attitude of tinkering in which the structuralist Lévi-Strauss (1968, p. 29) saw the instrument of any progress. The "bricoleur" collects objects upon which he stumbles without knowing what to assemble from them. From these collected things, he produces useful objects (Jacob 1983, pp. 50–55). Thus the "bricolage" is characteristically purpose-oriented even if the purpose is not absolutely dominant. Therefore it is not a game but rather a creative way of achieving scientific findings.

In synthetic biology as bioengineering field the purpose- and application-oriented approach is dominant. Thus, the major mode in synthetic biology is the creative but purpose-oriented bricolage, and that is no game.

3 Critique of "Play" as a Label in Synthetic Biology

3.1 Covering up High Hazards

Playful-creative scientific approaches were applied as one method to accomplish scientific findings long before synthetic biology emerged and are still used today in other fields, too. On the other hand, play has in science normally not been explicitly conceived as a form of gaining knowledge. Therefore, the *expressly desired* playful component can be seen as a specific characteristic of synthetic biology.

However, the simplicity with which the techniques of synthetic biology can be grasped, even by undergraduates and laymen, let them appear as playfully simple and riskless. This reveals that the label play with its connotation as "innocent" is of importance. It conveys an image of playful nonhazardous research. This perspective of "innocent playing" is in clear contrast to the potential hazards of synthetic biology in the real world. As a matter of fact, the risks of synthetic biology for humans and the environment are not sufficiently investigated and can therefore not be reliably estimated or excluded at present.

Therefore, creative elements are relevant in science and also in synthetic biology to achieve scientific findings. However, the label "play" conveys an inadequate presentation of synthetic biology to the public. This could contribute to mildening a potentially adverse attitude towards synthetic biology in society. Furthermore, the potential hazards of synthetic biology for humans and the environment may fall from view and the responsibility dimension might thus get lost out of sight. This could be highly problematic because potentially high-risk biotechnologies like synthetic biology should be accompanied by ethical reflections, for example the issue of responsibility, while "playing games". Therefore, synthetic biologists should avoid downplaying potential hazards and rather adopt an attitude of conscious responsibility that entails transparency (cf. Grunwald 2012, pp. 96–99; Engels 2003, p. 43).

3.2 Orientation Towards Academic and Financial Profit

According to Staudinger (1984, pp. 36–38), the money used for scientific research arises from the public authority and reveals one possible limitation of freedom in view of the playful character of sciences. The money lender, i.e., society, is allowed to codetermine what is to be done with the benefits. This touches upon aspects of science policy and strategy, when the public gives money and in return expects the fulfillment of needs for health, energy, protection of the environment, and much more. Furthermore, Staudinger underlines that due to the negative development in sciences, dubious craving for recognition and name-dropping is dominant. Science becomes a mere academic gimmick. The public charter, the invested money that is allocated to the sciences, is being abused.

In fact, the majority of research areas in synthetic biology are still in the state of—apparently playful—basic research, notwithstanding the fact that in the long term the applications are of significant interest. Specific corporations founded by the researchers themselves shall bring these applications to the market. Like no other scientist in synthetic biology, Craig Venter is in the public eye, presenting himself as a researcher "playing with Lego Bricks" and as a tough-minded economist bringing his research to application in a profitable way at the same time. As a researcher, he may have raising new funds in view. As an economist, he glances at the profitability of applications on the bioengineering market.

The economic obligation of researchers in the new biotechnologies, is apparent. Insofar, synthetic biology is liable to an increasing commercialization, although, as I have described above, neither technological and commercial orientation nor playfulness are novel components of or in the life sciences.

Also in the iGEM competition, two important aspects become apparent: On the one hand the basic idea of play and the open source concept of the parts registry. On the other hand, there are economic aspects, when BioBricks are patent-registered and when the teams have to find industrial sponsoring partners. In addition, some projects are brought on the market when the iGEM competition is finished (cf. Wagner and Morath 2012, pp. 134–135).

In the context of the increasing orientation of researchers towards values like prestige and financial profit, it appears that the playful character of synthetic biology receives a particular meaning. An analogy to the financial world may illustrate that: gambling is not done just for the sake of playing, but to a significant extent for financial profit. Besides the desire to play, there is the goal to gain economic benefits. The playful character of gambling places the dominant greed of financial enrichment closer to the sphere of harmless play and thus covers it up.

In the new biotechnologies, scientists can profit in several ways from their own scientific findings in the "game". This reveals that in the context of synthetic biology, notably with regard to the iGEM competition, the playful character is a perspective that—implicitly or consciously—pretends a non-purpose-oriented direction. Like every scientific field synthetic biology contains playful and creative elements as a way to achieve scientific findings. However, in synthetic biology the major mode is the purpose-oriented bricolage, and that is no game. Thus, synthetic biology in general is no area of purposeless play but of responsibility. In synthetic biology "playing games" is primarily a label that may serve the economization in the sciences. It is therefore connected to an inadequate public exhibition of the research of synthetic biology. Thus the orientation towards academic and financial profit may be lost out of sight.

4 Conclusion

The perspective "playing games" in synthetic biology, particularly with regard to the iGEM competition, may support the increasing interest of young researchers and is relevant because it may lead to new scientific findings. At the same time it is connected with an inadequate representation of the research of synthetic biology, not least in and for the public.

The expressly desired playful image of synthetic biology is problematic because it evokes associations of innocence and of purposelessness. However, in application-oriented synthetic biology, "bricolage" is the major mode, and this is not a game but rather a creative way to achieve scientific findings. Therefore, application-oriented synthetic biology is *not* an area of *innocent* and *purposeless* play. *Risks* of synthetic biology for humans and the environment cannot be reliably estimated or excluded at present. Further, scientists can *profit* in several ways from their own scientific findings in the new biotechnologies. Thus the label "play" in synthetic biology can serve the economization in the sciences and could be used to downplay the potential hazardousness of this new biotechnology. This is at odds with the responsibility dimension necessary for an appropriate ethical deliberation.

References

Benner SA (2012) Aesthetics in synthesis and synthetic biology. Curr Opin Chem Biol 16:581–585. doi:10.1016/j.cbpa.2012.11.004

Bennett G, Gilman N, Stavrianakis A et al (2009) From synthetic biology to biohacking. Are we prepared? Nat Biotechnol 27(12):1109–1111. doi:10.1038/nbt1209-1109

Billerbeck S, Panke S (2012) Synthetische Biologie – Biotechnologie als eine Ingenieurwissenschaft. In: Boldt J, Müller O, Maio G (eds) Leben schaffen? Philosophische und ethische Reflexionen zur Synthetischen Biologie. Mentis, Paderborn, pp 19–40

Charisius H, Friebe R, Karberg S (2012) Wir Genbastler. Frankfurter Allgemeine Sonntagszeitung, pp 57–60, 29 Apr 2012

Engelhard M (2011) Die synthetische Biologie geht über die klassische Gentechnik hinaus. In: Dabrock P, Bölker M, Braun M, Ried J (eds) Was ist Leben – im Zeitalter seiner technischen Machbarkeit? Beiträge zur Ethik der Synthetischen Biologie. Alber, Freiburg i.Br., pp 43–59

Engels EM (2003) Die Rolle der Bioethik für Politik und Forschungsförderung – Meine Erfahrungen im Nationalen Ethikrat. In: Haf H (ed) Ethik in den Wissenschaften. Beiträge einer Ringvorlesung der Universität Kassel. Kassel University Press, Kassel, pp 43–59

Feynman RP, Leighton R, Hutchings E (1985) Surely you're joking, Mr. Feynman! Adventures of a curious character. W.W. Norton, New York

Fritsche O (2013) Die neue Schöpfung. Wie Gen-Ingenieure unser Leben revolutionieren. Rowohlt, Reinbek bei Hamburg

Genspace (2014) Genspace. New York City's community biolab. http://genspace.org. Accessed 12 Nov 2014

Grunwald A (2012) Synthetische Biologie: Verantwortungszuschreibung und Demokratie. In: Boldt J, Müller O, Maio G (eds) Leben schaffen? Philosophische und ethische Reflexionen zur Synthetischen Biologie. Mentis, Paderborn, pp 81–102

Grupe O (1982) Bewegung, Spiel und Leistung im Sport. Grundthemen der Sportanthropologie. Karl Hofmann, Schorndorf

Grupe O (2001) Spiel/Spiele/Spielen. In: Grupe O, Mieth D (eds) Lexikon der Ethik im Sport. Karl Hofmann, Schorndorf, pp 466–469

Huizinga J (1949) Homo ludens. A study of the play-element in culture. Routledge & Kegan Paul, London

iGEM-Team UNIK Copenhagen (2014) LEGO—is it good or bad? http://2014.igem.org/Team: UNIK_Copenhagen/Lego_is_it_good_or_bad. Accessed 12 Nov 2014

Jacob F (1983) Das Spiel der Möglichkeiten. Von den offenen Geschichten des Lebens. Piper, Müchen

Köchy K (2012a) Was ist Synthetische Biologie? In: Köchy K, Hümpel A (eds) Synthetische Biologie. Entwicklung einer neuen Ingenieurbiologie? Themenband der interdisziplinären Arbeitsgruppe Gentechnologiebericht. Forum W(30), Dornburg, pp 33–49

Köchy K (2012b) Philosophische Implikationen der Synthetischen Biologie. In: Köchy K, Hümpel A (eds) Synthetische Biologie. Entwicklung einer neuen Ingenieurbiologie? Themenband der interdisziplinären Arbeitsgruppe Gentechnologiebericht. Forum W(30), Dornburg, pp 137–161

Kuiken T (2013) DIYbio: low risk, high potential. Citizen scientists can inspire innovation and advance science education—and they are Proving Adept at Self-Policing. Scientist 27(3):26. http://www.the-scientist.com/?articles.view/articleNo/34443/title/DIYbio–Low-Risk–High-Potential/. Accessed 12 Nov 2014

Künsting W (1990) Spiel und Wissenschaft: Versuch einer Synthese naturwissenschaftlicher und geisteswissenschaftlicher Anschauungen zur Funktion des Spiels. Academia-Verlag Richarz, Sankt Augustin

Lévi-Strauss C (1968) Das wilde Denken. Suhrkamp, Frankfurt a.M

Lorenz K (1983) Der Abbau des Menschlichen. Piper, München

Montada L (1988) Verantwortlichkeitsattribution und ihre Wirkung im Sport. In: Schwenkmezger P (ed) Sportpsychologische Diagnostik, Intervention und Verantwortung. bps-Verlag, Köln, pp 13–39

Nature Editorials (2010) Garage biology. Nature 467(7316):634. doi:10.1038/467634a

Potthast T (2009) Paradigm shifts versus fashion shifts? Systems and synthetic biology as new epistemic entities in understanding and making 'life'. EMBO Rep 10(S1):42–45. doi:10.1038/embor.2009.130

Schrauwers A, Poolman B (2013) Synthetische Biologie – der Mensch als Schöpfer?. Springer, Berlin

Staudinger H (1984) Forschung – ein Spiel? In: Ströker E (ed) Ethik der Wissenschaften? W. Fink, F. Schöningh, München, pp 27–42

Wagner H, Morath V (2012) iGEM – Eine studentische Ideenwerkstätte der Synthetischen Biologie. In: Köchy K, Hümpel A (eds) Synthetische Biologie – Entwicklung einer neuen Ingenieurbiologie? Themenband der interdisziplinären Arbeitsgruppe Gentechnologiebericht. Forum W(30), Dornburg, pp 134–135

Metaphors of Life: Reflections on Metaphors in the Debate on Synthetic Biology

Daniel Falkner

1 Introduction

> If we view life as a machine, then we can also make it: this is the revolutionary nature of synthetic biology. Until recently, biotechnologists focused on modifying the DNA of existing organisms (genetic modification). Synthetic biologists go one step further. They want to design new life and construct this from scratch. (de Vriend et al. 2007, p. 2)

This is how the Rathenau Institute introduces a letter "Synthetic biology: constructing life", which was addressed to the Dutch parliament in 2007, at a very early stage in the development of the research field of synthetic biology.[1] The "revolutionary nature of synthetic biology" here means both a paradigm shift—from the reading to the writing of DNA, from trial and error to programming software, from the modification to the design and construction of living organisms—and a "revolutionary power of converging technologies" to influence and to drive scientific and technological developments (de Vriend et al. 2007, p. 2). This claim of a revolution in paradigms and in progress often comes together with a terminology which contains the remarkable metaphorical concepts of *life as a machine* and of *constructing, designing,* and *programming life* (Boldt et al. 2009; Köchy 2012).

The early prophecy of a revolution in life science and biotechnology, worded in metaphors of machines and computers, seemed to come true when in May 2010 J. Craig Venter announced the world's first synthetic cell and brought synthetic biology into the awareness of public perception (Gibson et al. 2010). Along

---

[1]For a detailed description of the Rathenau Institute's initiatives to facilitate early engagement with synthetic biology, see Rerimassie (this volume).

D. Falkner (✉)
SYNMIKRO LOEWE-Zentrum für Synthetische Mikrobiologie, Marburg, Germany
e-mail: falkner@staff.uni-marburg.de

© Springer International Publishing Switzerland 2016
K. Hagen et al. (eds.), *Ambivalences of Creating Life*, Ethics of Science and Technology Assessment 45, DOI 10.1007/978-3-319-21088-9_13

with the presentation of this milestone in the young history of synthetic biology a remarkably computer metaphor was introduced to describe the process of creating and activating the synthetic genome:

> I describe DNA as the software of life and when we activate a synthetic genome in a recipient cell I describe it as booting up a genome, the same way we talk about booting up a software in a computer. (Venter 2012)

Venter not only used the same metaphorical frames as the letter from the Rathenau Institute five years earlier—he also interpreted synthetic biology as a kind of revolution in science and claimed the start of the "Digital Age of Biology" (Venter 2012).

In the present chapter I want to analyze the impression that arises from these (and other) citations: that there seems to be a connection between the paradigm shift in the epistemological approach, the technological development, the societal discourse and the metaphors that are used to describe, explain and argue the new field of synthetic biology and its revolutionary nature. My hypothesis is that metaphors play a constitutive and mostly underestimated role in science in general, in the modern life sciences and bio-technologies in particular, and also in the accompanying ethical debate. The current discussion on synthetic biology can be seen as a prime example for the different ways metaphors enter into an area of conflict between science, technology, society and ethics. In a first step I take a look at the ethical debate on synthetic biology and analyze the ways in which metaphors have been addressed (2). Then, due to a lack of a theory of metaphor within the synthetic biology debate, I give a short excursion into the history and theory of metaphor (3) and start to develop an analytical frame to determine and decipher the specific role and functions of metaphors in the intersection of science, technology and society (4). This analytical frame is then applied to the metaphor of the *genetic code* which is the common reference point and driving force in a reconstructed story from Erwin Schrödinger to Craig Venter (5). This leads to a reassessment of synthetic biology between science and art but also to a focus on the obscure and ideological dimension of the metaphorical speech about the revolutionary nature of synthetic biology (6). The last section sums up the results and presents three functions of metaphors that allow three perspectives of reflection and critique on metaphors in synthetic biology (7).

2 Metaphors of Life. How Metaphors Enter the Debate on Synthetic Biology

The frequent use of metaphors in the field of synthetic biology is observed and addressed by various authors of the accompanying Ethical, Legal and Social Implications (ELSI) research. Amelie Cserer and Alexandra Seiringer notice that metaphors of synthetic biology are to a large extent borrowed from the field of technical artefacts and industrial products (Cserer and Seiringer 2009). The complexity and newness of synthetic biology is explained in mainly mechanistic and industrial analogies which could lead to disturbing social consequences: "[...] the mechanistic and industrial metaphors give the impression that the creation of life by the Synthetic Biology technologies end up in artefacts, which are as easy to control as a car or an inkjet printer." (Cserer and Seiringer 2009, p. 34)

Kirsten Brukamp comes to a similar conclusion. She classifies the metaphors and unusual expressions in synthetic biology into different topics: engineering, construction and architecture, electro-technics, information theory, computer science, design, and theology (Brukamp 2011, pp. 70–71). She also suggests an evaluation of the usage of this terminology: A new descriptive vocabulary, such as "engineer biology", would be acceptable and sometimes even necessary. But metaphors such as "to program cells" are already highly problematic, because they could be understood as provocative and thereby could lead to an escalation of the debate. However the use of implicit valuation, such as "to awaken synthetic life", should be entirely avoided, because it often leads to misunderstandings and causes moral irritation and conflicts. Brukamp concludes that metaphors and eye-catching terminology contribute to a hype about synthetic biology, but are on the whole inadequate and even wrong with respect to the factual state of research (Brukamp 2011, p. 73).

Boldt, Müller and Maio note that within the discussion of problematic ethical and anthropological implications of synthetic biology a critical analysis of metaphors must be undertaken in three steps (Boldt et al. 2009, p. 57): first, a critical inquiry of metaphors that reveals their semantic content and their historic cultural implications, second, a demonstration of the innovative and epistemic potential of metaphors; and third, an investigation of the reality-constituent function of metaphors from an ontological and an ethical point of view. With these methodological considerations, technomorphic metaphors such as *living machine* and *artificial cell* are introduced as expressions of an ontological constitution of a new world of objects ("Ontologisierung", Boldt et al. 2009, p. 55). In a careless and non-reflective use, this could imply an artificialization and reification of nature (Boldt et al. 2009, p. 60). Not only could this lead to problems with regard to the ontological and moral status of artificial organisms, it could also affect our concept and value of life:

All of this vocabulary identifies organisms with artifacts, an identification that, given the connection between 'life' and 'value,' may in the (very) long run lead to a weakening of society's respect for higher forms of life that are usually regarded as worthy of protection. (Boldt and Müller 2008, p. 388)

Jens Ried, Matthias Braun and Peter Dabrock reconstruct the socio-cultural background and motives of metaphors in relation to their use in the ethical debate on synthetic biology in "Unbehagen und kulturelles Gedächtnis" (Ried et al. 2011). Characteristic of the public debate are a lack of knowledge on what synthetic biology is and the feeling of discomfort as a response to visions and goals of synthetic biology and as an expression of vague concerns about unforeseeable safety and security risks. In reference to Freud's study on "Unbehagen in der Kultur" and Assmann's theory of cultural memory such uneasiness is attributed to a lack of meaningful figures or blocked interpretive frames ("gesperrte Deutungsmuster") on the basis of common metaphors (Ried et al. 2011, p. 356). Therefore, particularly religious metaphors like "playing God" and "creating life" could serve as implicit indicators that refer to a deep-rooted sphere of societal subconsciousness and cultural memory in which the basic cultural and anthropological limits are questioned by synthetic biology.

In conclusion, metaphors in the ethical debate on synthetic biology are seen as the reason and cause of moral uncertainty and irritation in society. The engineering paradigm and the instrumentalist approach to life are reflected in mechanistic and industrial metaphors such as "living machines", "engineering life" etc. They illustrate how synthetic biology touches culturally and normatively charged and deeply rooted distinctions of living and non-living matter. Because metaphors seem to conflate the categories of "life" and "machine" they are rated as inadequate for an ontological determination of the new entities of synthetic biology, and, due to an artificialisation and reification of nature, it is assumed that they could lead to problematic ethical consequences with regard to our concept of life. At last metaphors such as "playing God" and "creating life" are perceived as expressions of social discontent and moral irritation. In this view they are ciphers on the surface to a deeper hermeneutical dimension in the unconscious of the debate on synthetic biology.

3 Metaphors We Live By. A Short Excursion into the History and Theory of Metaphor

Most of the contributions mentioned above see a need for critical evaluation of the ethical implications of using metaphorical language. There seems to be agreement that there is a danger in using metaphors because they are not adequate and could cause trouble. I suggest that this view on metaphors does not correctly conceive the specific role of metaphors in scientific contexts and misses the actual ethical dimension of metaphors as driving forces and argumentative instances in communication between science and society. Although a need to analyze metaphors is declared, there is a lack of a theory of metaphor and of a methodological framework to analyze and criticize metaphors in the specific context of synthetic biology. Furthermore, the relation between scientific and technological developments and the ethical relevance and societal dimension of metaphors remains unclear.

Technological and scientific progress can cause social and ethical conflict, when new options emerge that can no longer be regulated by the established ethical concepts and terms of a society. Such "situations of normative uncertainty" require an ethical re-orientation for both the scientists and society (Grunwald 2008, p. 55). This is currently the case in synthetic biology. Most of the different approaches and research projects that gather under this umbrella term (Balmer and Martin 2008, p. 3) can be related to the topic of "creating life" (see Eichinger, this volume), which is present in all dimensions of the scientific process. In synthetic biology the scientific, technological and societal dimension of research can no longer be kept apart. In this sense synthetic biology is a *technoscience*, which means it is not only a scientific programme but also a cultural and societal phenomenon (see Müller, this volume). This also means that the ethical conflicts and moral irritations, evoked by the claim to create life in the lab, also appear on all levels of the epistemic process of theory building, of technological progress in research praxis, and on the level of societal discourse. However, to face these conflicts is also a task of communication. Ethical debates and social discourses are the places where these ethical conflicts appear as the subject of a social praxis of argumentation and reasoning, in which language plays an important role. Therefore, situations of moral uncertainty and missing ethical orientation call for adaptation of language to a changing and evolving world that we perceive, describe and interpret as a world in which we act and argue. This brings us back to metaphors.

At first glance, metaphors are linguistic devices to illustrate and paraphrase complex, abstract and unknown issues in terms of known concepts. This is already true for the definition of metaphor by Aristotle: to give to a thing a name that belongs to something else (Aristotle et al. 1920, 1457b 6–9, pp. 71–72; cf. Ricoeur 1978b, pp. 13–24). Later, metaphors were attributed and reduced to a mere rhetorical function of substitution, and in the traditional philosophy of science metaphors were ignored or even disregarded as improper or metaphysical language because of their vague and ambivalent character. But at least since the interaction theory of the Anglo-Saxon philosophers I.A. Richards and Max Black there has been a remarkable turn in the history of the theory of metaphor. They defined the metaphor no longer as a semantic transfer process or a paraphrasing substitution bound to the linguistic level of words, but as an interaction of metaphorical and literal meaning and an interplay of the metaphorical speech and its context of use. Since then many authors have insisted on taking metaphors and their use in science and communication more seriously. In their famous book "Metaphors we live by" George Lakoff and Mark Johnson exposed a fundamental cognitive function that goes far beyond a merely rhetorical function of metaphors and transferred the results of the interaction theory into cognitive linguistics (Lakoff and Johnson 2003). They argue that metaphors as entities on the surface of language are based on deep-rooted, embodied schematic concepts and thereby structure and organize our perception of the world. In short: Thinking in metaphors allows us to understand the world—and therefore to live and to act in it—by explaining new and unknown things in terms of already known, experienced concepts. This means that both scientific language and daily communication are fundamentally rooted in cognitive metaphorical transfer processes, which is why Lakoff and Johnson speak of "metaphors we live by".

4 Living Metaphors. The Role and Functions of Metaphors in the Intersection of Science, Technology and Society

There is a wide and vast range of theories and literature about metaphors in science in general and in biology and life sciences in particular. Max Black, Mary B. Hesse, and Evelyn Fox Keller are just a few authors who described the role and function of metaphors in scientific language (Black 1962; Hesse 1970; Keller 1996). The constitutive aspect of metaphors in common language and their foundation in basic cognitive schemes of perception are the latest results of a theory of cognitive linguistics by George Lakoff and Mark Johnson and many others (Fauconnier and Turner 2002; Kövecses 2010; Lakoff and Johnson 2003). Bernhard Debatin, Michael Pielenz and Martin Seel are representatives for approaches to the argumentative dimension of metaphors in reasoning (Debatin 1995; Pielenz 1993; Seel 1990).

The goal of the following attempt is to bring aspects of these authors' analyzes together in a critical perspective on metaphors in synthetic biology and thereby to develop an analytical frame, which can be applied as an instrument of critique and evaluation of discourses between science and society. My starting point is the critical hermeneutics of Paul Ricoeur, as presented in "The rule of metaphor" (Ricoeur 1978b). This theory of the "living metaphor" historically and systematically combines linguistic and philosophical traditions and concepts, from Aristotle to Max Black, in a very fruitful and productive way. He provides a semantic determination of structure and function of metaphor, clarifies the philosophical relation between metaphor and reality, and situates it as a dialectic principle of innovation and critique in the social praxis of human understanding and argumentation.

The essence of Ricoeur's thesis is captured in the figure of the *paradox of copula* and means language creativity and innovation on the basis of a "semantic twist at the level of sense" (Ricoeur 1978a, p.146). By contradicting the rules of literal language the metaphor releases a new, metaphorical meaning which refers to reality in the mode of the statement "being-as", and claims "is" and "is not" at the same time: "Being-as means being *and* not being. Such-and-such was and was not the case." (Ricoeur 1978b, p. 306) This dialectical tension within the semantic and referential structure of the metaphor is the origin of the possibility of creativity and innovation in language. Above all, Ricoeur aims at an innovative and critical dimension of a theory of metaphorical reference, when he designates the metaphor as the heuristic function of a *redescription of reality* (Ricoeur 1978b, Introduction). With this concept of metaphorical redescription, borrowed from Mary B. Hesse, he includes the interaction theory of Max Black and adopts his analysis of models in science to his own metaphor theory:

> The central argument is that, with respect to the relation to reality, metaphor is to poetic language what the model is to scientific language. Now in scientific language, the model is essentially a heuristic instrument that seeks, by means of fiction, to break down an inadequate interpretation and to lay the way for a new, more adequate interpretation. (Ricoeur 1978b, p. 240)

According to Black, models are "sustained and systematic metaphor[s]" and "sometimes not epiphenomena of research, but play a distinctive and irreplaceable part in scientific investigation" (Black 1962, p. 236). Hesse takes up this trail and reformulates Blacks interaction theory of metaphor in terms of Wittgenstein's family resemblances (Arbib and Hesse 1986, pp. 151–153; cf. Hesse 1988). Her main thesis is that the deductive model of scientific explanation must be modified, acknowledging the underlying metaphoricity of language, and complemented by an approach of theoretical explanation as *metaphoric redescription* of the explanandum (Hesse 1970, pp. 157–177). Such a perspective on the innovative potential and imaginative power of metaphors avoids the dualism of reality and fiction and supports a critique of an objectivist, positivist ideal of scientific progress. This leads to the acknowledgement of an epistemic normativity of metaphors and models in science, which lies in the function of "redescription" and the mode of "seeing as" (Arbib and Hesse 1986, pp. 149–150). Revolutions in scientific progress and paradigm shifts now appear driven by the dynamics of a metaphorical redescription: "Scientific revolutions are, in fact, metaphoric revolutions, and theoretical explanation should be seen as metaphoric redescription of the domain of phenomena." (Arbib and Hesse 1986, p. 156). In terms of Ricoeur, the metaphorical reference appears as a *"critical* instance, directed against our conventional concept of reality" in philosophical discourses and leads to an extended concept of truth (Ricoeur 1978b, p. 305).

The place where a metaphor can act as such a revolutionary and critical instance is at last communication, i.e. the social practice of reasoning and arguing (Debatin 1995, p. 323). The metaphor can thus be seen as a starting point and instrument of critique of nominalistic theories of language and meaning. To use metaphors as such instruments of reflection and critique in a rational way requires, as Debatin points out, a method of "reflexive metaphorization" which means to reveal underlaying metaphors *as* metaphors in the processes of scientific theory building and philosophical discourses (Debatin 1995, pp. 163–168). In this sense Martin Seel refers to metaphor as a "Trojan tournament horse" for and against the "fortress of a systematic theory of meaning" (Seel 1990, p. 237). This critical function of metaphors leads to an affirmative emphasis on the communicative and argumentative power of metaphors. The metaphor is a truth-apt agent in argumentative reasoning, not by expressing statements about facts in a figurative, non-literal meaning, but by opening up new perspectives. Metaphorical speech puts things into a new light and introduces a new way to talk about reality. At the same time, the new perspective, introduced by the metaphor, is reflected *as* perspective *through* the metaphor. This is the dialectical and reflexive structure of the metaphor (Debatin 1995, p 338; cf. also Zimmer 2003, pp. 27–37). To understand metaphorically means to see something *as* something and establishes this perspective as an autonomous truth claim within a social practice of reasoning where different perspectives, background concepts, and normative beliefs could come into an "debate about truth" (Wellmer 2007; cf. also Wellmer 2004, pp. 166–173 and pp. 250–252).

With this background the processes of scientific theory construction and practical discourses occur as points of intersection for applying an ethical perspective on metaphors between science and society. Hence, three characteristic functions of metaphors can be deduced:

1. An *innovative function* and the epistemic normativity of metaphors as the condition and driving forces of scientific inquiry, paradigm shifts, technological progress and political/social/ethical discourses
2. A reflexive *critical function* of metaphors to introduce a new description language, open up new perspectives, and thereby correct and replace old, established concepts on the level of theory building and philosophical discourses on truth (see something *as* something)
3. An *argumentative function* of metaphors as truth-apt statements in the social praxis of reasoning, involving a specific rationality and "logic of plausibility" and bearing the potential of innovation and progress as well as ideological disturbance.

5 A Metaphor Comes to Life. The Story of the Genetic Code and Venter's Digitalization of Life

Metaphors from the fields of information theory and computer science are, alongside mechanistic and industrial metaphors, very prominent in synthetic biology. This is no surprise, because in most of the approaches of synthetic biology the technical progress in DNA sequencing and computational methods, play an essential and crucial role (Bölker 2011, pp. 28–30; de Lorenzo and Danchin 2008). One of the most prominent projects of synthetic biology, the "creation of a bacterial cell controlled by a chemically synthesized genome" (Gibson et al. 2010), would be unthinkable without enormous computing power and the efficiency of next-generation sequencing. One could say the synthesis of life is preceded by a digitization of life. Before the artificial cell can be brought to life, a huge amount of data has to be handled, the cellular processes must be represented as digital code, and virtually simulated and designed in the computer.[2] This is the point Craig Venter is aiming at when he speaks of "DNA as software of life", and makes claims of a "Digital Age of Biology", which is the metaphorical rhetoric in his lecture "What is life? A 21st century perspective", given in Dublin in 2012 (Venter 2012).

At the same location, seventy years earlier, Erwin Schrödinger gave a homonymic series of lectures and published these in the book "What is life? The Physical Aspect of the Living Cell" in 1944 (Schrödinger 2012). With this pioneering work Schrödinger introduced the metaphor of the "genetic code" and

---

[2]The following argument applies only to approaches which focus on DNA and are based on the differentiation of the somatic and the genetic level. Some areas of synthetic biology, such as bottom up protocell research, are not concerned with DNA. Although they work with computational methods and concepts, too, the metaphor of the genetic code script does not play a role for these research projects.

thereby significantly influenced the further history of modern molecular biology and genetics (Fischer and Mainzer 1990). As a leading physicist of his time, who witnessed and strongly influenced the paradigm shift from classical Newtonian physics to quantum theory, Schrödinger looked at the phenomenon of living cells. He asked from a "naive physicist's" point of view how the statistically unlikely case of life can be explained in physical terms and is even possible under the second law of thermodynamics. To solve this riddle, he searched for models and analogies such as the *aperdiodic crystal*, the *Laplacedemonian*, or "some kind of codescript":

> It is these chromosomes [...] that contain in some kind of code script the entire pattern of the individual's future development and of its functioning in the mature state. (Schrödinger 2012, p. 21)

This metaphor of a coded script should explain two conditions which are specific for the phenomenon of life: first, the stability of the heritage factor against environmental microphysical forces; second, the mutability as a condition to enable selective evolution (Blumenberg 1986, p. 372). For this purpose Schrödinger embedded his code script metaphor into other metaphorical concepts, such as *law code* and *architecture*:

> The chromosome structures are at the same time instrumental in bringing about the development they foreshadow. They are law-code and executive power - or, to use another simile, they are architect's plan and builder's craft - in one. (Schrödinger 2012, p. 22)

In the end it was the interpretation as *Morse code* that had a deep impact on the following history of molecular biology, biochemistry and genetics:

> For illustration, think of the Morse code. The two different signs of dot and dash in well-ordered groups of not more than four allow thirty different specifications. (Schrödinger 2012, p. 61)

The philosopher Hans Blumenberg has described this episode within his metaphorological history of the legibility of the world (Blumenberg 1986, pp. 372–409). The genetic code, descendent from the old script metaphor of the *book of nature*, here fulfills the function of closing the gap between metaphor and model, i.e. a transition from initially struggling with different metaphorical explanation models to a hypothetic scheme that drives scientific research and initiates a paradigm shift. According to Blumenberg, biochemistry and genetics were successful not least because Schrödinger's metaphorical idea was taken literally (Blumenberg 1986, pp. 376–379). The approaches and visions of synthetic biology are now the latest highlight of this successful story of modern life sciences—and, thereby, of the genetic code metaphor. Venter referred to this in his anniversary lecture and placed himself in an ancestral story of discovering the genetic code (Venter 2012).

But not only in the history of science can a line be drawn from Schrödinger to Venter. The story from the discovery of the gene to the synthesis of a bacterial genome is also the success story from reading the genetic code to writing DNA in digital code of bits and bytes. Evelyn Fox Keller and Lily E. Kay have both pointed out that it is not only the metaphor of the book of nature that sets the background concept for reading and writing the genetic code: the technological and practical dimensions of research, the information discourse in computer science and cybernetics, and the political and social context of the cold war and the human genome project all strongly influenced the history of biology in the middle of the 20th century—and are contained, one could say "encoded", in the metaphor of the genetic code (Kay 2000; Keller 2002).

With this perspective it becomes clear that the metaphor of reading and writing a binary code of life is not only for illustrative and eye-catching purposes, but rather it is an expression of the theoretical approach and epistemic foundation with which Venter describes and understands the development of biology and his own research. He directly refers to the *Morse code* metaphor and brings it up to date by reformulating it as *digital computer code*:

> I view DNA as an analogue coding molecule, and when we sequence the DNA, we are converting that analogue code into digital code; the 1s and 0 s in the computer are very similar to the dots and dashes of Schroedinger's metaphor. I call this process 'digitalizing biology'. (Venter 2012)

Venter does not just see DNA as the software of life, he actually and literally writes this code. In this sense, DNA as digital computer code is not only an illustrating and heuristic description, but a practical instruction for its own realization. The digital world of artificial life in the computer becomes physical reality. "We can digitize life, and we generate life from the digital world." (Venter 2012) Now the thesis of a potential ontologization and artificialization of life, mentioned above, takes a remarkable turn: The metaphor of the genetic code comes to life in Venter's creation of the first artificial organism *in a true sense of the word*. The metaphor of the genetic code is obviously more than *just* a metaphor. It is a redescription of reality in the sense that the cellular processes of replication under the conditions of stability and mutability of the genetic factors can not only be *explained* as based on "some kind of codescript", as Schrödinger assumed, but also *initiated* and *created* as digital code. This shift from the domain of explanation to the domain of phenomena —which can be seen analogous to the shift from *analysis* to *synthesis* that is claimed as the revolutionary nature of synthetic biology—had a striking impact on the epistemic and technical principles of research and the scientific and societal discourse in the modern life sciences and biotechnologies.

6 Between Living Art and Artificial Life. An Ethical Perspective on the Metaphors in Synthetic Biology

This perspective on one of the most prominent and much-noticed projects of synthetic biology may lead to a reassessment of synthetic biology with regard to scientific theory. If metaphors are constituent elements and driving forces of innovation and creativity on all levels of scientific inquiry, and Venter's creation of a bacterial genome could be seen as the "realization" of the metaphor of the genetic code, then one could say that Venter is a kind of artist and creative mind. Horst

Bredekamp und Hans-Jörg Rheinberger see synthetic biology in the context of the historical debate between science and art, and thereby between artificial life and living art (Bredekamp and Rheinberger 2012). From this perspective, the world's first synthetic cell is a spectacular result of hard scientific work as well as a piece of designed technology. Venter not only reconstructed the DNA sequence of an existing bacterial genome, he also rewrote the code and inserted watermark sequences including an e-mail address, the names of all authors, and three quotations from famous scientists and writers (cf. JCVI 2010). Venter's masterpiece was created, signed and presented like a piece of art and in this sense dissolves the distinction between aesthetic representation and scientific description of the living.

Of course, this interpretation may be a little bit overambitious and exaggerated, but it fits in this picture that, according to Bredekamp and Rheinberger, Venter presents his own work as a painting and himself as an artist. They even interpret this event as one of the most remarkable upheavals in the history of the life sciences and therefore speak of a "century painting". However, I think that the example of the metaphors of "genetic code", "DNA as software of life" and other computer metaphors in the context of synthetic biology illustrate that they are more than just rhetorical ornaments and an eye-catching strategy (which they *also* are). They have crucially influenced the history of modern biology and life sciences, made possible new epistemic and technical approaches and changed the way we speak of life.

What does such a classification of one synthetic biology event, located between science and art, mean for the *ethical* evaluation of metaphors in synthetic biology in general? The example of the genetic code metaphor has shown a close connection and interplay between metaphors, scientific proceeding and technical developments that can be understood as innovative redescription and critical reflection, on the level of scientific explanatory language, the level of technological developments and the level of societal discourses. However, this says nothing about the ethical risk and danger of a change of our self-conception or the concept of life. It would have to be shown whether synthetic organisms must be attributed a moral status and whether the human self-conception is affected in a *morally relevant* way. This is no new issue with respect to the engineering approach to life in synthetic biology, but refers to a long tradition of well-known debates around the concept of life in the history of science and philosophy (see Steizinger, this volume).

I want to draw attention to another ethical dimension which is implied in the communicative function of metaphors and their role in discourses. Pielenz describes an evocative function of metaphors as topical inference rules and claims a specific logic of the plausibility of metaphors in reasoning (Pielenz 1993). This means, analogous to the epistemological normativity in scientific inquiry, metaphors in argumentations evolve a specific metaphorical rationality and evoke associative ideas and contexts of actions in terms of plausible and probable arguments. Herein is the innovative function of metaphorical redescription, in opening up new perspectives on issues in the mode of "seeing-as".

But this logic of plausibility also has the potential of supporting questionable ideology and intentional rhetorical disturbance. Speaking of virtualization and digitizing the world often means the implicit promise of a free, democratic and connected world. The possibility of digital networking all over the world suggests free exchange of information as well as the abolition of hierarchical power structures. It is the phantasmic picture and utopian promise of a decentralized and real democratic world community, organized and realized in the internet (Fröhlich 1996). This is of course and particularly true for the scientific world too. The "digital age of biology" proclaimed by Craig Venter means not only that "life is based on DNA software", which offers new ways of creative design for science, but also a simplification and acceleration of exchange of information and access to scientific knowledge:

> Scientists send digital code to each other instead of sending genes or proteins. [...W]e can send digital DNA code at close to the speed of light and convert the digital information into proteins, viruses and living cells. (Venter 2012)

The revolutionary visions of synthetic biology go even further and promise to solve nearly every urgent problem of humankind:

> Synthetic life will enable us to understand all life on this planet and to enable new industries to produce food, energy, water and medicine as we add 1 billion new humans to earth every 12 years. (Venter 2012)

But there is also a dark side to this picture. This dream of a "new age of scientific swarm intelligence" (Moos 2014, p. 17) is at least the old Baconian dream of enlightenment, of humanizing society by science. But the historical lessons learned from the dialectical nature of enlightenment should make us skeptical. As Dirk Vaihinger reminds us in relation to the promises of the digital revolution, it is quickly forgotten that the financial basis of an economically profitable venture is still the powerful context for efficient selection and processing of large amounts of data (Vaihinger 1997, p. 31). This could relativize the supposed subversive possibilities of hackers and all experimental private users, and condense the theory of new media into a concrete ideology. The image of the brilliant, subversive biohacker, who does synthetic biology in his own garage lab, and thereby helps to solve almost any burning problem of humanity, is surely unrealistic, considering the current and expected state of research in synthetic biology. It is also in contrast to the reality of an aggressive, monopolizing patent policy by the established biotechnology companies, the maintenance of socio-economic imbalances, and the unequal distribution of benefits and costs of biotechnological developments (see ETC Group 2007).

7 Conclusions. Three Functions—Three Perspectives

I have drawn from Ricoeur's theory of the living metaphor three functions of metaphors in the intersection field of science, technology and society. This analytical frame, applied to the story of the genetic code and its revitalization by Craig Venter, has led to three perspectives:

1. Metaphors have innovative and epistemological value. Metaphors in this perspective appear as driving forces in the research process on all levels of theory building, practical norms of science, technological development, and the accompanying political and social discourses, as the metaphor of the genetic code—from Schrödinger to Venter—illustrates.
2. Metaphors are critical instances in scientific description language and the philosophy of science discourse on truth. Therefore, metaphors can open up new perspectives and introduce new ways of seeing things. In our case, this leads to a new perspective and reassessment of synthetic biology in the context of the historical debates between science and art: synthetic biology can be conceived as an endeavor between artificial life and living art.
3. Metaphors have a communicative dimension and argumentative power. With a specific rationality and logic of the plausible, metaphors can rule a debate in a positive, creative way as well as in negative, distorting ways. To make this distinction is a task for ethical critique and leads to revealing some of the promises of synthetic biology, transported in and by metaphors, as ideological and research funding policy strategies.

I want to end this reflection on metaphors in science in general and in synthetic biology in particular with a comment which takes up this last point—the argumentative power and ideological danger of metaphors. Whenever progress is evoked with a revolutionary claim, it is advisable to be skeptical and to look closely at the visions, prophecies and promises that were claimed in the name of a better future and the humanization of society. Promises and visions accompanied by metaphors can help to provide legitimacy and attention, which may be important factors in research policy. However, the example of the genetic code has shown that metaphors are more than *just* metaphors and eye-catchers. They influence both the social discourse and the epistemic and technical processes in scientific inquiry. Therefore, metaphors must be recognized as an irreplaceable condition and driving-force of progress and innovation. Yet, given the risk of the ideological transfiguration of metaphors, it is also a matter for ethical critique to reflect such metaphors *as* metaphors, reveal their positive and negative implications and claim communication responsibility with respect to the implicit and explicit usage of metaphors. The ethical responsibility of researchers includes, besides the fundamental norms of the scientific ethos (Merton 1968), the communicative task of a sincere presentation of their own scientific work and research activities to society as a whole. Academic freedom and the fundamental rights of a democratic social system depend intimately on each other (Özmen 2012, pp. 126–132). From this perspective an ethical dimension of synthetic biology comes to the fore, which is in the fundamental relationship of responsibility and trust between science and society (cf. EGE 2009, p. 37). This relationship seemed to have been plunged into a "crisis of confidence" (Mittelstraß 2006, p. 9). The actual ethical question in the debate on synthetic biology can be seen as manifestation of this crisis. It is then no longer: "What is life?", but rather: "How do we want to live together and what role should synthetic biology play in our society?" A part of the answer probably lies in the metaphors we use to talk about life and synthetic biology.

References

Arbib MA, Hesse MB (1986) The construction of reality. Cambridge University Press, Cambridge

Aristotle, Bywater I, Murray G (1920) Aristotle on the art of poetry. Clarendon Press, Oxford

Balmer A, Martin P (2008) Synthetic biology: social and ethical challenges. Institute for Science and Society, University of Nottingham, England

Black M (1962) Models and metaphors. Studies in language and philosophy. Cornell University Press, Ithaca

Blumenberg H (1986) Die Lesbarkeit der Welt. Suhrkamp, Frankfurt a. M

Boldt J, Müller O (2008) Newtons of the leaves of grass. Nat Biotech 26(4):387–389. doi:10.1038/nbt0408-387

Boldt J, Müller O, Maio G (2009) Synthetische Biologie. Eine ethisch-philosophische Analyse. Beiträge zur Ethik und Biotechnologie, vol 5. Bundesamt für Bauten und Logistik BBL, Bern

Bölker M (2011) Revolution der Biologie? In: Dabrock P, Bölker M, Braun M, Ried J (eds) Was ist Leben - im Zeitalter seiner technischen Machbarkeit? Beiträge zur Ethik der Synthetsichen Biologie, 1st edn. Alber, K, Freiburg, München, pp 27–42

Bredekamp H, Rheinberger H (2012) Die neue Dimension des Unheimlichen. In: Köchy K, Hümpel A (eds) Synthetische Biologie. Entwicklung einer neuen Ingenieurbiologie? Themenband der interdisziplinären Arbeitsgruppe Gentechnologiebericht, Berlin-Brandenburgische Akademie der Wissenschaften, pp 162–163

Brukamp K (2011) Lebenswelten formen. Synthetische Biologie zwischen Molekularbiologie und Ingenieurtechnologie. In: Dabrock P, Bölker M, Braun M, Ried J (eds) Was ist Leben - im Zeitalter seiner technischen Machbarkeit? Beiträge zur Ethik der Synthetischen Biologie. Alber, Freiburg i. Br

Cserer A, Seiringer A (2009) Pictures of synthetic biology. Syst Synth Biol 3(1–4):27–35. doi:10.1007/s11693-009-9038-3

de Lorenzo V, Danchin A (2008) Synthetic biology: discovering new worlds and new words. The new and not so new aspects of this emerging research field. EMBO Rep 9(9):822–827. doi:10.1038/embor.2008.159

de Vriend H, Walhout B, van Est R (2007) Constructing life—The world of synthetic biology. Rathenau Instituut, The Hague

Debatin B (1995) Die Rationalität der Metapher. Eine sprachphilosophische und kommunikationstheoretische Untersuchung. Dissertation, Technische Univ. Berlin

EGE (European Group on Ethics in Science and New Technologies to the European Commission) (2009) Ethics of synthetic biology. Opinion 25, Brüssel

Fauconnier G, Turner M (2002) The way we think. Conceptual blending and the mind's hidden complexities. Basic Books, New York

Fischer EP, Mainzer K (eds) (1990) Die Frage nach dem Leben. Piper, München

Fröhlich G (1996) Netz-Euphorien. Zur Kritik digitaler und sozialer Netz(werk-)metaphern. In: Schramm A (ed) Philosophie in Österreich 1996. Graz, 28. Februar - 2. März 1996. Hölder-Pichler-Tempsky, Wien

Gibson DG, Glass JI, Lartigue C et al (2010) Creation of a bacterial cell controlled by a chemically synthesized genome. Science 329(5987):52–56. doi:10.1126/science.1190719

ETC Group (2007) Extreme genetic engineering. An introduction to synthetic biology. www.etcgroup.org. Accessed 20 June 2015

Grunwald A (2008) Auf dem Weg in eine nanotechnologische Zukunft. Philosophisch-ethische Fragen, Orig.-Ausg. Angewandte Ethik, vol 10. Alber, Freiburg i. Br., München

Hesse MB (1970) Models and anologies in science, 2nd edn. Notre Dame University Press, Indiana

Hesse MB (1988) Die kognitiven Ansprüche der Metaphern. In: van Noppen JP (ed) Erinnern, um Neues zu sagen. Die Bedeutung der Metapher für die religiöse Sprache, Athenäum, Frankfurt am Main, pp 128–148

JCVI (J. Craig Venter Institute) (2010) First self-replicating, synthetic bacterial cell constructed by J. Craig Venter Institute Researchers. http://www.jcvi.org/cms/press/press-releases/full-text/article/first-self-replicating-synthetic-bacterial-cell-constructed-by-j-craig-venter-institute-researcher/home/. Accessed 11 May 2015

Kay LE (2000) Who wrote the book of life? A history of the genetic code. Writing science. Stanford University Press, Stanford

Keller EF (1996) Refiguring life. Metaphors of twentieth-century biology. Columbia University Press, New York

Keller EF (2002) The century of the gene. Harvard University Press, Cambridge

Köchy K (2012) Philosophische Implikationen der Synthetischen Biologie. In: Köchy K, Hümpel A (eds) Synthetische Biologie. Entwicklung einer neuen Ingenieurbiologie? Themenband der interdisziplinären Arbeitsgruppe Gentechnologiebericht, Dornburg, pp 137–180

Kövecses Z (2010) Metaphor. A practical introduction, 2nd edn. Oxford University Press, Oxford

Lakoff G, Johnson M (2003) Metaphors we live by. University of Chicago Press, Chicago

Merton RK (1968) Social theory and social structure. Free Press, New York

Mittelstraß J (2006) Taking it on trust. In: Ernst Schering Foundation (ed) Trust in Science—A dialogue with society. Berlin, pp 5–11

Moos T (2014) There is no such thing as artificial life. Notes on the ethics of synthetic biology. systembiologie.de (8):16–17

Özmen E (2012) Die normativen Grundlagen der Wissenschaftsfreiheit. In: Voigt F (ed) Freiheit der Wissenschaft. Beiträge zu ihrer Bedeutung, Normativität und Funktion. de Gruyter, Berlin, Boston

Pielenz M (1993) Argumentation und Metapher. Narr, Tübingen

Ricoeur P (1978a) The metaphorical process as cognition, imagination, and feeling. Critical Inquiry 5(1):143–159

Ricoeur P (1978b) The rule of metaphor. Multi-disciplinary studies of the creation of meaning in language. Transl. by Robert Czerny with Kathleen McLaughlin and John Costello. Routledge & Kegan Paul, London

Ried J, Braun M, Dabrock P (2011) Unbehagen und kulturelles Gedächtnis. Beobachtungen zur gesellschaftlichen Deutungsunsicherheit gegenüber Synthetischer Biologie. In: Dabrock P, Bölker M, Braun M, Ried J (eds) Was ist Leben - im Zeitalter seiner technischen Machbarkeit? Beiträge zur Ethik der Synthetischen Biologie. Freiburg i. Br, Alber, pp 345–369

Schrödinger E (2012) What is Life? With Mind and Matter and Autobiographical Sketches. Cambridge University Press (Canto Classics), New York

Seel M (1990) Am Beispiel der Metapher. Zum Verhältnis von buchstäblicher und figürlicher Rede. In: Forum für Philosophie Bad Homburg (ed) Intentionalität und Verstehen. Suhrkamp, Frankfurt a Main, pp 237–272

Vaihinger D (1997) Virtualität und Realität - Die Fiktionalisierung der Wirklichkeit und die unendliche Information. In: Krapp H, Wägenbauer T (eds) Künstliche Paradiese, virtuelle Realitäten. Künstliche Räume in Literatur-, Sozial- und Naturwissenschaften. Fink, München

Venter JC (2012) What is Life? A 21st century perspective. On the 70th Anniversary of Schroedinger's Lecture at Trinity College by J. Craig Venter. http://edge.org/conversation/what-is-life. Accessed 05 Nov 2014

Wellmer A (2004) Sprachphilosophie. Eine Vorlesung, Suhrkamp, Frankfurt am Main

Wellmer A (2007) Der Streit um die Wahrheit. Pragmatismus ohne regulative Ideen. In: Wellmer A (ed) Wie Worte Sinn machen. Aufsätze zur Sprachphilosophie. Suhrkamp, Frankfurt am Main

Zimmer J (2003) Metapher, 2nd edn. Edition panta rei, vol 5. Transcript, Bielefeld

Debasement of Life? A Critical Review of Some Conceptual and Ethical Objections to Synthetic Biology

Tobias Eichinger

1 The Productional Paradigm of Creating Life

One of the expressions most commonly used in media reporting on synthetic biology as well as in diverse forms of self-presentation of scientists is the formula of *creating new life forms* or the *creation of life*. This is often coupled with the adjective *artificial*, so that synthetic biology is often presented as a new techno-science that deals with the *artificial creation of life* in the lab. Let us illustrate that characterization with some examples from the field. Martin Fussenegger writes the goal of synthetic biology is "to create and engineer functional biological designer devices and systems" (What's in a name? 2009). Quite similarly, Anthony Forster and George Church state that "creating bacteria" is part of the agenda of synthetic biology (Forster and Church 2007). Victor de Lorenzo and Antoine Danchin talk about "the creation of new organisms" and the goal "to recreate a cell" (Lorenzo and Danchin 2008). And the paper from the Venter Institute that attracted so much attention in 2010 was entitled "Creation of a Bacterial Cell Controlled by a Chemically Synthesized Genome" (Gibson et al. 2010). Philosophers and ethicists who observe the field are adopting these expressions and are discussing one of the key issues of synthetic biology under the umbrella of the term of *life creation*: Mark Bedau refers to the activity in synthetic biology as "creating genuinely new forms of life" (Bedau 2011) and Joachim Boldt and Oliver Müller emphasize the crucial fact that synthetic biology "can create new life forms" (Boldt and Müller 2008).

T. Eichinger (✉)
Institute of Biomedical Ethics and History of Medicine, University of Zurich, Zurich, Switzerland
e-mail: eichinger@ethik.uzh.ch

© Springer International Publishing Switzerland 2016
K. Hagen et al. (eds.), *Ambivalences of Creating Life*, Ethics of Science and Technology Assessment 45, DOI 10.1007/978-3-319-21088-9_14

Thus, synthetic biology is not only concerned with the living in terms of disassembling, analyzing and maybe modifying it, but with *creating* life. Especially that more fundamental claim makes a great media stir, causes and focuses manifold hopes and visions, fears and vehement critique at the same time.

One of the objections that are advanced against the aim of creating life from an ethical perspective refers to conceptual implications for the whole emerging field of research. This critical view refers above all to the technical concept of creating and the engineer-driven paradigm of production. As an example of this main feature, consider a paper by the *Bio Fab Group* in which a whole range of expressions characterize synthetic biology in a very technical and explicit engineering way. The authors use terms like "to construct", "to produce", "to reengineer", "to generate", "to build", "manufacturing" and "designing" for the activities in synthetic biology (Baker et al. 2006). Thereby, the new section of biology that calls itself 'synthetic' is linked to the conceptual field of technical production and engineering. This conceptual field is shaped by the paradigm of means-end-relations and instrumental rationality. That means that under the paradigm of engineering, the activity in question is primarily targeted at an end that lies outside of itself. The activities of building, constructing, manufacturing and producing are paramount examples for that paradigm. The end that sets the whole activity in motion and lies outside of it, is a product. The product as the result of a production process is characterized by its concrete objectivity that makes it independent of its production history and its producer. It is an artificial thing which is constructed, built, produced, etc. to serve its purpose. The purpose of a product and its production process is (usually) a human purpose, the product typically would not have come into existence without human action, and consequently extends the range of nature and natural objects.

These more general features of the pattern of production indicate that the productional paradigm is predominant for synthetic biology in conceptual and practical regards. They fit in with a very common definition of synthetic biology that stresses "the design and construction of new biological systems *not found in nature*" (Schmidt 2009a). Moreover, in the majority of cases, the products of synthetic biology are not produced for their own sake, but for a certain instrumental purpose. An exception—and it seems to be a big and relevant exception—is foundational research. Here the main aim is to increase biological knowledge, to make progress in understanding life, that is discovering the preconditions of life's emergence in the past or exploring the minimal functional requirements of life. Another exception of the purpose-driven productional paradigm is the field of bioart, where the agenda and activity of synthetic biologists are items for artistic reflection and playful work. Beyond these two fields, synthetic biology largely is concerned with creating life forms in a productional manner.[1]

---

[1]Presumably in the near future great efforts will be made promoting industrial application of synthetic biology.

2 The Critique of the Concept and Handling of Life in Synthetic Biology

Ethical criticism based on the fact that synthetic biology deals with life concerns two aspects: the conceptual dimension of the paradigm of life in synthetic biology and the more practical or consequentialistic implications of the synthetic biologist's approach to life and the living.

2.1 A Misguided Concept of Life: The Reductionism Argument

As synthetic biology is a subdiscipline of biology, *the* natural science of life, it is by definition dealing with life. What makes it special and new is its productional or engineering approach. And it is no wonder that this technically dominated approach implicates and shapes the objects of its activity: life, life forms, living systems and so on. This controversial understanding of life in synthetic biology is different from the implicit concepts of life in other fields of biology. Even more importantly, there seems to be a considerable difference between the synthetic biology conception of life and conceptions of life outside the natural sciences, not to mention in the public and media.

The very concept of life in synthetic biology arouses criticism and contradiction. It is feared that the leading science-oriented paradigm of synthetic biology in dealing with life forms could have negative consequences on the concept and the understanding of life and the living in general. According to the criticism which addresses a kind of ontological question on a descriptive level, the methodological strategy of synthetic biology. Thus, synthetic biology's methodology which is strongly shaped by a pragmatical engineer's approach, represents and reinforces epistemically a misguided reductionistic conception of life.

A very fundamental version of this objection highlights the fact that every arrangement and observation of living phenomena in an experimental manner—in an artificial setting like a laboratory—could necessarily reveal only a very special and limited view of and insight into life. What experimental research in the lab never could show are insights according to sample discovering the 'real logic' or 'true principles' of life. Revolutionary and final answers to the question what life 'as such' actually is could never be found in a scientific laboratory (if indeed such an insight enterprise is possible).

Criticism of this type rejects every comprehensive claim that synthetic biologists make when they publish—or are quoted with—phrases like 'life is not more than...'. For this skeptical position, it is clear and almost self-explaining that life is always more than any reductionistic or focused view on some aspects of what it could contain. What reinforces the objection from this holistic-life-position is the idea of separating, disassembling and modularizing living systems in small and smallest possible parts—an idea that influences biology in general all along.[2] The tool kit pattern, the metaphor of life as a tool box and the concept of the *BioBricks* illustrate this approach of modularization and standardization of life in a concise way with a certain symbolic value. And in addition to that, this view is also incompatible with a concept of life as some kind of complex, whole and undividable entity. Above all, concentrating on minimal preconditions of life's functioning as an epistemic way to figure out what life is—as in top-down-synthetic-biology—is therefore misguided from the start.

As part of this argument, sometimes the 'mystery', 'self-will' or the 'wisdom' of life is mentioned. Here life often is understood as an elusive, holistic entity which could necessarily never be explained completely—even less by means of hard science, there always will be an unachievable rest. More specific for synthetic biology, that view which sometimes recalls vitalistic motives, could be confirmed by the problem of *biosafety* issues. The fact that synthetic biology has to face difficulties of unexpected and—to a certain degree—unpredictable outcome, is tied back to this peculiar remain of life that defies control. For some critics, ignoring life's specific intangibility by claiming to exercise total controllability is condemned to failure. From this point of view, this kind of failure underlies the conceptional misapprehension in the understanding of life.

2.2 A Problematic Handling of the Living: The Instrumentalization Argument

The technological paradigm determining synthetic biology of controlling, constructing and creating the living is not only criticised in a descriptive or ontological way as fundamental misguided research approach; critics also fear questionable effects in normative regards. As mentioned above, according to the dominant premise in the field of synthetic biology, living beings could be taken apart in minimal functional components out of which completely novel living systems could be reassembled from the scratch independently from natural standards. Besides the descriptive objections the engineering approach of modularization and standardization, of partition and building, of deconstructing and reconstructing has obviously practical implications concern the ways of doing synthetic biology and give

---

[2]Cf. the chapter by Andreas Christiansen in this volume.

occasion to criticism on a prescriptive level. Doing science under the label of synthetic biology implies always a certain mode of working with living entities. This mode of handling of the living which is strongly shaped by the technical, productional and instrumental paradigm that determines synthetic biology, leads for some to ethical problems.

The disputable norms of acting are consequences of the understanding of the living. By seeing life as a result of a technological process of production which is activated for arbitrary purposes, a manipulative and instrumental approach to the outcome of the production is predefined. Inevitably, the products of synthetic biology—which are alive by definition—would thereby be subject to a handling according to inanimate artifacts that doesn't acknowledge any value for its own sake. For critics of synthetic biology (and biotechnology in general) such an attitude of a technical-instrumental use is not compatible with living objects in principle because it has no regard for any intrinsic value the living may have. For them, even the designing and engineering of microorganisms from the scratch could mean a devaluation of these basal life forms.

But this apprehension of instrumentalization and debasement of life and natural values go beyond the particular living systems. In further consequence, critics fear negative retroaction and a certain ethical impact on how life in general is understood and valued (cf. Boldt 2013). In a temporal respect, this fear concerns effects in the present as well as in the future. As a current impact it is assumed that if low or rudimentary forms of living organisms are treated like nonliving material without any intrinsic value, as they belong to the sphere of the living, also the status of other living objects—that are not only bacteria or yeast—could diminish as a side effect. This would imply a weakening of the respect for higher forms of life that are usually and so far regarded as worthy of respect and protection. Very similar, the prospective fear contains a kind of slippery slope argument. If man is practicing the instrumentalization and exploitation of low life forms and gets used to it, the moral barrier to include higher organisms into the paradigm of technical-instrumental engineering and production decreases. In the course of a process of habituation to the usage of the living, even life forms that are commonly worth protecting would step-by-step lose their state of intrinsic value and the morally dangerous effect of a general brutalization is expected.

Some proponents of that line of slippery slope argumentation predict as further consequence, that this development would finally lead, by implication, to changes in the conception and estimation of a very special kind of a very high life form—the moral appreciation and status of humans. Some fear certain dubious effects on the self-conception of man if synthetic biology expands a technical-instrumental handling to higher forms of life. This argument gets a special touch as the creative moment of drafting and designing in synthetic biology is emphasized. According to this the shift from ,traditional' biotechnology and genetic engineering to synthetic biology is interpreted as a shift that goes along with an alarming transition in the human self-conception. This idea contains a shift from the anthropological term of the *homo faber* that is 'only' manipulating existing organisms to a *homo creator* that reinvents nature and is therefore at risk of overestimating his comprehension of nature (Boldt 2013).

3 Assessment and Conclusion

How valid are these strategies of criticism? From a strict and critical point of view these objections turn out to be not very plausible or convincing arguments to identify synthetic biology as a novel and peculiar alarming biotechnology and to advice increased caution.

3.1 The Reductionism Argument

It is doubtful if one can derive from the methodological reductionism in synthetic biology a comprehensive and conceptional claim of explaining life 'as such'. The reductionist view on life here is introduced and established for specific purposes, for purposes of building and engineering, not for apprehending the essence of life or for explaining life in its full range and complexity. There are at least two facts that could count as an indication for that. First, the importance of biosafety issues in synthetic biology is probably not denied by any serious scientist in the field (cf. exemplarily Schmidt 2009b). This illustrates that there is a strong awareness concerning the limits of control of living entities. And second, there are scientists in synthetic biology who are suggesting a necessary broadening of the range of their own activity by the term of *tinkering*. Steven Benner and colleagues do so with the term of "tinkering biology" (Benner et al. 2011) or Petra Schwille, who published a programmatic paper for bottom-up-synthetic-biology entitled "Engineering in a Tinkerer's World" (Schwille 2011). Thereby, the methodological procedure is to be supplemented with the principle of trial-and-error, which means more a kind of reacting and cooperating with the living material and its contingency, more than determining and commanding it. By that, the tinkering synthetic biologists are also acknowledging certain limits of prediction that emerge by working with living objects in principle. In this sense, the designing, synthesizing and engineering of artificial life forms could only be successful if it reckons on a certain own internal dynamics of the living. As for example, Matthias Heinemann and Sven Panke are considering "a fundamental difference between engineering biology and engineering in other natural sciences such as chemistry or physics". According to these synthetic biologists, the difference lies in the fact that "biological systems have the capacity to replicate and to evolve" (Heinemann and Panke 2006). Regarding this dimension of a kind of natural originality and obstinacy or self-will, one could rather come to the opposite conclusion: by doing synthetic biology, the respect and estimation of the living is rather reinforced than weakened.

3.2   The Instrumentalization Argument

Here one has to differentiate two levels of argumentation. First, the assumption that the purposeful designing and engineering of living systems implies that these low life forms are treated as mere nonliving material. Second, the fear that this kind of objectifying handling leads to negative and brutalizing side effects. The first assumption seems to be only an assertion that lacks the proof of its necessity. Only because synthetic biology follows a strong technical character of engineering and production, it doesn't need to devaluate its objects and treat them like nonliving material. But even if one may concede a certain tendency of such an approach, one would have to show why this should be a normative risk specific to synthetic biology that does not already exist in the case of conventional biotechnologies or genetic engineering. Similarly the effect of brutalization, of an expansion of a devaluating and careless handling from lower life forms to higher stages seems to be more an exaggerated fear than a probable and realistic consequence (as it is a peculiar weakness of slippery slope arguments in general). Why should the systematic design and fabrication and an instrumentalized usage of yeast cells and bacteria lead to a problematic devaluation of higher life forms or even should have any consequences on the self-understanding of man? In case of these lower forms of life, the commonality with plants, animals and particularly humans is only comprehensible on an abstract level. We could understand or accept that bacteria are part of the same dimension—being alive—as we are, only if we 'learn' biological facts. There are no perceptible properties that connect us with yeast cells. In this regard, 'living machines' as an output of synthetic biology are in ethical regards actually more machines than living entities for us. From a morally relevant perspective these objects are much further away from higher organisms, that it is absolutely not convincing that a certain way of handling bacteria in the petri dish would jump over to our attitude to mammals for example. Not to mention our own self-understanding and self-evaluation. Most notably, it obscure why this should occur of all things here in the case of synthetic biology, if this has not happened after decades and centuries of factory farming and industrial meat production. Here chicken, pigs and cows—quite higher organisms as bacteria—are produced in an industrial manner and are treated solely as means to human ends and not as living beings that do possess intrinsic value.

And with regard to negative developments in the future, that could result from a current practice, but also could not result, the claim for an initial prohibition or moratorium seems to reveal an unfounded pessimism. That point of view shows little confidence in the possibility of a cautious and permanent monitoring of new and open processes in science and society. As numerous projects of accompanying research for ethical, legal and social aspects of new biotechnologies in general and synthetic biology in special show, there exists at least the possibility of such a continuous control within a highly differentiated and reflective scientific culture.

So as a result of a critical appraisal of the outlined types of conceptual criticism and ethical objections against the concept of life and the handling of the living in synthetic biology, it is to state that these concerns are hardly sustainable as beating arguments. Nor could they serve as a basis for general regulations and policy making. They rather should be understood as the expression of a conceptual uncertainty and a corresponding uneasiness resulting from the transgression and blurring of terminological and ontological boundaries that seemed unalterable so far. These boundaries comprise the distinction between technology and nature, between products and living beings, between natural and artificial, etc. Thereby it is to emphasize that this uneasiness has a potentially big societal impact, because it does exist (not only among ethicists and biocentrists). Moreover, regarding its theoretical and conceptual impact, that uneasiness should lead us perhaps rather to question the conceptual desire or demand on clear and definite distinctions and boundaries—especially if we are dealing with thresholds between nature and technology, as it is to a considerable degree the case with synthetic biology.

Acknowledgment I am grateful to Christian Illies for numerous helpful remarks and inspiring comments.

References

Baker D, Church G, Collins J et al (2006) Engineering life: building a FAB for biology. Sci Am 294(6):44–51

Bedau M (2011) The intrinsic scientific value of reprogramming life. Hastings Cent Rep 41 (4):29–31

Benner S, Yang Z, Chen F (2011) Synthetic biology, tinkering biology, and artificial biology. What are we learning? CR Chim 14:372–387

Boldt J (2013) Life as a technological product: philosophical and ethical aspects of synthetic biology. Biol Theory 8:391–401

Boldt J, Müller O (2008) Newtons of the leaves of grass. Nat Biotechnol 26(4):387–389

Forster A, Church G (2007) Synthetic biology projects in vitro. Genome Res 17(1):1–6

Gibson D, Glass J, Lartigue C et al (2010) Creation of a bacterial cell controlled by a chemically synthesized genome. Science 329:39–50

Heinemann M, Panke S (2006) Synthetic biology—putting engineering into biology. Bioinformatics 22(22):2790–2799

Lorenzo V, Danchin A (2008) Synthetic biology: discovering new worlds and new words. The new and not so new aspects of this emerging research field. EMBO Rep 9(9):822–827

Schmidt M (2009a) Introduction. In: Schmidt M, Kelle A, Ganguli A, Vriend H (eds) Synthetic biology. The technoscience and its societal consequences. Springer, Dordrecht, pp 1–4

Schmidt M (2009b) Do I understand what I can create? Biosafety issues in synthetic biology. In: Schmidt M, Kelle A, Ganguli A, Vriend H (eds) synthetic biology. The technoscience and its societal consequences. Springer, Dordrecht, pp 81–100

Schwille P (2011) Bottom-up synthetic biology: engineering in a tinkerer's world. Science 333:1252–1254

What's in a name? (2009) Nat Biotech 27(12):1071–1073

Engineers of Life? A Critical Examination of the Concept of Life in the Debate on Synthetic Biology

Johannes Steizinger

1 The Concept of Life. Its Return in the Debate on Synthetic Biology

The long and complex history of the concept of life has reached a paradoxical point: On the one hand, the term life is used ubiquitously. An array of disciplines under the umbrella term 'life sciences' dominate the theoretical discourse of our times. In fields such as medicine, pharmacology and agriculture, numerous technological applications are changing our daily world. These applications can be understood as an indicator of the far-reaching implications that the scientific discourses on life have for society and culture. In the view of some, we are living in a "culture of life", which moves away "from the ideals of the Enlightenment towards an idea of individual perfectibility and enhancement" (Knorr Cetina 2005, p. 76).

On the other hand, there is no precise and generally valid definition of life. This is not least because in current biology the status of the concept of life is controversial. In June 2007 an editorial article in the journal Nature claimed: "It would be a service to more than synthetic biology if we might now be permitted to dismiss the idea that life is a precise scientific concept" (Editorial 2007, p. 1032). Moreover, scientists assure us that "the impossibility of a sharp distinction between animate and inanimate would not create difficulties for the biology in its everyday scientific practice" (Budisa 2012, p. 101; see also Toepfer 2011, pp. 467, 468).

Thus, for many scientists the possibility of a precise biological definition of life is not important. They regard life as a "fuzzy concept" and are satisfied with the notion that biology allows a plurality of approaches to life (see Witt 2012, p. 37). Some scientists, like Dominique Homberger, even claim that biologists have an intuitive knowledge of the border between inanimate matter and living beings, but

J. Steizinger (✉)
Department of Philosophy, University of Vienna, Vienna, Austria
e-mail: johannes.steizinger@univie.ac.at

© Springer International Publishing Switzerland 2016
K. Hagen et al. (eds.), *Ambivalences of Creating Life*, Ethics of Science and Technology Assessment 45, DOI 10.1007/978-3-319-21088-9_15

are not able to explain the phenomenon of life physically (see Homberger 1998). Similarly, the philosopher Jean Gayon thinks that life could disappear as a scientific concept and remain only as a "folk concept" for our everyday practice. He claims: "When this point will be reached, life will be no longer a concept for the natural sciences, but just a convenient word in practice, in the world we inhabit. 'Life' will be a folk concept. Its specialists will be no longer chemists, biologists, and robo- ticists; life will be a subject for psychology, cognitive science and anthropology." (Gayon 2010, p. 243). But Gayon does not only deny that there can be a scientific definition of life in the strong sense. He assumes also that "the recognition of 'life' has always been and remains primarily an intuitive process, for the scientists as for the layperson. However we should not expect, then, to be able to draw a definition from this original experience." (Gayon 2010, p. 231). Against this background, it is not surprising that some critics assert that the concept of life is only used as a buzzword to create attention in a world in which the selling of a scientific result is as important as the result itself.

The concept of life has no better a reputation in current philosophy than in science. Traditionally, the philosophical concept of life points to a realm which cannot be captured completely by thinking. Wilhelm Dilthey, one of the most important philosophers of life, claimed around 1890: "The expression 'life' for- mulates what is most familiar and most intimate to everyone, yet at the same time something most obscure indeed totally inscrutable. What life is remains an insol- uble riddle. All reflection, inquiry, and thought arise from this inscrutable [source]" (Dilthey 2010b, p. 72). Dilthey was part of the development of a philosophy of life in the 19th century. The term philosophy of life (*Lebensphilosophie* in German) groups together highly different authors (e.g., Friedrich Nietzsche, Wilhelm Dilthey, Georg Simmel, Henri Bergson or Ludwig Klages), who are united more by their impact than by their doctrines. Most of them were driven by a critique of the one-sided emphasis on reason and rationality in both idealistic philosophy and science. Therefore, it is not surprising that there is a tension between systematic philosophy and philosophy of life. Moreover, since some philosophers of life were entangled in the theoretical foundation of National Socialism (see e.g., Lebovic 2013)—a part of the history of philosophy, which has still to be investigated—the philosophy of life tradition largely disappeared after 1945. While there have been a few attempts to renew philosophical reflection on life (see Fellmann 1993; Worms 2013), the concept of life is mostly used as a critical concept in political philosophy (see Agamben 1998; Esposito 2013). There is little systematic work on the concept of life in contemporary philosophy.

Against this background, synthetic biology and the discourse on its scientific and societal consequences is clearly an exception. Here, the concept of life is not only used as buzzword (a) but also discussed theoretically (b) and plays a crucial role in the debate about the ontological, epistemological and ethical dimensions of synthetic biology (c). In what follows, I will briefly outline these different aspects of its use[1]:

---

[1]For a discussion of life as metaphor in the context of synthetic biology see Falkner, this volume.

(a) Some protagonists of synthetic biology like Craig Venter consider the 'creation of life' as the central aim of their research. Moreover, 'the creation of artificial life' is advertised as "the most sensational success of synthetic biology with the promise to provide solutions to our energy, health, environmental and nutritional problems" (Budisa 2012, p. 103). Briefly speaking, a lot of hopes and concerns which are connected with the production of synthetic biological systems focus on the formula 'creation of life'. Public press and mass media have readily accepted this self-advertisement and have reported on scientific developments in synthetic biology from the beginning. Anna Deplazes-Zemp and Nikola Biller-Andorno remark correctly that this use of the expression 'creation of life' is based on the ambiguity in the concept of life: "Headlines such as 'Life 2.0', 'Engineering life: building a FAB for biology' or 'Synthetic life' illustrate this tendency [that synthetic biology would lead to 'synthetic life'; J.S.]—such titles would not produce the same effect, if 'life' was purely a scientific concept." (Deplazes-Zemp and Biller-Andorno 2012, p. 959).

(b) A lot of scholars—philosophers as well as scientists—in synthetic biology assure us that they want to contribute to the basic understanding of life (see Deplazes-Zemp 2012, pp. 762, 763). Due to experimental results like the "synthetic cell" (Gibson et al. 2010) created by Craig Venter and his colleagues, debate on the question 'what is life?' has reignited. Mark Bedau, for instance, emphasizes that "we now have an unprecedented opportunity to learn about life. Having complete control over the information in a genome provides a fantastic opportunity to probe the remaining secrets of how it works" (Bedau 2010, p. 422). Others, like Arthur Caplan, conclude that we have already learnt enough to end an old and for a while forgotten debate:

> Venter and his colleagues have shown that the material world can be manipulated to produce what we recognize as life. In doing so they bring to an end a debate about the nature of life that has lasted thousands of years. Their achievement undermines a fundamental belief about the nature of life that is likely to prove as momentous to our view of ourselves and our place in Universe as the discoveries of Galileo, Copernicus, Darwin and Einstein. (Caplan 2010, p. 423)

Craig Venter claims that his synthetic genomics approach will provide a reductionist explanation of life (Cho et al. 1999; Deplazes-Zemp 2012, p. 763). Since, as Michel Morange puts it, "life is on the way to being 'naturalized'", it thus seems "fully accessible to scientific enquiry" (Morange 2010, p. 181).

As one would expect, these claims have invited objections. Take, for example, Deplazes-Zemp and Biller-Andorno, who answer directly to Caplan:

> Synthetic biology, even with the production of a living protocell, could not bring an end to this debate [about the nature of life; J.S.]. Those who argue that life is more than merely a scientific phenomenon would say that a synthetic organism, if it is considered to be alive, also has features that cannot be captured by the life sciences. [...] Biocentrists argue that a synthetic organism has moral value, and other philosophers claim that a synthetic organism is an autonomous system with subjectivity and a self. (Deplazes-Zemp and Biller-Andorno 2012, p. 962)

As these few examples already show, there is a diverse debate[2] about the concept of life in synthetic biology, which is far away from a precise and universally accepted answer to the question 'what is life?'.

(c) The concept of life plays a crucial role in discussions about the societal dimensions of synthetic biology. This is because the special ethical relevance of synthetic biology is supposed to be explained by the conviction that synthetic biology "entails a confrontation with life" (Ruiz-Mirazo and Moreno 2013, p. 378). Therefore, some regard the concept of life as the focal point of the ethical, legal and political questions raised by the development of synthetic biology (e.g., Dabrock et al. 2011b, p. 14). The relevance of the concept of life for the debate about the societal impact of synthetic biology is closely connected with its other meanings. Within the concept of life the ethical aspects are intrinsically linked with the epistemological prerequisites and the ontological consequences of synthetic biology.

In the next section I will examine this point of intersection, and analyse some of the issues which are discussed in terms of the concept of life. I will trace some typical arguments in the debate on synthetic biology. My analysis is based on the following assumption: If we take the idea that there is no precise and generally valid definition of the concept of life seriously, its use raises a question rather than a solution, contrary to what is often suggested.

2 Engineers of Life? Current Issues in the Debate on Synthetic Biology

2.1 Fabrication of Life? The Epistemological Question

The umbrella term synthetic biology groups together a set of different scientific and methodological disciplines, which share a constructive approach to their object (see Acevedo-Rocha in this volume; Billerbeck and Panke 2012; Bölker 2011). In this respect, synthetic biology can be seen as a new form and development of biotechnology. In contrast to other biotechnologies, synthetic biology systematically introduces engineering concepts and methodologies like standardization, modularization and hierarchical organisation (see Boldt 2013, pp. 391, 392; Deplazes-Zemp 2012, p. 772). Moreover, as both practitioners and theoreticians of synthetic biology emphasise, synthetic biology research has a creative aim: Novel products with useful functions should be designed in a rational manner (see Boldt 2013, p. 392; Bölker 2011, pp. 35–39). In the best case, the human designed biological systems cannot be found in nature.

---

[2]Further examples include Boldt et al. (2012), Bedau et al. (2010), Dabrock et al. (2011a), Hacker and Hecker (2012), Witt (2012).

Briefly speaking, with synthetic biology the engineer enters biology and gets an epistemological model of biological research. Since in contrast to other biotechnologies synthetic biology is not only the application of theoretical knowledge, the fabrication of biological systems should also lead to a better understanding of their composition and functioning. The phrase "knowledge through fabrication" (Ruiz-Mirazo and Moreno 2013, p. 377) summarizes the methodological approach of the various forms of synthetic biology (see also Köchy 2012a, pp. 160, 161). To sum up their attitude, several scientists refer to Richard Feynman's saying: "What I cannot create I do not understand" (see Deplazes-Zemp 2012, p. 762; Ruiz-Mirazo and Moreno 2013, p. 377; Weiss 2011, p. 17).

This *bon mot* condenses a long tradition of modern scientific thinking. Many studies mention the prehistory of the epistemological imperative of synthetic biology. It goes back at least to the 17th Century, in which—after 300 years of progress in the production of mechanical gadgets and devices—a general scientific research strategy was established in the sciences.[3] René Descartes and Julien Offray de La Mettrie exemplify a scientific attitude which identifies the explanation of a natural phenomenon with a demonstration of how it can be generated by the action or activity of a mechanism (see Ruiz-Mirazo and Moreno 2013, p. 376; see also Deplazes-Zemp 2012, pp. 762, 773). Thus, a constructive approach is essential for modern scientific thinking. From its early stages theory and practice are entangled. Synthetic biology introduces this epistemological principle into a new realm, the realm of biology. The ground breaking aim of synthetic biology "is to learn more about the living by means of re-construction or fabricating it" (Ruiz-Mirazo and Moreno 2013, p. 377).

In the debate on synthetic biology, it is exactly this combination—the engineering attitude and life forms as its objects—which mostly stands for both the potentials and problems of its way to gain and apply scientific knowledge. For instance, Deplazes-Zemp argues:

> [...] that the notions of 'new life-forms' in synthetic biology, the way that synthetic biologists want to contribute to the understanding of life, and how they want to modify life by a rational design reveal a conception of life that differs from that of traditional biotechnology. As a result, synthetic biology adds a new facet to the multifarious notion of life. For certain ethical positions this production- and design-oriented conception of life may raise concerns. (Deplazes-Zemp 2012, p. 758)

Further, ethicists like Joachim Boldt completely reject the "conception of life as a toolbox" (Deplazes-Zemp 2012). Boldt adopts Hannah Arendt's distinction between "action" and "work" (he calls it "fabrication")[4] and criticises synthetic

---

[3]Martin Weiss even argues with reference to Martin Heidegger that the association of knowledge with the notion of building characterizes the occidental philosophical tradition since Plato (see Weiss 2011, p. 179; this volume).

[4]As Boldt mentions, Arendt originally distinguishes three types of human activity: "labor", "work" and "action". Boldt drops "labor" with the pragmatic argument that "for the purpose of this article [...] fabrication and action are the two types of human activity that are of special interest" (Boldt 2013, p. 393).

biology because it is an "implementation of the ideals of fabrication in the realm of the living" (Boldt 2013, p. 398). For Boldt, the notion of fabricating life encapsulates the problematic assumptions and implications of synthetic biology: It reduces life forms to "complex conglomerations governed by regularities that apply to physical and chemical matter" (ibid.) and is unable to explain their inherent value (ibid. p. 397). I return to this below.

2.2 Living Beings or Artefacts? The Ontological Question

The criticisms of synthetic biology as 'fabrication of life' presuppose that its objects have a particular ontological status and deduce ethical consequences from that status: Its objects are regarded as new variants of life and have to be treated *as* forms of life. But there is a philosophical controversy about the ontological status of the products of synthetic biology (see, e.g., Gehring 2010; Brenner 2012; Beuttler 2011). This is not least because synthetic biology challenges the well-established distinction between nature and technology. The ontological relevance of this distinction can be traced back to Aristotle who classifies in his *Physics* "all the things that are" (Aristotle 2004, p. 49) into two forms: He claims that "some are by nature" (ibid.) and "others through other causes" (ibid.). Natural objects have the "source of motion and rest, either in place, or by growth and shrinkage, or by alteration" (ibid.) in themselves. The "other things" are produced and "none of them has in itself the source of its making" (ibid.).

This influential distinction between "physis" and "techne" is still used as a starting point to define life: Living beings are identified with natural objects whose definition is extended by, for example, the notion of autopoietical organisation (see, e.g., Brenner 2012, pp. 106–108). And, they are distinguished from artefacts which are fabricated by human beings and exist only in relation to their use. From such a dualistic perspective, a constructive approach to the realm of the living seems to be impossible. In fact, philosophers like Andreas Brenner, claim that synthetic biology is a misleading concept (see Brenner 2012, p. 118), because living beings cannot by definition be produced. Here, the notion of synthetic life is rejected as contradictio in adiecto. Since life emerges out of itself, human technology can only produce artefacts.

Such arguments are problematic, because they miss the significance of synthetic biology regarding the relation of nature and technology. Kristian Köchy emphasizes that nature and technology are closely related, although the well-established distinction suggests an opposition. Moreover, he shows convincingly that new technological possibilities change not only our concept of technology, but also our concept of nature (see Köchy 2012a, p. 159). Synthetic biology is clearly an example for the blurring of the demarcation between nature and technology, which can be interpreted in different ways.

Some scientists as well as philosophers regard synthetic biology as the ultimate proof of a technological understanding of nature and as the closer of the mechanistic world view of modern science (see, e.g., Morange 2009; Venter 2013). They put forward a reductionist concept of life and claim that the further development of synthetic biology will enable us to explain the animate part of nature by constructing it. Michel Morange claims, for example, that the "rise of synthetic biology is a return to the 'old' traditions: one can claim that a system has been fully described only when it has been possible to reconstruct it. [...] the achievement of the distant goal of constructing an artificial living cell will be the ultimate proof that life has been fully explained." (Morange 2009, p. 52). The notion of constructing life suggests to understand the products of synthetic biology as living machines.[5] What qualifies the synthetic biological systems as machines is not only their way of production, but also their rational design and their function. Anna Deplazes-Zemp remarks: "When synthetic biologists speak of their products as machines they imply these entities have lost their independence and are thus controllable." (Deplazes-Zemp 2012, p. 767).

But it is controversial, if the notion of synthetic biological systems as living machines captures the combination of natural and technological properties which you find in the existing results. Products like the "synthetic cell" made by the JCVI or Biobricks—the standardised biological parts which are used to engineer novel biological devices and systems—present a mixture of natural and technological properties on several levels. According to Köchy, in current research natural systems are used as material and models for both the products and the production process (see Köchy 2012b, 147–149).[6] Moreover, he shows that the technological approaches of synthetic biology are always framed by the requirements of complex biological systems (see Köchy 2012a, 165). Köchy defines these requirements as the natural prerequisites of the synthetic products and shows that their increasing complexity intensifies the mixture of natural and technological modes of production (see ibid. 172). Likewise, this complexity sets limits to the possibility of planning and controlling the process of production. According to Köchy, in synthetic biology the production process has to be conceived as a form of directed self-organisation (see ibid. 171; Köchy 2012b, 157). Thus, applying engineering principles in biology causes changes in the concept of technology and its relation to nature. These changes also affect the background beliefs which coin the self-understanding of synthetic biologists. Some refer to their research activity as "tinkering" (see, e.g., Benner et al. 2011), a concept which is in tension to essential features of the

---

[5]For a discussion of the complex history of the analogy between living beings and machines see Köchy (2012b, pp. 150–157).

[6]Köchy indicates that the reference to natural systems is also important for the proclaimed future of synthetic biology and refers to the report of the NEST High-Level Expert Group for the European Commission which claims: "[...] synthetic biology aims to go one step further by building, i.e. synthesizing, novel biological systems from scratch using the design principles observed in nature [...]" (European Commission 2005, p. 11).

mechanistic paradigm, especially the idea to design biological system "in a rational and systematic way" (European Commission 2005, p. 5).

Some authors in the debate about the ontological status of the objects of synthetic biology claim that their mixed character can only be understood by a hermeneutic concept of life which connects nature and culture. From such a perspective, synthetic biology systems are life forms, because they are both natural objects and technological products. Here, the concept of life should achieve in theory what synthetic biology performs in practice: a dialectic of nature and technology. Ulrich Beuttler argues, for example, that on the one hand, synthetic biological systems like artificial cells would still be similar to natural life forms, insofar as they are constructed as self-maintaining systems, which serve their purpose in an independent way (see Beuttler 2011, p. 292). According to Beuttler, reductionist explanations cannot capture this form of selfhood. But in contrast to certain autopoietical theories (e.g., Brenner 2012) he also emphasizes that the self-reliance of a living being is not based on the form of its emergence: "Life moves, preserves, organises and develops itself, but it does not create itself." (Beuttler 2011, p. 297; translated by J.S.). On the other hand, the entities of synthetic biology are as technological products part of human culture. As Beuttler emphasizes, his concept of life avoids the fatal alternative nature or technology, because it encompasses the sphere of culture without blurring all distinctions. He claims that in his theory life as culture and life as nature are united and distinguished likewise (see ibid. 300, 301). Regardless of the details and validity of Beuttler's view, such approaches are interesting, because they show that synthetic biology can be interpreted in a non-reductionist way and, thus, indicates the limits of a mere biological concept of life.

2.3 The Value of Synthetic Biological Systems? The Ethical Question

These epistemological and ontological considerations are closely related to ethical questions. The assumption that the products of synthetic biology are human-designed life forms extends the scope and depth of the relevant ethical considerations. Ethicists, like Joachim Boldt, emphasize that the ethical questions which are raised by synthetic biology are an inherent part of the research activity, and not just problems of technology assessment (see Boldt 2012, pp. 189, 190). Here, the definition of the objects of synthetic biology as life forms is used to think of this research in terms of interactions between different life forms, i.e. human and non-human life. In other words, it adds an existential dimension to the practice of synthetic biology. From this perspective, the attitude of the researchers towards their objects becomes an important issue (see, e.g., Boldt 2013, pp. 397–400). Others argue that the artificial production of living beings is a new challenge regarding our responsibility for nature and its future form (see, e.g., Aurenque 2011, pp. 342–344).

But the concept of life also motivates immediate ethical claims. There are several versions of the argument: If synthetic biological systems are life forms, they have to be treated as life forms. All these arguments rest on the conviction that life has an inherent value and, thus, ethical relevance—a claim, which is intuitively right, but nevertheless controversial (see Toepfer 2014). Take, for example, Boldt's criticism of current synthetic biology, which is based on an axiological concept of life. He declares self-activity to be the bearer of the normative content of the concept of life. This cybernetic concept of life—life as self-activity and communicative interaction with an environment (see ibid.)—is combined with the holistic potential of the concept of life: The boundaries between the different life forms are sublated (see Boldt 2013, p. 399) and all life forms are conceived as being engaged in seeking a common good. The good is a "practical notion of truth" (Boldt 2013, p. 398), because:

> The entity to which one relates is conceived of as taking part in the search for the good in which one is immersed oneself. Hence, on this view one does not have an a priori right to discard the interests and behavior of the entity, but is supposed to commence action towards it in order to get to know the entity and accommodate its interests, if this appears reasonable. The inherent value of the entity is a result of conceiving of the entity as a proto-subject. Thus, the observer is compelled to respect its ways of behaving and turns from observer into companion. (Boldt 2013, p. 397)

Another claim for the "attitude of respect for nature" (Taylor 1986, p. 59) rests on the "biocentric outlook on nature" (ibid. p. 99). Biocentrists, like Paul Taylor, hold "that all organisms are teleological centers of life in the sense that each is a unique individual pursuing its own good in its own way" (ibid. p. 100). As moral agents, human beings are obliged to acknowledge the "inherent worth" of "entities that have a good of their own" (ibid. p. 75). Deplazes-Zemp remarks rightly that for biocentrists "the production of synthetic organisms would [...] imply a moral responsibility towards the produced organism" (Deplazes-Zemp 2012, p. 770). Already in 1986, Taylor emphasizes the special relevance of an "ethics of the bioculture" which "is concerned with the human treatment of animals and plants in artificially created environments that are completely under human control" (Taylor 1986, p. 53). He argues that "it becomes a major responsibility of moral agents in this domain of ethics to work out a balance between effectiveness in producing human benefits, on the one hand, and proper restraint in the control and manipulation of living things, on the other" (ibid. pp. 57, 58).

There are profound arguments which can be raised against both the cybernetic and the biocentric starting points. Georg Toepfer remarks, for example, that neither differentiates between the immanent normativity of every organic life form and the normativity which is posited by individual reflection and in distance to the organic presuppositions of life (see Toepfer 2014). But the purposes that guarantee the self-preservation of a system are not necessarily ethically valuable. There can be a difference between teleological purpose and the ethical good. Thus, both the cybernetic and the biocentric position are problematic because of their equivocation on the concept of purpose. In addition, they do not provide any possibility of deducing the reflexive normativity and can be accused of "immanentism" (ibid.).

3 The Concept of Life in Philosophical and Scientific Discourses Around 1900. A Key Constellation

I have shown that the concept of life plays a crucial role in current debates on the philosophy of synthetic biology. I will now turn to the history of the concept of life. All of the issues I have discussed, have a long history in philosophical and scientific reflections on life. They are what we might call tropes in the modern discourse on life. The latter starts around 1800 and culminates in the late 19th century. But in the context of synthetic biology the turn of the 20th century is of special interest, because it was around this time that a strictly biological belief in the possibility of creating life arose. Scientists like Emil Fischer or Jacques Loeb concluded independently that the artificial production of life should be possible (see Budisa 2012, p. 106). For Fischer—as Nediljko Budisa emphasizes—the chemical synthesis of life seemed to be "an achievable goal":

> Fischer believed that modifications, design and creation of organisms with chemical methods is a kind of beginning of a grand future project: he expected new forms of life with novel/alternative chemical compositions created by synthetic means to have fundamental advantages over the known living organisms with great potentials to gain technological benefit for society. (Budisa 2012, p. 107).

Loeb was also convinced that the artificial production of living beings would be possible in the future. In his opinion, only technological problems explained the failure of contemporary attempts to synthesize life and he saw no reason why the artificial production of living organism should be impossible in principle (see Loeb 2008, p. 258). Loeb's lecture *Das Leben* (*Life*), which was delivered in 1911 at the first conference of the monists in Hamburg, explained the theoretical framework behind his belief in the possibility of "a practical, useful and controlled design of 'synthetic life'" (Budisa 2012, p. 106). He developed a rigid "philosophy of reductionist experimentation that sees living organisms as chemical machines" (ibid. p. 106). Moreover, Loeb claimed that if life could be explained completely by physics and chemistry, we could also build our social and ethical life on a scientific foundation (see Loeb 2008, p. 255). The engineering of natural life would enable us to engineer social life as well. His lecture ends with reflections on the natural basis of human ethics.

The social, political and ethical implications of Loeb's strict naturalism indicate two important features of the scientific and philosophical discourses on life around 1900: Firstly, the reduction of the social sphere to the scientific method points to the reach of the technical imperative of the engineer. Scientists like Loeb or Fischer saw themselves as engineers of life. Moreover, as Petra Gehring shows convincingly, the practical orientation of the social sciences led to a technological understanding of society, which would enable its reform in the name of life (see Gehring 2009, pp. 123, 134).[7]

Secondly, Loeb's integration of the social sphere into his naturalistic concept of life, together with the practical direction of his scientific approach, indicates the claim to comprehensiveness in the discourses on life around 1900. There were different approaches to life, which were connected to one another mainly because of their all-inclusive concept of life. Regardless whether the conceptualisation of life was based on nature (e.g., Loeb or Ernst Haeckel), culture (e.g., Georg Simmel) or history (e.g., Wilhelm Dilthey) it was supposed to comprise the whole of reality. Thus, new epistemological models arose that were mainly connected in that they strove to transcend the opposition between the natural sciences and the humanities. These methodological claims to comprehensiveness were based on the all-inclusiveness of the central concept of life. Its all-inclusiveness provided the possibility of undermining the epistemic dichotomy—so it was at least claimed.

Wilhelm Dilthey's philosophy of life is a surprising example of the attempt to overcome the dualism of nature and spirit (*Geist* in German) within the paradigm of life. From the moment life entered the systematic part of his philosophy, a new imperative started to guide his philosophy of science, with "life" providing the nexus that brought together all sciences without reducing any one science to any other. In particular, his writings from the 1890s show that this approach is based on the definition of life as a process of articulation. This idea also grounds Dilthey's attempt to develop a concept of mind out of biological structures.[8] The life of the spirit is considered as a more subtle and complex embodiment of structural characters, which are valid for all processes of life—for instance, thinking is, according to Dilthey, an interpolation between stimulus and response (see Dilthey 1981, p. 13). Dilthey posits continuity between the different life forms. Moreover, the relation between "higher" and "lower" life forms is defined as development, which is characterised by an increasing differentiation and delineation. In other words, life in itself develops continually more manifold and complex forms. The following passage articulates Dilthey's aim of grounding structures of meaning in biological processes:

---

[7]Gehring analyses writings of Arthur Ruppin, Albert Hesse and Wilhelm Schallmayer on the significance of the theory of evolution for social and political issues as examples of this perspective of intervention. There are also other examples of the importance of an engineering approach in the social thinking of this time: in 1881 Julius Post published his contributions to "Social-Engineering" (*Arbeit statt Almosen. Beiträge zur Social-Technik; Labour rather than Charity. Contributions to Social-Engineering*). In 1899, Paul Natorp introduced the concept of social engineering in his *Sozialpädagogik.Theorie der Willenserziehung auf der Grundlage der Gemeinschaft (Social pedagogy. A theory of the cultivation of the will on the basis of community)*. In 1904, Albert Kellner presented the *Gesellschaft für Ethische Kultur (Society for Ethical Culture)*, his Social-Engineer (*Der Sozial-Ingenieur*) and reports about different approaches to enhancing working conditions in the United States and in Great Britain (see Neef 2012, pp. 257–259).

[8]This part of Dilthey's philosophy was and still is neglected. An exception, on which I base my own interpretation, are the works of Matthias Jung (see Jung 1996, 2003, 2008).

Wherever psychic inwardness emerges, namely, in the entire animal and human world, the structure and articulation of life is the same. But what is still completely lacking in the lower forms of life is that manifold of discrete sensations and feelings on which psychology as a rule is built. The primordial nucleus of inner life is always and everywhere the progression from an impression stemming from the milieu of the living creature to the movement that adapts the relationship of the milieu of the living creature. There is no more original connection than this in all inner life. [...] Seen from inside, the development of the living creature into higher forms involves an articulation; life articulates itself. And to the inner articulation there corresponds an external articulation of the animate, organic body in a series of stages. Intermediaries between the impression and the executed movement multiply. Both the initial impressions and the final response assume more complex forms. But everything happens on the basis of a schema common to animal and human life. And precisely in this great, encompassing nexus and its relation to our intellectual inner life there lies the convincing, unimpeachable proof that thought appears as part of life, is linked to it, and serves its sphere. A biological perspective is necessary in order to be convincing about the structure of life. (Dilthey 2010b, pp. 70, 71)

This glance at Dilthey's hermeneutics of life shows that he was attempting to integrate biological concepts into his theory of mind without deducing the latter from a biological basis. In this respect, his philosophy is an example of the comprehensive claims of the discourses on life around 1900. Moreover, Dilthey's theoretical considerations on human nature always have a practical slant. The "practical nature of human beings" (Dilthey 2010b, p. 14) is the starting point of his epistemology. Therefore Dilthey developed an anthropology of knowledge in which the nexus of life and value plays a crucial role. He claims that the main work of life is to recognize what is really valuable for us (see Dilthey 2004, p. 88). Even knowledge of reality serves to elaborate life-values. It is this aim which gives human actions their teleological aspect. Moreover, Dilthey wanted to integrate humanity into the purposiveness of nature, and defined the individual as the reference point of purposiveness. Dilthey envisaged the human life-unit as a subject of self-preservation. The "course of a life" is defined as a "unity", which constitutes a "complete and self-enclosed, clearly delineated process" (Dilthey 2002, pp. 92, 93). The lived experiences "belong to a nexus that persists as permanent amidst all sorts of changes throughout the entire course of life" (ibid. p. 102). On the level of culture, the continual process of individuation is exactly this 'permanent persistence'[9] of a single nexus in the course of life. The process of individuation has a biological foundation: Dilthey conceives individuation as a descendant of the self-preservation instinct. He thinks that the human life-unit is ultimately nothing but "a bundle of drives" (Dilthey 2010a, p. 14). And life-experience, in which the value of things is proved, can be seen as an enhancement of self-identity by reflective awareness (see Dilthey 1990, p. 409). This process also enables the objectification of life values. Its highest form is philosophy, which develops a

---

[9]Manfred Sommer notes correctly that this phrase is an "emphatic pleonasm" (see Sommer 1984, p. 61). But his judgment that Dilthey's life-unit is isolated in itself is false. For Dilthey the life-unit is always exposed to "the pressure of the outside world" (Dilthey 2010a, p. 25). This difference is "first experienced in impulse and resistance" (ibid. p. 23) and develops into social interaction.

system of values and analyses their genesis and validity—a process from which non-human life-forms are nevertheless excluded by definition.

It is important to acknowledge that it is not just the philosophical discourse on life that is characterised by the practical turn, which entangles the concept of life with the concept of value. As Gehring emphasises:

> The oscillating concept of 'life-value' is a hinge of the discourse on life. Life is value and life – as valuating authority – signalises likewise what really has value [...]. On the one hand 'life-value' is (almost) a pleonasm, but on the other hand the certainty of the 'value' of life constitutes what turns the mere discourse on life into a dispositif, in other words an effective guideline also for institutional actions outside science.[10] (Gehring 2009, p. 134; translated by J.S.)

Gehring convincingly argues that this action-theoretical and existential concept of reality emerged with this all-inclusive concept of life. The idea is that reality should be reshaped in the name of life and values (see Gehring 2009, p. 132). Already Friedrich Nietzsche refers in his *Untimely Mediations* to the "master builder of the future" ("Baumeister der Zukunft", Nietzsche 1874, p. 294; translated by J.S.), who shapes a new future by the power of their will. In a posthumous fragment Nietzsche also uses the phrase "master builder of life" ("Baumeister des Lebens", Nietzsche 1980a, p. 631; translated by J.S.). But Nietzsche points above all to a problem in the axiological concept of life. His vitalistic theory of history emphasises the destructive features of innovation. For Nietzsche, the creation of something new always presupposes the destruction of the established. Precisely in this sense, he conceives of history as life. For Nietzsche, "life always lives at the expense of other lives"[11] (Nietzsche 1980b, p. 167; translated by J.S.) since "in its basic functions" life "essentially [...] harms, oppresses, exploits, and destroys, and cannot be conceived at all without this character."[12] (Nietzsche 1887, p. 312; translated by J.S.) It is exactly this essential connection between the preservation and destruction of life which is an argument against a cybernetic starting point for justifying an ethical dimension of the concept of life (see Toepfer 2011, pp. 457, 458, 2014). This argument holds in the realm of biology, because Nietzsche's view is proven by biological observation.

---

[10]See the German original: "Ein Gelenkstück des Lebensdiskurses ist der in sich changierende Begriff 'Lebenswert'. Leben *ist* Wert und Leben zeichnet zugleich—als wertende Instanz—'das wirklich' Wertvolle aus [...]. Einerseits ist 'Lebenswert' (fast) ein Pleonasmus, andererseits liegt in jener Gewissheit vom 'Wert' des Lebens das, was aus dem bloßen *Diskurs* des Lebens ein *Dispositiv* macht, also eine praxiswirksamen Leitstrahl auch für außerwissenschaftliches institutionelles Handeln." (Gehring 2009, p. 134).

[11]See the German original: "Leben lebt immer auf Unkosten andern Lebens". (Nietzsche 1980b, p. 167).

[12]See the whole sentence in the German original: "*An sich* von Recht und Unrecht reden entbehrt alles Sinns, *an sich* kann natürlich ein Verletzen, Vergewaltigen, Ausbeuten, Vernichten nichts 'Unrechtes' sein, insofern das Leben *essentiell*, nämlich in seinen Grundfunktionen verletzend, vergewaltigend, ausbeutend, vernichtend fungirt und gar nicht gedacht werden kann ohne diesen Charakter." (Nietzsche 1887, p. 312).

4 Life Again? Conclusions

This short examination of an important part of the history of the concept of life suggests a surprising result. In the light of this history, synthetic biology leads to well-known debates, arguments, notions and questions—especially in the philosophical and ethical discourse on this new approach in biological research and technology. The concept of "life as a toolbox", which was shaped by the interpretation of the research activity of synthetic biology, has already been formulated in another context. It does not add "a new facet to the multifarious concept 'life'" as Deplazes-Zemp claims (2012, p. 772). There is a long tradition of discussing and answering the questions about life which synthetic biology raises. Therefore, the questions themselves have to be analysed critically. Do they capture the epistemological, ontological and ethical consequences of synthetic biology? And does the concept of life provide solutions to actual problems, or at least help to understand this new approach in biology, together with its societal impact? To answer these questions I return to the current issues:

1. Our discussion of the current issues has shown that the phrase "fabrication of life" encapsulates many of the hopes and concerns connected with synthetic biology. Emphasis on this notion's historical roots permits a more sober judgement on synthetic biology and the expectations which are connected with this new technology.

2. But should synthetic biological systems be defined as living? In my opinion, the concept of life is too ambiguous and controversial to be useful for capturing the actual practice of synthetic biology. Therefore, synthetic biology should follow other branches of biology in being reluctant to use the term. This would be helpful for both synthetic biology itself, and philosophical debates about it. Yet, from another angle, it is to be welcomed that because of this new constructive approach in biology, established dualisms such as that of nature and technology are being questioned again. In these debates, the concept of life could play a crucial role since, because of its ambiguity—especially its "double meaning as a material object [...] and as a value of certain spiritual or even moral dimensions" (Budisa 2012, p. 101)—the concept of life could mediate between heterogeneous spheres like the natural sciences and the humanities, theory and practice, descriptive contexts and normative contexts. Following this view, synthetic biology could be considered as the latest example of the entanglement of theory and practice in modern science. It presents an essential feature of the life sciences, because they are characterized by their close relation to practical application. Moreover, their significance is based on their practical impact, which already shapes our daily life. Thus, for a critical understanding of the life sciences it is all the more important to consider scientific research as a social practice. The concept of life could indicate this entanglement of science and society. Furthermore, it emphasises the existential dimension of every cultural practice.

3. History tells us that the concept of life is strongly associated with the concept of value. There are various versions of the equation "life is value". Nevertheless there are strong arguments against its ethical relevance, especially in the context of synthetic biology.[13] First, both concepts (life and value) are highly ambiguous and problematic. It would need careful elaboration to establish an ethics of life which is based on the intrinsic value of life. Second, in the history of the concept of life there have been arguments for and against its ethical relevance. Careful examination of the history of ethics would reveal that it is highly controversial whether life as such has an intrinsic value. Generally, life is regarded only as a prerequisite for the realization of ethically valuable qualities (see Toepfer 2014). Moreover, the historical findings clearly show that, for example, the cybernetic version of an ethics of life is not tenable. Third, the question remains, what counts as life? To use life as an ethical concept presupposes a precise and generally valid definition of life, which is not available so far—whether in biology or in philosophy.

But perhaps this fact does not matter because whether the objects of synthetic biology are conceived as living beings or not isn't important for the ethical dimension of synthetic biology. Synthetic biology is definitely a further step in the mastery of nature; in others words, in its cultivation. It could even lead to "a parallel biological world" (Budisa 2012, p. 115). It is no accident that Budisa compares the route to such a world with "a road that follows a course that has to be optimally designed" (ibid). This "directed evolution" (ibid.) is distinguished from "natural evolution" which is described as "a contingent historical process, like the flow of the river" (ibid). This metaphor clearly presents synthetic biology as cultivating natural evolution. Not least because of this radical possibility, synthetic biology has to be seen as a new technology which raises old philosophical questions: Which attitude towards nature is reasonable? Is the "human intellect" (ibid.) a good "directing principle" (ibid.) for natural processes? How far should we change the face of the world? Can we take the responsibility for the consequences of such deep interventions, especially if we think of future generations? What can we know about the potential consequences? In the context of such questions, the special ethical challenge of synthetic biology becomes apparent. As several authors convincingly argue, its risks are not captured by traditional technology assessment (e.g., Engelhard 2011). Complex biological systems are difficult to predict (see, e.g., Köchy 2012a, pp. 172, 173). The more they diverge from nature, the less models for comparison are available. Therefore, in my opinion, the questions of biosafety and biosecurity are urgent. For a valid technology assessment it is certainly important to know, as concretely as possible, how synthetic biological systems behave. But the question whether this activity has to be conceived as life or not is a

---

[13]See also Eichinger, this volume.

philosophical one. And in this context, Hegel's famous claim that "the owl of Minerva begins its flight only with the onset of dusk" is probably true:

When philosophy paints its grey in grey, a shape of life has grown old, and it cannot be rejuvenated, but only recognized, by the grey in grey of philosophy. (Hegel 2003, p. 23)

References

Agamben G (1998) Homo Sacer: Sovereign power and bare life (trans. D Heller-Roazen). Stanford University Press, Stanford

Aristotle (2004) Physics (trans. J Sachs). In: Sachs J (ed) Aristotle's Physics. A Guided Study, 4th edn. Rutgers University Press, New Brunswick/London

Aurenque D (2011) Natur, Leben, Herstellung: Worin liegt die ethische Herausforderung der Synthetischen Biologie? In: Dabrock P et al (eds) Was ist Leben – im Zeitalter seiner technischen Machbarkeit. Karl Alber, Freiburg, pp 327–344

Bedau M (2010) The power and the pitfalls. Nature 465:422

Bedau M, Church G, Rasmussen S et al (2010) Life after the synthetic cell. Nature 465:422–424

Benner SA, Yang Z, Chen F (2011) Synthetic biology, tinkering biology, and artificial biology. What are we learning? C R Chim 14(4):372–387

Beuttler U (2011) Strukturelemente und Wert des Lebens – theologisch-hermeneutische und – ethische Überlegungen zum Lebensbegriff der Synthetischen Biologie. In: Dabrock P, Bölker M, Braun M, Ried J (eds) Was ist Leben – im Zeitalter seiner technischen Machbarkeit Karl Alber, Freiburg, pp 277–305

Billerbeck S, Panke S (2012) Synthetische Biologie – Biotechnologie als eine Ingenieurswissenschaft. In: Bold J, Müller O, Maio G (eds) Leben schaffen? Philosophische und ethische Reflexionen zur Synthetischen Biologie, Mentis, Paderborn, pp 19–40

Boldt J (2012) "Leben" in der Synthetischen Biologie: Zwischen gesetzförmiger Erklärung und hermeneutischem Verstehen. In: Boldt J, Müller O, Maio G (eds) Leben schaffen? Philosophische und ethische Reflexionen zur Synthetischen Biologie, Mentis, Paderborn, pp 177–191

Boldt J (2013) Life as a technological product: philosophical and ethical aspects of synthetic biology. Biol Theory 8:391–401

Boldt J, Müller O, Maio G (eds) (2012) Leben schaffen? Philosophische und ethische Reflexionen zur Synthetischen Biologie, Mentis, Paderborn

Bölker M (2011) Revolution der Biologie? Ein Überblick über Voraussetzungen, Ansätze und Ziele der Synthetischen Biologie. In: Dabrock P, Bölker M, Braun M, Ried J (eds) Was ist Leben – im Zeitalter seiner technischen Machbarkeit, Karl Alber, Freiburg, pp 27–41

Brenner A (2012) Leben leben und Leben machen. Die Synthetische Biologie als Herausforderung für die Frage nach dem Lebensbegriff. In: Bold J, Müller O, Maio G (eds) Leben schaffen? Philosophische und ethische Reflexionen zur Synthetischen Biologie, Mentis, Paderborn, pp 105–120

Budisa (2012) A brief history of the "Life Synthesis". In: Hacker J, Hecker M (eds) Was ist Leben? Nova Acta Leopoldina, vol 116:394. Wissenschaftliche Verlagsgesellschaft, Stuttgart, pp 99–118

Caplan A (2010) The end of Vitalism. Nature 465:423

Cho MK, Magnus D, Caplan AL et al (1999) Policy forum: genetics. Ethical considerations in synthesizing a minimal genome. Science 286(2087):2089–2090

Dabrock P, Bölker M, Braun M, Ried J (eds) (2011a) Was ist Leben – im Zeitalter seiner technischen Machbarkeit? Beiträge zur Ethik der Synthetischen Biologie, Karl Alber, Freiburg

Dabrock P, Bölker M, Braun M, Ried J (2011b) Einleitung. In: Dabrock P, Bölker M, Braun M, Ried J (eds) Was ist Leben – im Zeitalter seiner technischen Machbarkeit, Freiburg, pp 11–24

Deplazes-Zemp A (2012) The conception of life in synthetic biology. Sci Eng Ethics 18:757–774

Deplazes-Zemp A, Biller-Andorno N (2012) Explaining life. EMBO Rep 13(11):959–962

Dilthey W (1981) System der Ethik. In: Gesammelte Schriften, vol X, 4th edn. Vandenhoeck & Ruprecht, Göttingen

Dilthey W (1990) Das Wesen der Philosophie. Gesammelte Schriften, vol V, 5th edn. Vandenhoeck & Ruprecht, Göttingen, pp 339–416

Dilthey W (2002) The formation of the historical world in the human sciences. In: Makkreel RA, Rodi F (eds) Selected works, vol III. Princeton University Press, Princeton/Oxford

Dilthey W (2004) Logik und Werk. Späte Vorlesungen, Entwürfe und Fragmente zur Strukturpsychologie, Logik und Wertlehre (ca. 1904–1911). In: Gesammelte Schriften, vol XXIV. Vandenhoeck & Ruprecht, Göttingen

Dilthey W (2010a) The origin of our belief in the reality of the external world and its justification. In: Makkreel RA, Rodi F (eds) Selected works, vol II. Princeton University Press, Princeton/Oxford, pp 8–57

Dilthey W (2010b) Life and cognition. In: Makkreel RA, Rodi F (eds) Selected works, vol II. Princeton University Press, Princeton/Oxford, pp 58–114

Editorial (2007) Meanings of "life". Synthetic biology provides a welcome antidote to chronic vitalism. Nature 447:1031–1032

European Commission (2005) Synthetic biology. Applying engineering to biology. Report of a NEST High-Level Expert Group. Luxembourg, Office for Official Publications of the European Communities

Engelhard M (2011) Die synthetische Biologie geht über die klassische Gentechnik hinaus. In: Dabrock P, Bölker M, Braun M, Ried J (eds) Was ist Leben – im Zeitalter seiner technischen Machbarkeit. Karl Alber, Freiburg, pp 43–59

Esposito R (2013) Terms of the political. Community, immunity, biopolitics (trans. NS Welch). Fordham University Press

Fellmann F (1993) Lebensphilosophie. Elemente einer Theorie der Selbsterfahrung, Rowohlt, Reinbeck

Gayon J (2010) Defining life: synthesis and conclusions. Orig Life Evol Biosph 40:231–244

Gehring P (2009) Wert, Wirklichkeit, Macht. Lebenswissenschaften um 1900. Allgemeine Zeitschrift für Philosophie 34(1):117–135

Gehring P (2010) Das Leben im Bakterium. Wenn es kein Leben ist, was ist es dann?. Frankfurter Allgemeine Zeitung 26 May 2010. http://www.faz.net/aktuell/feuilleton/forschung-und-lehre/das-leben-im-bakterium-wenn-es-kein-leben-ist-was-ist-es-dann-1983342.html?printPagedArticle=true. Accessed 24 Apr 2015

Gibson DG et al (2010) Creation of a bacterial cell controlled by a chemically synthesized genome. Science 329:52–56

Hacker J, Hecker M (eds) (2012) Was ist Leben? Nova Acta Leopoldina, vol 116:394. Wissenschaftliche Verlagsgesellschaft, Stuttgart

Hegel GWF (2003) Elements of the philosophy of the right (trans: Nisbet HB). 8th edn. Cambridge University Press

Homberger DG (1998) Was ist Biologie? In: Dally A (ed) Was wissen Biologen schon vom Leben? Loccumer Protokolle, Loccum, pp 11–28

Jung M (1996) Dilthey zur Einführung. Junius, Hamburg

Jung M (2003) "Das Leben artikuliert sich". Diltheys performativer Begriff der Bedeutung. Revue Internationale de Philosophie 226(4):439–454

Jung M (2008) Wihelm Diltheys handlungstheoretische Begründung der hermeneutischen Wende. In: Kühne-Bertram G, Rodi F (eds) Dilthey und die hermeneutische Wende in der Philosophie. Wirkungsgeschichtliche Studien, Vandenhoeck & Ruprecht, Göttingen, pp 257–272

Kellner A (1904) Der Sozial-Ingenieur. Ethische Kultur 12:27–29

Köchy K (2012a) Zum Verhältnis von Natur und Technik in der Synthetischen Biologie. In: Boldt J, Müller O, Maio G (eds) Leben schaffen? Philosophische und ethische Reflexionen zur Synthetischen Biologie, Mentis, Paderborn, pp 155–175

Köchy K (2012b) Philosophische Implikationen der Synthetischen Biologie. In: Köchy K, Hümpel A (eds) Synthetische Biologie. Entwicklung einer neuen Ingenieurbiologie? Themenband der interdisziplinären Arbeitsgruppe "Gentechnologiebericht", Wissenschaftlicher Verlag, Dornburg, pp 137–161

Knorr Cetina K (2005) The rise of a culture of life. EMBO Rep 6(1):76–80

Lebovic N (2013) The philosophy of life and death. Ludwig Klages and the Rise of a Nazi Biopolitics. Palgrave MacMillan, New York

Loeb J (2008) Das Leben. In: Then C (ed) Dolly ist tot: Biotechnologie am Wendepunkt. Rotpunktverlag, Zürich, pp 255–280

Morange M (2009) A new revolution? The place of systems biology and synthetic biology in the history of biology. EMBO Rep 10(1):50–53

Morange M (2010) The resurrection of life. Orig Life Evol Biosph 40:179–182

Natorp P (1899) Sozialpädagogik. Theorie der Willenserziehung auf der Grundlage der Gemeinschaft. Frommann, Stuttgart

Neef K (2012) Die Entstehung der Soziologie aus der Sozialreform: Eine Fachgeschichte. Campus-Verlag, Frankfurt

Nietzsche F (1874) Von Nutzen und Nachtheil der Historie für das Leben. In: Sämtliche Werke, vol 1. dtv/de Gruyter, München/Berlin/New York, 1980, pp 243–334

Nietzsche F (1887) Zur Genealogie der Moral. In: Sämtliche Werke, vol 5. dtv/de Gruyter, München/Berlin/New York, 1980, pp 245–412

Nietzsche F (1980a) Nachgelassene Fragmente 1880–1882. In: Sämtliche Werke, vol 9. dtv/de Gruyter, München/Berlin/New York, 1980

Nietzsche F (1980b) Nachgelassene Fragmente 1885–1887. In: Sämtliche Werke, vol 12. dtv/de Gruyter, München/Berlin/New York, 1980

Post J (1881) Arbeit statt Almosen. Beiträge zur Social-Technik, Roussell, Bremen

Ruiz-Mirazo K, Moreno A (2013) Synthetic biology: challenging life in order to grasp, use and extend it. Biol Theory 8:376–382

Sommer M (1984) Leben aus Erlebnissen. Dilthey und Mach. In: Orth EW (ed) Dilthey und der Wandel des Philosophiebegriffs seit dem 19. Jahrhundert. Studien zu Dilthey und Brentano, Mach, Nietzsche, Twardowski, Husserl, Heidegger. Karl Alber, Freiburg, pp 55–79

Taylor PW (1986) Respect for nature. A theory of environmental ethics. Princeton University Press, Princeton

Toepfer G (2011) Leben. In: Metzler JB (ed) Historisches Wörterbuch der Biologie. Geschichte und Theorie der biologischen Grundbegriffe, vol 2. Stuttgart/Weimar, pp 420–483

Toepfer G (2014) Die Unbegrifflichkeit von "Leben" in der Begrifflichkeit der Ethik. Welche Rolle die Rede von "Leben" in der Ethik spielt und warum sie nicht zentral ist. Jahrbuch für Wissenschaft und Ethik 18:199–234

Venter JC (2013) Life at the speed of light: from the double helix to the dawn of digital life. Viking, New York

Weiss MG (2011) Verstehen, was wir herstellen können? Martin Heidegger und die Synthetische Biologie. In: Dabrock P, Bölker M, Braun M, Ried J (eds) Was ist Leben – im Zeitalter seiner technischen Machbarkeit. Freiburg, Karl Alber, pp 173–193

Witt E (2012) Konzepte und Konstruktionen des Lebenden. Philosophische und biologische Aspekte einer künstlichen Herstellung von Mikroorganismen. Karl Alber, Freiburg

Worms F (2013) Über Leben. Merve, Berlin

Synthetic Biology and the Argument from Continuity with Established Technologies

Andreas Christiansen

1 Introduction

In efforts to defend synthetic biology against critics, it is common to see references to earlier, established technologies and practices that synthetic biology is based upon and/or an extension of. These technologies include basic domestication and cultivation of plants and animals, systematic selective breeding, and newer techniques such as directed evolution and recombinant genetic modification. The precise point of such references, particularly in public debate, is not always perfectly clear. Here, I reconstruct them as making the (tacit) claim that since synthetic biology is in some respect just like some technology that we think is unproblematic, we should think synthetic biology unproblematic as well. My aim in this chapter is to assess the argument and determine the extent to which it is successful.

I will first present a couple of examples of the line of argument as it appears in public discussion, as well as three examples from the philosophical literature. I will attempt to reconstruct these as instances of a type of argument that I call the Continuity Argument. The Continuity Argument, in summary, says that some fact about synthetic biology cannot be a reason to be critical of it unless we should also be critical of established technologies of which that fact holds.[1] For example, the fact that synthetic biology constitutes human manipulation of the genome of organisms cannot be a reason to be critical of synthetic biology unless we should also be critical of domestication in general, since domestication is also a form of

---

[1] I use "being critical of" as a generic term covering any negative reactions to synthetic biology. I'll return to this point below.

A. Christiansen (✉)
Department of Media, Cognition and Communication, University of Copenhagen, Copenhagen, Denmark
e-mail: rms998@hum.ku.dk

© Springer International Publishing Switzerland 2016
K. Hagen et al. (eds.), *Ambivalences of Creating Life*, Ethics of Science and Technology Assessment 45, DOI 10.1007/978-3-319-21088-9_16

such manipulation. It is, then, an argument that *defends* synthetic biology against a *suggested reason to be critical* of it. It is thus a *type* of argument, rather than a single argument; there can be many different instances of the Continuity Argument. Since the Continuity Argument is an answer to a claim that there exists some *reason* to be critical of synthetic biology, much of my reconstructive effort will make use of the notion of a reason and how reasons work.

Having reconstructed the Continuity Argument, I present three issues that may undermine, to some extent, the soundness of the argument. These are (i) that the argument doesn't show that we should stop being critical of synthetic biology rather than start being critical of the established technologies; (ii) that the argument doesn't allow that the degree to which a fact about synthetic biology holds may make a difference to how that fact works as a reason; and (iii) that the argument doesn't distinguish between what reasons we have and what we should do all things considered. I then discuss a specific, commonly expressed reason to be critical of synthetic biology, namely that it involves human intentional manipulation of organisms' DNA, in light of these issues. Finally, I conclude with some reflections about what thinking about continuity can teach us.

2 The Continuity Argument

As mentioned, the line of thinking I have in mind appears in both the academic literature and in discussions in society at large. In this section I will first present two examples from the public sphere. I then present three examples of the argument found in the philosophical literature on synthetic biology. Finally, I construct a general form of the argument.

2.1 *Continuity Arguments in Public Debate*

Example 1
At the technology-friendly info-tainment website cracked.com, an image was posted showing a dog, a chicken, a pig and a cob of corn juxtaposed, respectively, with a wolf, a junglefowl, a boar, and an ear of teosinte—that is, with their wild ancestors. Accompanying the image was the following text:

> Hate how common GMOs seem to be nowadays? Just remember... by one means or another, humans have *always* loved screwing with nature. And we made all the really big changes AGES ago. (Cracked.com 2014)

Example 2
In an editorial arguing against the labelling of GMO foods, Scientific American makes the following claim:

> We have been tinkering with our food's DNA since the dawn of agriculture. By selectively breeding plants and animals with the most desirable traits, our predecessors transformed organisms' genomes, turning a scraggly grass into plump-kerneled corn, for example. For the past 20 years Americans have been eating plants in which scientists have used modern tools to insert a gene here or tweak a gene there, helping the crops tolerate drought and resist herbicides. (Scientific American 2013)

These examples bring out some of the features of the Continuity Argument that I described in the introduction. Cracked and Scientific American are arguing *against* some attitude or action—namely "hating how common GMOs are becoming" and labelling GM foods, respectively. They do so by pointing out that a fact about GMOs, namely that they are the products of human alteration of organisms' genomes—of our "screwing with nature" and "tinkering" with DNA—is also a fact about earlier practices. Assuming that this fact is supposed to be relevant for their case, what Cracked and Scientific American seem to be saying is that *this* fact is not a good reason for hating or labelling GMOs, since we don't hate or label non-GM organisms that are also products of human alterations. Supposedly, someone has made the opposite claim, that the fact that GMOs are products of human tinkering *is* a reason for labelling/hating them, although neither Cracked nor Scientific American provide sources for this.

2.2 Continuity Arguments in the Philosophical Literature

2.2.1 Douglas, Powell and Savulescu

Douglas et al. (2013) are interested in the question of whether the creation of artificial organisms—which they define as organisms constructed from chemically simple, non-living materials[2]—is "morally significant". It is in defining what they mean by "morally significant" that they put forward a Continuity Argument. In order for the creation or artificial organisms to be morally significant, according to Douglas et al., it has to be the case that (a) there are moral reasons not to create such organisms (or factors that weaken reasons to create them),[3] and (b) these reasons are

---

[2]I take this to be sufficiently close to the program of synthetic biology that I will not distinguish between synthetic biology and the creation of artificial organisms.

[3]As Douglas et al. recognize, the moral significance of artificial life might be *positive*, i.e. there might be moral reasons *for* creating artificial organisms. Since the main thrust of the debate about synthetic life has concerned whether it is *negatively* morally significant, Douglas et al. do not further discuss this possibility. For my purposes, given the fact that Continuity Arguments are typically defences against criticism of synthetic biology, it is mainly the (possible) negative moral significance that is at issue. But note that Continuity Arguments could be made with the opposite valence—i.e. arguments that dismissed a reason that was alleged to count in favour of synthetic biology (or another technology) by noting that the same reason would hold for a technology we were disposed to reject.

specific to the creation of artificial organisms. The latter condition means that the reasons in question must not also apply to certain "contrasting practices", namely all other ways of generating new organisms—from sexual reproduction through selective breeding to directed evolution and recombinant genetic engineering.

Already at this stage of their argument, Douglas et al. rule out the fact that synthetic organisms are made to the specification of rational agents (i.e. humans) as a possible reason for critical attitudes to synthetic biology. They do so on specifically continuity-based grounds, arguing that since this fact is true of widespread practices going back to the beginnings of agriculture, it would be "surprising" if it were a reason for being critical of synthetic biology per se.

2.2.2 Holtug

Holtug's (2009) point of departure is somewhat different. He is not concerned with synthetic biology per se, but rather with "creating new life forms". The latter includes other methods of "genetic engineering", such as recombinant techniques. This means that Holtug cannot use recombinant genetic engineering as the established technology with which synthetic biology is compared. Since, as noted, Continuity Arguments are defences against criticisms of synthetic biology, this could be either a gain or a loss depending on the specific criticism one is addressing. If the critic accepts recombinant genetic engineering, then it makes sense to take the Douglas et al. route. If the opponent is equally critical of recombinant technology, Holtug's route is more useful. As it happens, I think most critics of synthetic biology are also critical of recombinant genetic engineering, and so I think Holtug's route is typically the superior one. However, some authors argue that synthetic biology, but not recombinant genetic engineering, is problematic in some sense. For example Boldt and Müller (2008)—who Douglas et al. cite in motivating their focus on synthetic biology per se—argue that synthetic biology's move from "manipulation" to "creation" is an ethically significant step.

Holtug introduces his Continuity Argument in assessing one interpretation of a classic objection to genetic engineering, namely that it constitutes our 'playing God'. The interpretation he targets says that "genetic engineering is unnatural, and therefore wrong" (Holtug 2009, p. 237). Unnaturalness is thus supposed to be a reason for finding a practice morally wrong. Holtug then questions whether there is a (non-ad hoc) sense of 'unnatural' that (a) would make it true that there is a reason of the relevant kind, and (b) does not also apply to selective breeding, since "even proponents of the "playing God" objection would want to allow this technology". He thus suggests that, since the critics of genetic engineering do not think that selective breeding should be disallowed (i.e. do not think it is wrong), any sense of 'unnatural' that applies to selective breeding cannot be used by them as a reason to find genetic engineering wrong.

2.2.3 Preston

Preston (2013) argues in great detail that synthetic biology is a "red herring" in discussions about the distinction between organisms and artefacts (a discussion that synthetic biology and technologies like it are sometimes claimed to give rise to). Her main focus is on theoretical issues, not on the ethical argument that is the main interest of both Douglas et al. and of Holtug. She is primarily concerned with showing that a range of changes that synthetic biology is claimed to cause with respect to the distinction between organism and artefact actually occurred much earlier, with the advent of agriculture. However, she several times hints at the fact that this is supposed to be part of a (more) complete Continuity Argument. Those against whom she directs her arguments are making the case that the changes mentioned are ethically salient in some way. They argue that the fact that synthetic biology entails some specific conception of the distinction between organism and artefact is a reason for being critical of it. What Preston tries to show is precisely that since this relationship between organism and artefact is present already in agriculture, it cannot work as such a reason (at least unless it also is a reason to be critical of agriculture as such). It is worth noting, however, that Preston does not argue that the fact of continuity means that all the ethical worries surrounding synthetic biology disappear, but merely that a prominent diagnosis and 'treatment' of the problem looks unhelpful (see esp. Preston 2013, p. 658).

2.3 The General Form

As the examples illustrate, the Continuity Argument amounts to the claim that some proposed reason to be critical of synthetic biology is not in fact a reason unless it is also a reason to be critical of established practices to which that reason would also apply. Often, the tacit or explicit assumption is added that we should not, in fact, be critical of the established practice, and hence not of synthetic biology either. I now want to state this argument more precisely, in order to see clearly what its premises and conclusions are. Before doing that, however, since the argument makes claims about reasons, it might be useful to look briefly at the nature of reasons.

2.3.1 Reasons

At its core, a reason is a kind of relation; it is a relation between (at least) a fact or consideration on one side and an action or attitude on the other, such that the fact or consideration counts in favour of the action or attitude. There are different ways of spelling the relation out in detail. In particular, there are differences of opinion as to how many relata are in the relation (see Scanlon 2014, p. 31, n. 21 for an overview). There are certainly two core relata, namely a fact or consideration (which I will call r)

and the action or attitude a that r is a reason for. Scanlon (2014, p. 31) and Cuneo (2007, p. 65) add a person, x, and a set of circumstances, c, as relata. On their conception a reason is a relation of the type $R(r, x, c, a)$: r is a reason for x in c to do action or take attitude a. Some philosophers, e.g. Skorupski (2010, pp. 36–37) add even more relata (in his case a time, t, and a degree of strength, d).

One of the terms of the relation, r, is typically referred to as the 'reason'. To avoid ambiguity, I will use the capitalized 'Reason' to refer to whole relations, and 'reason' to refer to a fact that functions as an r-term in some Reason relation. In thinking about (lower-case) reasons, it is important to remember that they only get to be reasons by being part of a Reason relation. Hence the other terms of the relation must be there as well (we can set aside the question of what the correct list of relata is—whatever relata turn out to be in that correct list must all be present). The non-core relata (i.e. the agent x, the circumstance c, the time t etc.) can sometimes be left unspecified because the Reason is supposed to be true for all values of those terms (i.e. for all agents in all circumstances at all times, r is a reason to take a). However, this cannot be done for the core relata, i.e. the reason r and the action or attitude a. It makes no sense to claim that some consideration is a reason for *all* actions and attitudes. And similarly, there is no action or attitude such that *every* consideration is a reason for it.

In the rest of this chapter, I will focus exclusively on the r and a-terms, and simply assume that proponents of the Continuity Argument and their opponents do not disagree on the circumstances under which and the persons for whom r is supposed to be a reason to take a.

2.3.2 The Argument

The Continuity Argument attempts to show that a *specific* fact, r, is *not* a good reason for taking a *specific* action or attitude, a—or, in other words, that this specific Reason relation, R, does not hold. Since the Reason under scrutiny has been suggested by critics of synthetic biology, we can assume that a is some action or attitude that is *negative* towards it. It could be very strongly negative, such as *banning* synthetic biology, or it could be less negative, such as creating a certain safeguard system, or labelling products that contain synthetic organisms, or not supporting synthetic biology research financially. Whenever I use a, it will be implied that the action or attitude is a negative one in this broad sense.

The r-term will be a proposition that is true about synthetic biology. Of course, people sometimes suggest Reasons where r is not true of synthetic biology, but in such cases the Continuity Argument is redundant anyway. The fact r will also be true of some established technology, e.g. domestication or selective breeding. I will use 'E' to refer to some such technology.

As the name of the Continuity Argument suggests, the idea is that E and synthetic biology are *continuous*. This is not a term with a clear meaning in the literature. For the examples I have used, i.e. domestication, selective breeding, recombinant genetic modification and synthetic biology, it seems obvious to see each new technology as based upon and developed from the last—that is, as taking the guiding ideas of the predecessor technology and pushing them further. However, the sense of continuity that is directly important to my argument is only continuity in terms of r, i.e. the fact that r holds for both synthetic biology and E.[4] Strictly speaking, then, the fact that synthetic biology is continuous with E in former sense is beside the point. We might just as well compare synthetic biology with very different technologies, depending on what the r under consideration suggests; for instance, beer brewing is invoked to dismiss criticisms based on the fact that some applications of synthetic biology makes use of fermentation products. In the case I take up in Sect. 4, the two senses of continuity align. The r discussed there is precisely the thing that synthetic biology has in common with those technologies with which it is continuous in the sense of being based on and extensions of—namely that it consists in human manipulation of organisms' genome.

The core of the argument, then, is this: Critics have suggested that the fact r is a reason for taking some negative action or attitude a to synthetic biology. In other words, the Reason $R(r, a)$ is claimed to hold with respect to synthetic biology. This r is also a fact about an established technology E. So R should hold with respect to E as well. But if R holds with respect to E, then we ought to take a towards E. The converse of this is that if we ought not to take a towards E, then R does not hold with respect to E. Since R does not hold with respect to synthetic biology unless it holds with respect to E, we can conclude that if we ought not to take a to E, then R does not hold of synthetic biology. In schematic form, the argument is this (I abbreviate synthetic biology 'SB' in the schematic form of the argument):

(1) r is true about SB and a fact about E.
(2) If (1) is true, then if R is true with respect to SB, then R is true with respect to E.
(3) If R is true with respect to E, then we ought to take a with respect to E.
(4) From (3) it follows that, if we ought not to take a with respect to E, then R is not true of E.
(5) Therefore from (2) and (4) it follows that, if we ought not to take a with respect to E, then R is not true of SB.

It is easy to see that the examples from the philosophical literature that I described above are instances of this argument. In the case of Douglas et al., the claim is this: The only fact that could function as an r-term with respect to synthetic biology is that synthetic organisms are derived from chemically simple, non-living materials. This is because all other (plausible) facts are also facts about

---

[4]This means that, in principle, Continuity Arguments could be made wherein the established technology is as new as, or even newer than, synthetic biology—although I am not aware of any such arguments.

"contrasting" (that is, established) practices, including traditional genetic modification techniques. So Douglas et al. propose a Continuity Argument that hits a range of potential r-terms, rather than a specific one. In their paper, they then apply this Continuity Argument (among other arguments) in rejecting several well-known reasons proposed by synthetic biology critics (namely the 'playing God' worry in various forms, the worry about reductionism, and worries concerning the interests of organisms).

For Holtug, there is a specific r-term, namely unnaturalness, which he takes to be an interpretation of the playing God worry. Likewise, there is a specific a-term, finding genetic engineering morally wrong, and a specific E, namely selective breeding. So Holtug's is a paradigm example of a Continuity Argument; its conclusion is that if we do not find selective breeding morally wrong, then unnaturalness cannot be a reason for finding genetic engineering morally wrong. Holtug adds the assumption, mentioned above, that "even proponents of the "playing God" objection" do not find selective breeding morally wrong, which leads to the conclusion that unnaturalness is not in fact a reason for finding genetic engineering morally wrong.

Finally, Preston attempts to show that agriculture constitutes a breakdown of the organism/artefact distinction. This does not in itself amount to making a Continuity Argument. Rather, what Preston does is to provide detailed support for premise (1) of the Continuity Argument as stated above, i.e. the claim that r is a fact about E as well as synthetic biology. However, as mentioned, her discussion assumes that the (dis)continuity between synthetic biology and established practice is ethically salient. In fact, acceptance of the Continuity Argument is common ground between Preston and her opponents, who also assume that the ethical status of synthetic biology is (at least partly) determined by whether it is continuous with established technologies or not. Otherwise, their efforts in proving the *dis*continuity of synthetic biology and established practices would be pointless.

3 Issues with the Continuity Argument

Returning to the Continuity Argument in its general form, there are several problematic issues to take note of. First, the conclusion of the argument is not that we ought not take a towards synthetic biology, and neither can that conclusion be straightforwardly inferred from (5). It takes the extra assumption (like the one Holtug makes) that we should not, in fact, take a towards E. Second, there are conditions under which premise (2) is not true. These are conditions where there is some potentially ethically salient difference in degree with respect to r (e.g. if r is the fact that technologies are risky, one technology might be more risky than another). Third, there are conditions under which (3) is not true. These are conditions in which *another* Reason outweighs R with respect to E, and thus makes it the case that we ought not to take a with respect to E all things considered. I will now go through each of these issues in turn.

3.1 Modus Ponens Versus Modus Tollens

The conclusion of the Continuity Argument as I have formulated it is this:

(5) if we ought not to take a with respect to E, then R is not true of E.

As noted, this is not the conclusion that proponents of the Continuity Argument are ultimately looking for—nor is it that interesting for a neutral spectator to the debate who wants to know what she should do or think about synthetic biology. What is sought is the conclusion that r is not a reason to take a to synthetic biology. Getting there takes one more step.

The conclusion as stated in (5) is a conditional statement of the form *if A, then B.* A and B here stand for claims—so an example of a fleshed out version could be *if* it is always wrong to kill, *then* the death penalty is wrong. When faced with a conditional statement, two routes are possible. First, one may claim that A is true and infer, *modus ponens*, that B is true. Or one may claim that B is false and infer, *modus tollens*, that A is false. In the killing/death penalty example, one may affirm that killing is always wrong and infer that the death penalty is wrong (*modus ponens*). Or one may deny that the death penalty is wrong and infer that killing is not always wrong (*modus tollens*).

This means that from the conclusion of the Continuity Argument—'if we ought not to take a with respect to E, then R is not true of SB'—two routes are possible: (i) claiming that we should not take a to E, and inferring that R is not true with respect to synthetic biology; or (ii) claiming that R is true with respect to synthetic biology, and inferring that we ought to take a to E as well. In other words, we can just as easily use a Continuity Argument where a is, say, banning a technology and r is the fact that the technology involves manipulation of DNA to show that we should in fact ban E, as to show that R doesn't apply to synthetic biology. In other words, for all this version of the Continuity Argument tells us the correct conclusion may be that we ought to ban agriculture.

Whether we should make the *modus ponens* or the *modus tollens* inference depends on how plausible the respective claims are. In the version of the Continuity Argument sketched just above, the first claim—that agriculture should not be banned—is a point of view that few of us can imagine giving up on. After all, agriculture is *the* way in which humanity feeds itself. So it is fairly plausible that the *modus ponens* inference is the one most would be inclined to make. But some environmental philosophers have been willing to give up on agriculture, even while recognizing its importance for humanity. For example, the Norwegian philosopher and 'Deep Ecologist' Arne Naess famously argued that the aim of securing the flourishing of non-human life and ecosystems was sufficient reason to make a sizeable reduction of the human population "a high priority" of policy in industrial societies (Naess 2005, p. 44).

The main point of interest here, however, lies not in such substantive issues. The point is rather that a Continuity Argument is not sufficient to show that we ought not take a with respect to synthetic biology. The mere fact that established technologies exhibit the same characteristic that you find problematic about synthetic biology does not show that you are wrong to find that characteristic problematic. Perhaps you just needed the clearer case of synthetic biology to realize how problematic that feature of an established technology, e.g. agriculture, actually was. What the Continuity Argument (if sound) *can* show is the implicit commitments that your insistence that r is a reason to be critical of synthetic biology carries with it with respect to other technologies and practices of which r is also true. And these commitments can potentially be quite extensive.

3.2 Matters of Degree

I said above that synthetic biology might simply provide a *clearer case* of r than the cases one has previously considered—and that this could make someone realize that a well-established technology that had this feature was problematic for the same reason. One thing 'a clearer case of r' could mean is that r is in some sense present *to a larger degree* in synthetic biology than in E. Such difference in degree might not only make someone see E as problematic because of r—it might also justify the judgment that r is a reason to take a in the case of synthetic biology, but not in the case of E. In other words r to degree d_2 could be a reason to take a, while r to degree d_1 is not (where $d_1 < d_2$).

A case in which matters of degree clearly make a difference is *expensiveness*. Suppose that we agree that a restaurant visit is expensive if it costs more than 100 € per person. Clearly, expensiveness might provide a reason r to take a certain a, e.g. not to go to the restaurant. Still, it is not irrelevant *how* expensive a restaurant is. It is quite coherent to think that in the case of *Chez Jacques*, costing 100 €, its expensiveness is not a reason not to go, while thinking in the case of *The Lean Pig*, costing 500 €, expensiveness is a reason not to go. A Continuity Argument for this case would not work, since taking r, expensiveness, to be a reason not to go to *The Lean Pig*, does not commit us to taking expensiveness to be a reason not to go to *Chez Jacques*. In other words, premise (2)—'if r is a fact about *CJ* and a fact about *LP*, then if R is true with respect to *CJ*, then R is true with respect to *LP*'—of that Continuity Argument would not be true.

Interestingly, Beth Preston explicitly rejects that degree makes a difference. As mentioned in Sect. 2.2, the case she is interested in is the proposed r that the advent of synthetic organisms would blur the boundary between artefact and organism. On domesticated plants and animals and synthetic organisms, she writes: "They are all biological artifacts—the differences are merely a quantitative matter of how much of their function and structure is under our control" (Preston 2013, p. 658). Of course, Preston is not simply making a mistake here in not taking the difference in degree to undermine her argument. Not every case is like the case of *Chez Jacques* and *The Lean Pig*. But the argument has to be made that the case one has in mind is not like the restaurants case. There are three possible strategies one might take in making such an argument. I'll first present those strategies and then briefly discuss how Preston deals with the issue.

The first strategy is the most straightforward one, namely arguing that the degree of r does not in fact differ between synthetic biology and E. In a restaurants case, if both restaurants cost 200 €, then expensiveness *by itself* cannot be a reason not to go to one, but not a reason not to go to the other.[5]

The second strategy is to argue that there is a limit of r above which it provides a reason to take a, and that both E and synthetic biology are on the same side of this limit with respect to the level of r. The limit could be a fuzzy one, as long as it is agreed that both E and synthetic biology are on the same side of the limit. As an example, suppose you are deciding where to buy a house, and do not want to live somewhere where a natural disaster might occur. Presumably, any location has *some* chance of being hit by a natural disaster, but you only are only concerned to avoid a location where the chance is *high*, i.e. above a certain level. If two houses are both above the level, the chance of a natural disaster occurring is a reason to avoid buying either of the two houses, but it cannot be a reason to avoid buying one house but not a reason to avoid buying the other.

The third strategy denies that r (at least within this specific R) works like expensiveness does in the restaurants example. That is, r is a reason to take a whenever it is present, no matter to what degree, or r might be a feature that essentially does not admit of degrees. An example of the first kind might be rights: the fact that a is a violation of someone's rights is often thought to be a reason not to do a no matter how small a violation it is. Examples of the second kind are legion; the fact that one's dog died is a reason to be sad, and the dog is either dead or it isn't—it cannot be more or less dead.

Preston can be interpreted as using either of these strategies. She discusses matters of degree in more detail in the course of arguing against the claim that synthetic biology is problematic for ontological reasons. The idea she is opposing is that the fact that synthetic biology gives rise to entities that are neither purely artificial nor purely natural—i.e. synthetic organisms—is a reason to be wary of SB. Preston does not deny that synthetic organisms are ontologically 'blurry'. Instead, she argues that everyone accepts that the distinction between natural and artificial is not a strict one, but is a matter of degree. She then argues that the "qualitative

---

[5]It should be noted, however, that the structure of reasons can be quite complex. Suppose that *Chez Jacques* is a three-star Michelin restaurant, while *The Lean Pig* is a run-of-the mill brasserie. In that case, the judgment that *The Lean Pig* is too expensive (i.e. that its expensiveness is a reason not to go) while *Chez Jacques* is not is intuitively quite plausible, even if they both cost the same (e.g. 200 €). One way of interpreting this is that the restaurants' respective status provides reason for judging one (*The Lean Pig*) to be expensive, while the other was not ('*Chez Jacques* is actually *cheap* for a three-star Michelin', someone might say). So expensiveness becomes a property that is relative to the product. Alternatively, we might hold on to the idea that anything above 100 € is expensive, but then argue that the fact that *Chez Jacques* is a three-star restaurant *defeats* or *attenuates* the reason-giving force of expensiveness. Here, the different status of the restaurants provides reasons for thinking certain things about how other features they have—here, their expensiveness—count in favour of or against going.

break" at which "ontologically problematic" entities come into existence occurs with the advent of agriculture. This is because agriculture entails domestication, and because domesticated plants and animals are ontologically blurry in the relevant way. Such organisms are simultaneously somewhat natural and somewhat artificial —natural *qua* organisms, artificial in virtue of being essentially shaped by human intervention. She then recognizes that there is a difference between agriculture and synthetic biology in that the latter is more "refined" since it uses better knowledge and more precise techniques for its interventions (Preston 2013, pp. 650, 651).

Preston could here be interpreted as using any of the three strategies. She at least thinks that synthetic biology and domestication lie on the same side of a relevant limit—the "qualitative break". That break is marked by our determining the nature of *species* of organisms, i.e. our shaping organismic DNA systematically. This interpretation matches the second strategy above. On a stronger interpretation, matching the first strategy, there is no further distinction in degree of the relevant sort. Hence synthetic biology and domestication share the same level of blurriness. On a still stronger interpretation, blurriness itself doesn't even admit of degrees, either in general[6] or insofar as blurriness is supposed to be a reason to be wary. That matches the third strategy. Which interpretation is best depends on whether Preston thinks ontological blurriness and 'refinement' are the same thing—i.e. whether being a result of more refined intervention (such as synthetic biology) moves an entity further into the territory of blurriness. I will not try to answer this question, but I will say more about the sort of argument Preston is discussing in Sect. 4.

3.3 Pro Tanto *Versus All Things Considered*

In Sect. 3.1 I discussed the problem of deciding between *modus ponens* and *modus tollens* given that the conclusion of the Continuity Argument is a conditional (if… then) statement. In some cases, I suggested, the plausibility of one of the claims— that we ought not take a with respect to E—was so high that most would be prepared to accept the *modus ponens* version and hence arrive at the desired further conclusion that r is not a reason to take a with respect to synthetic biology. This would be the case, for example, where the proposed a is a ban and the established technology an indispensible technology like agriculture. So one might be tempted to accept, at least, that there are no r-terms that (i) apply to agriculture as well as synthetic biology, and (ii) would count in favour of a ban on synthetic biology.

However, this would be too hasty a judgment. The antecedent—that we ought not to take a with respect to E—is an *all things considered* judgment. But a Reason does not only apply where we ought, all things considered, take the a in question. To the contrary, most reasons are *pro tanto* (or, synonymously, *contributory*)

---

[6]In other words, the boundary between blurriness and non-blurriness is *not* a blurry boundary.

reasons. The expression '*pro tanto*' is Latin for "for so much", or more colloquially "for what it's worth". Its usage in ethics is due to Shelly Kagan. On Kagan's definition, a *pro tanto* reason "has genuine weight, but nonetheless may be outweighed by other considerations" (Kagan 1989, p. 17). Being *in fact* outweighed is not a condition for being *pro tanto*. A reason for a is 'outweighed' when there is some other reason r' that counts in favour of not-a. It might do so by counting in favour of not taking a directly, as when my being ill is a reason not to go to the office on Saturday that outweighs my reason to do so (e.g. that I have a deadline Monday). Alternatively, outweighing might take the form of r' counting in favour of some a' that is incompatible with a, as when the fact that the weather is good is a reason to go to the beach that outweighs my reasons to go to the office on Saturday. In a case where r is outweighed, I ought not, all things considered, take a (or, as some put it, I have *most reason* to do not-a).

Since reasons are *pro tanto*, we cannot conclude that a reason to take a does not exist simply from the fact that we ought not take a all things considered. Drawing that conclusion would be mistaking *pro tanto* for *prima facie* reasons—the latter being reasons that seem, on the first look, to genuinely count in favour of a, but might later turn out not to do so. It is a mistake to think that a reason's being outweighed means that it was no reason at all. Consider the case of heroin use. Presumably, being high on heroin is highly pleasant, and this counts in favour of doing heroin. For most of us, however, other reasons, e.g. the risk of addiction or death, outweigh the pleasantness. Still, pleasantness *is* a reason, and it may come to the fore if, for some reason, the reasons not to do heroin lose some of their weight (say, if you were in the last stages of a terminal illness).

The important point is that the Continuity Argument is supposed to be about (*pro tanto*) reasons. If r is merely outweighed with respect to E, then premise (3) is false, since in that case R might be true with respect to E even though we ought not take a to E. The proponent of the Continuity Argument needs to *show* that if R applied to E, then we should take a to E. In other words, they owe an argument that shows that r could not simply be outweighed in this case.

4 The Continuity Argument and Human Intervention

I want now to apply the insights of Sect. 3 to a set of versions of the Continuity Argument, namely those where the proposed r-term is the fact that synthetic organisms are the products of human intervention. By their being 'products of human intervention' I mean that these organisms' genotype is, at least partly, determined by more or less deliberate human action. This is a common basis for criticism of synthetic biology, and it is addressed by all the authors described in Sect. 2. I have already touched upon Preston's arguments regarding the issue (in Sect. 3.2). My examples from public debate both concerned human intervention —"screwing with nature" (Cracked.com 2014) and "tinkering with [...] DNA" (Scientific American 2013). Holtug discusses unnaturalness, but I think what he has in mind is close to human intervention—modified organisms are unnatural because they are the products of human intervention. And as noted, Douglas et al. reject the fact that synthetic organisms are "made to specification" as a possible r-term early in their paper. They further discuss an interpretation of the 'playing God' objection according to which the problem is "exceptionally grandiose and hubristic attitudes" that are embodied in a human "drive to mastery"—i.e. a desire to intervene where one should not (Douglas et al. 2013, p. 690).

So how does a Continuity Argument aimed at human intervention stand with respect to the issues discussed in the previous section? First of all, it is clearly possible that the *pro tanto*/all things considered distinction could be relevant here, since it could be so in practically all cases. But more than that, the fact that the E in question is often agriculture, and that agriculture has the feature of being necessary for our survival, means that any r will *typically* or even *always* be outweighed with respect to E. I see no reason why critics using human intervention as an r-term could not view the intervention that agriculture embodies as sad, but unavoidable (they might even lament that humans ever started being agriculturalists, without thereby committing themselves to the view that we ought to abolish agriculture). When E is a less foundational technology, as it is for both Douglas et al. and Holtug, it becomes less obvious that there are strong reasons for being in favour of E (that is, in favour of not-a, since a is some negative attitude or action). But sometimes, these technologies are argued to be a necessary condition for feeding a population of the size that exists today (to the extent that it *is* actually adequately fed), or for feeding the global population at some time in the future (Borlaug 2000). If these claims are true, the counterweighing reasons in favour of selective breeding, or genetically modified crops, might be (almost) as strong as those in favour of agriculture.

Above all, the *pro tanto*/all things considered distinction brings to the fore the need to talk and think about what (*pro tanto*) Reasons there are directly, and to be cautious about using all things considered judgments. However, doing so might exacerbate the difficulty in deciding between *modus ponens* and *modus tollens*. Correctly interpreted, what we know from a Continuity Argument is that if r is not a reason to take a to E, then r is not a reason to take a to synthetic biology either. That is, in this case, if human intervention is not a reason to take a to the E in question, the human intervention is not a reason to take a to synthetic biology either. But often, our commitment to the claim that a Reason applies to a case is weaker than our commitment to all things considered judgments, in part because some Reasons may make very little difference. If that is the case, it is easier for the opponent of the Continuity Argument to simply bite the bullet, accept the *modus tollens* inference and conclude that we do have some reason, namely the fact that E embodies human intervention, to take a to E—for example, as mentioned, that intervention *does* count in favour of ending the practice of agriculture. Of course, the critic cannot use this strategy to defend *all* Reasons that she takes to apply to synthetic biology— rationality requires that there must be *some* Reason that applies in the synthetic biology case and not in the case of E if one wants different all things considered judgments about the two cases.

Finally, could matters of degree be important? As discussed briefly in Sect. 3.2 on Preston, there is an ambiguity here. On the one hand, intervention could be an either/or situation—an organism's DNA either is or is not (partly) determined by human intervention. On the other hand, there is a sense in which interventions are a matter of degree. Preston, recall, allowed that there is a difference in how "refined" our intervention is, where being more refined amounts to using better knowledge and more precise techniques for shaping organisms' genetic makeup. Higher refinement means that our interventions become more targeted; they determine as little of the genetic makeup of the organism as is necessary for achieving the goal of the interventions. In addition to refinement, we might think of the *scope* of intervention as another dimension in which degree distinctions can be made. An intervention is larger in scope when more of the organism's traits are there because of the intervention. Notice that there is a tension between refinement and scope insofar as less refined interventions tend to be larger in scope because they change traits that they did not intend to change. This means that a higher degree of intervention on one dimension may alleviate worries based on higher degrees of intervention on the other dimension. For example, a common worry in the debate concerning genetically modified food is that the modifications might alter other parts of the genome than those responsible for the traits we want (e.g. drought resistance), and therefore create unforeseen harmful traits. Increasing the *refinement* of the intervention may here alleviate the worry precisely by decreasing the *scope* of intervention.

I think it is reasonable to see the biotechnological development as a movement towards higher refinement of intervention. While it is difficult to say anything definite about how domestication got started, a plausible suggestion is that the first changes that human beings caused in the DNA of organisms were side effects of our eating plants with desirable qualities and consequently planting their seeds by accident (see Bulleit 2005 for an argument to this effect). If this is true, early domestication exhibited a zero degree of refinement, since neither knowledge nor technique was employed. Relatively soon after, more refined interventions emerged in the form of intentional selection of crops and animals with desirable traits for planting and breeding—basic selective breeding. In the 18th century, selective breeding methods were further improved by the use of scientific methods (e.g. exact measurements of wool or meat yield from sheep and cattle)—so much so that these developments served as inspiration for the ground-breaking work on heredity and evolution of Mendel and Darwin (the latter of whom coined the term 'selective breeding'). Selective breeding methods have since developed further. The biotechnology revolution represents yet another increase in refinement by making it possible to make very fine-grained selections, down to the level of individual genes. And at least some of the research programmes falling under the heading of synthetic biology embody the completion of the progression towards greater refinement: The creation of organisms with all and only those genes that human beings have decided they should have.

The question now is whether this increase in degree matters for our ethical evaluation of synthetic biology—that is, whether synthetic biology's level of refinement of intervention provides a reason for taking some a that domestication's or selective breeding's level of refinement does not. Preston argues explicitly that the difference in degree of control between bog-standard agriculture and synthetic biology does not matter, since agriculture was the "tipping point" in all relevant respects. As she herself points out, her arguments are especially aimed at those who imagine that rolling back of modern biotechnologies (and, for that matter, modern, industrialized agricultural techniques) would change humanity's way of relating to nature from one based on "control" to one based on "trust" and "nurturing" (Preston 2013, p. 658). For such people, it seems, any degree of control—that is, of intervention in my sense—is troubling.

But not all critics of synthetic biology need hold such a view. I think there are two plausible ways of understanding intervention-based worries that would make the difference in degree relevant. First, the problem might be related to the increased powers that more refined techniques of intervention bring us. A straightforward worry might be, for example, that synthetic biology will be used mainly with the interests of business in mind, and not promote (or perhaps even contradict) the interests of animals, the environment and 'ordinary people'. Of course, existing biotechnology may already be used in this way (think of chickens bred to be so meaty that they cannot carry their own weight). On that basis, one might argue that each new power merely provides a reason (or even demand) to use it carefully and right. But beyond this, one might use this line of reasoning as a way of grounding scepticism towards new technologies as such. If one is sceptical of the willingness and ability of scientists, business and the political system to use the powers that synthetic biology brings in the right way, it is reasonable to be critical of allowing synthetic biology in the first place, or accept it only on the condition that it is used only in certain ways.

Less straightforwardly, there is a connection between refinement and responsibility. A plausible conception of moral responsibility has it that an agent is morally responsible for an event to the extent that the occurrence of the event can be explained by the agent's motivations (Björnsson and Persson 2012). When the refinement of our interventions increase, the fact that an organism has a certain trait is more likely to be explained by the agent's motivations. In the case of some projects belonging to the umbrella of synthetic biology, *all* traits would be so explained. For example, if we were to create a sentient fully artificial organism, the fact that the organism is sentient is explained by the motivations of the designer. Consequently, the designer bears moral responsibility for sentience. And similarly for all other traits, be they good or bad. It is not exactly clear what reasons are provided by the fact that more refined interventions ground extra responsibilities, but I think it is plausible that some critics really worry about whether we—our society or humanity as such—are well-suited for taking on these extra responsibilities.

Second, there are views that explicitly view *too much* control, rather than control as such, as the problem. Control is objectionable only when it exceeds certain limits, either in the sense that specific things are controlled that ought not be controlled, or in the sense that it realizes or aims towards too much—perhaps ultimately *universal*—control. For example, G.A. Cohen argues that "[i]t is essential that some things should be taken as given: the attitude of universal mastery over everything is repugnant, and, at the limit, insane" (Cohen 2011, p. 207). This conservative or conservationist view is familiar, I think, from the debate about human impact on the environment, as well as from debates on bioethical issues such as human enhancement.

In our case, the argument made is that synthetic biology is tantamount to, or at least invites, the attitude of or desire for universal control. For instance, Boldt and Müller suggest that what gives rise to ethical concerns regarding synthetic biology is the move from manipulation to creation. Where earlier technologies aimed at "softening unpleasant edges of" and "adding extra value to" existing organisms, synthetic biology "brings into existence something that counts as valuable from our point of view" (Boldt and Müller 2008, p. 388). The latter, but not the former, invites a view of nature as "a blank space to be filled with whatever we wish". For worries of this type, the degree of intervention clearly makes a difference—a difference that makes sense of taking different attitudes to synthetic biology and other technologies. Of course, there may be other problems with these kinds of criticisms of synthetic biology, but those are not vulnerable to a Continuity Argument.

5 Conclusion

I have argued above that there exists a common line of argument, which I have dubbed the Continuity Argument, that is used in defence of synthetic biology. It consists in denying that some fact that is claimed to be a reason to be critical (in some specific way) of synthetic biology is in fact such a reason. The denial is supported by arguing that the same Reason relation—the fact r counts in favour of the critical action or attitude a—should hold with respect to some established technology E that we are not, or have not been, inclined to be critical of in the relevant way. I have argued that this line of argument has some limitations, especially with respect to three features: (i) It cannot determine whether to stop being critical of synthetic biology or to *start* being critical of E; (ii) It does not take into account the possibility that matters of degree of r make a difference; and (iii) It does not adequately distinguish between *pro tanto* reasons and reasons all things considered.

What can we learn from this if we are interested in the ethical assessment of synthetic biology? I think there are two general ways to go. The first is to hold on to the idea, which has been assumed throughout this chapter, that Continuity Arguments are, in fact, arguments—that is, that they are a set of premises intended to establish a conclusion. If we do this, my discussion of the argument reveals the circumstances under which the argument is and is not sound. It shows the user of the Continuity Argument what more, beyond the mere fact that r applies to E as well as synthetic biology, she needs to establish. Similarly, it shows the critic of synthetic biology what pitfalls she needs to avoid when her proposed r is shared by an established technology. If R is likely to be outweighed in the case of E but not synthetic biology, or if r differs in degree between synthetic biology and E, then it is likely that a Continuity Argument is not sound. This further suggests that technologies that are very similar to synthetic biology, such as recombinant genetic engineering, are better candidates for successful Continuity Arguments than those which are less similar, such as agriculture. Unfortunately this tends to diminish the strength of the Continuity Argument insofar as we are less certain that genetic engineering is ethically unproblematic than that agriculture is. Hence the jump from the demand that evaluation of synthetic biology and E be the same to the conclusion that synthetic biology is ethically unproblematic—i.e. from the establishment of the conditional conclusion (5) to the *modus ponens* inference from that—is more difficult to make.

The second way to go is to give up on the idea that Continuity Arguments are arguments in the sense described above. Instead, we might view them as heuristic devices that help us think more clearly about the issues surrounding synthetic biology. Noticing that domesticated plants and animals are products of human intervention, for example, may be no more than a way of making ourselves and others reflect on whether we really think that the fact of intervention is morally salient. On this interpretation, the level of similarity between synthetic biology and E is less important, since we are merely reminding ourselves of the kind of ethical role (which may include being ethically irrelevant) that r plays in other cases. The use of comparisons with established technologies, on this view, should be seen as playing the same role as thinking about historical analogies does for political decision makers. Here, no one would suggest that there is some kind of rational demand for aligning future actions with past ones, even past ones that one approves of. Instead, the past is a resource for understanding the current situation better, which enables a better decision about the current situation on its own merits. If this weaker way of understanding the Continuity Arguments is attractive, we should also heed the advice of those decision makers who have used history well: To read history widely, and not fixate on a single analogy. To do so would be to ignore sources of knowledge that could improve our decisions, whether about politics or synthetic biology.

References

Björnsson G, Persson K (2012) The explanatory component of moral responsibility. Noûs 46 (2):326–354

Boldt J, Müller O (2008) Newtons of the leaves of grass. Nat Biotechnol 26(4):387–389

Borlaug N (2000) The green revolution revisited and the road ahead. 30th Anniversary Nobel Lecture. http://www.nobelprize.org/nobel_prizes/peace/laureates/1970/borlaug-lecture.pdf. Accessed 12 Dec 2014

Bulleit RW (2005) Hunters, herders and hamburgers: the past and future of human-animal relationships. Columbia University Press, New York

Cracked.com (2014) 20 Annyoing 'Modern' Trends That Are Older Than You Think, #7. http://www.cracked.com/photoplasty_763_20-annoying-modern-trends-that-are-older-than-you-think_p3/. Accessed 5 June 2015

Cohen GA (2011) Rescuing conservatism. In: Wallace RJ, Kumar R, Freeman S (eds) Reasons and recognition: essays on the philosophy of T.M. Scanlon. Oxford University Press, Oxford

Cuneo T (2007) The normative web: an argument for moral realism. Oxford University Press, Oxford

Douglas T, Powell R, Savulescu J (2013) Is the creation of artificial life morally significant? Stud History Philos Biological Biomed Sci 44(4B):688–696

Holtug N (2009) Creating and patenting new life forms. In: Kuhse H, Singer P (eds) A companion to bioethics (2nd edn). Wiley-Blackwell, Oxford

Kagan S (1989) The limits of morality. Oxford University Press, Oxford

Naess A (2005) The deep ecology movement: some philosophical aspects. In: The Selected Works of Arne Naess, vol X: Deep Ecology of Wisdom. Springer, Dordrecht

Preston B (2013) Synthetic biology as red herring. Stud History Philos Biol Biomed Sci 44(4B):649–659

Scanlon TM (2014) Being realistic about reasons. Oxford University Press, Oxford

Scientific American (2013) Labels for GM foods are a bad idea. http://www.scientificamerican.com/article/labels-for-gmo-foods-are-a-bad-idea/. Accessed 23 Feb 2015

Skorupski J (2010) The domain of reasons. Oxford University Press, Oxford

Synthetic Biology Between Engineering and Natural Science. A Hermeneutic Methodology for Laboratory Research Practice

Michael Funk

1 Introduction

In this chapter, I investigate some basic methodological and epistemological issues of synthetic biology and its relations to a theory of engineering and laboratory research from a philosophical point of view. During the summer school on which this volume is based we tried to shed light on the differences and similarities between synthetic biology and biotechnology or genetic engineering—and more generally the relations between synthetic biology, engineering sciences and natural sciences. With regard to epistemological and methodological aspects of synthetic biology, a need for clarification was identified: How can we understand the practice and generation of knowledge in synthetic biology? Is synthetic biology engineering science or natural science? What is the relation between engineering and natural sciences? What is the methodology of synthetic biology? With this chapter, I want to contribute some possible answers to these questions. They cause new challenges because most philosophical reflections of science have been focused on natural sciences but not engineering sciences. Therefore I am going to develop a hermeneutic methodology of synthetic biology with respect to the works of Don Ihde, Bernhard Irrgang, Hans J. Münk, Hans Poser and others.

2 Three Methodological Turns

In the philosophy of science, physics was long seen as a paradigmatic ideal case of natural scientific research and thus shaped the classical philosophical approaches. Many epistemological investigations have been oriented on terminologies, theories

M. Funk (✉)
Philosophy of Technology, Technical University Dresden, Dresden, Germany
e-mail: funkmichael@posteo.de

© Springer International Publishing Switzerland 2016
K. Hagen et al. (eds.), *Ambivalences of Creating Life*, Ethics of Science and Technology Assessment 45, DOI 10.1007/978-3-319-21088-9_17

and experimental methodologies of physics—and not of biology or engineering (Poser 2004, pp. 175–184, 2012, pp. 312–333). During recent years, a number of philosophical investigations have also been emphasizing methodologies of engineering sciences (e.g. Banse et al. 2006; Irrgang 2010; Kornwachs 2012). But even though there is a growing number of efforts to understand the methodology of engineering, a general philosophical approach to the epistemology of engineering is still "work in progress" and a desideratum—especially with respect to the new general paradigm of biology (Poser 2004, pp. 175–178, 2012, pp. 312, 313, 336, 337).

The methodological turn from physics to biology in philosophy of science is one of three puzzle-pieces that form the background for the development of a hermeneutic theory of synthetic biology. Another puzzle-piece can be found in the development of genetic engineering and biotechnology since the 1970s in combination with a new technology-oriented form of laboratory research practice, which became the historical and technical condition for current synthetic biology. Synthetic biology is a hybrid of engineering and natural sciences which is constituted by a plurality of different methodologies (Irrgang 2004, pp. 289–294). This hybrid-status leads to epistemological problems that are more specific to synthetic biology—this seemed to be a general conclusion during the Summer School leading to this book—than the ethical questions. A similar hypothesis has also been elaborated by Hans J. Münk, who claims that the hybrid status causes the most difficult new questions of synthetic biology in philosophy. Evidenced by statements of protagonists of synthetic biology during the last years, another hypothesis—formulated by Irrgang (2003)—becomes more and more plausible: related to the rise of synthetic biology and the implementation of a general principle of engineering is a paradigmatic change of biology itself. Technological procedures and models become dominant in the laboratory practice of synthetic biology. Synthetic biology is biology shaped by an engineering-paradigm (Irrgang 2003, pp. 131–135, 2004, p. 285, 2012, pp. 347–353; Münk 2011, pp. 106–117). The second puzzle-piece is thus a methodological turn from classical natural sciences to engineering and laboratory research.

In contrast with the situation where technological procedures and models are dominant in the laboratory, classical experiments have often been described as tools for verification and falsification of scientific theories. But in fact even classical experiments are inherently forms of technical practice, not theory-dominated observations. The laboratory research practice of synthetic biology is the next technological step as it is based on the complete instrumental and technological embedding of empirical research in combination with constructive elements and systematic application of information- and computer-technologies (Irrgang 2003, pp. 69–78, 2004, pp. 285–294, 2012, pp. 348–353). Synthetic biology is the transformation of biology to a specific new form of technological practice, which is much more than experimental verification or falsification of logical theories. This is the third turn: from theory to practice (Irrgang 2003, pp. 129–133). In other words: knowing means acting, and if we want to understand synthetic biology, we need to understand the actions of synthetic biology.

At this point we can come to a first philosophical result. Epistemology and methodology of synthetic biology needs to implement three turns:

1. a turn from physics to *biology as new general paradigm*,
2. a turn from classical natural sciences and its experiments to an *engineering* and new *technological laboratory paradigm*, and
3. a turn from research theory to research *practice*.

Looking back into the history of 20th-century philosophy, we can see that the first turn from physics to biology was found in the works of Mathias Gutmann, Peter Janich, Michael Weingarten and others. The book *Wissenschaftstheorie der Biologie* (Janich and Weingarten 1999) is paradigmatic for the turn. Close to Ian Hacking, Bruno Latour, Steve Woolgar, Albert Borgmann, Carl Mitcham and others, Don Ihde developed the second and third methodological turns, to laboratory sciences and technical practice. Paradigmatic for these turns are his books *Technics and Praxis* (Ihde 1979), *Technology and the Lifeworld* (Ihde 1990) and *Instrumental Realism* (Ihde 1991), in which Ihde elaborated a lifeworld-based techno-scientific approach, focusing on the hybrid forms of instrumental practice in modern scientific laboratories. Following these turns, Bernhard Irrgang presented a *Technikhermeneutik* of genetic engineering, biotechnology and synthetic biology in his book *Von der Mendelgenetik zur synthetischen Biologie* (Irrgang 2003).

On the basis of these methodological turns and their philosophical representatives, I want to develop a hermeneutic heuristic of synthetic biology. As it is shaped by an engineering-paradigm such a heuristic must include basic characteristics of engineering methodology. Therefore, in the next step some fundamentals of a theory of engineering sciences are summarized.

3 Some Fundamentals of a Theory of Engineering Sciences

Understanding the methodology of synthetic biology means understanding the epistemology of engineering: engineering knowledge, technological practice and its relations to laboratory research procedures. In this process it is important to keep in mind that engineering—even classical physics-oriented engineering—is always more than just applied natural science (Poser 2004, p. 185, 2012, pp. 313–318). Laboratory research is more than just another form of experimentation, and any kind of experimentation is more than verification or falsification of scientific theories. With genetic engineering and biotechnology, genuine hybrids of technological practice and laboratory research have been generated. They are now being transformed into synthetic biology as a next step of the technoscientific development towards a practice of technoresearch (Irrgang 2004, p. 296, 2012, pp. 348–359).

It is important to note that on a general level, engineering is oriented at technological rules of procedure ("Verfahrensregeln"). Technical practice needs to be efficient and successful. Therefore, there are no isolated logical criteria of truth for the rules of procedure, their verification is pragmatic. This is one general difference between engineering and natural sciences: engineering needs to be pragmatically successful, whereas natural sciences want to generate theoretical and universal truth (Irrgang 2003, pp. 176–183, 2004, pp. 286–288). Not the artifacts, but the aims of engineering and natural sciences are different: natural sciences want to formulate theoretical laws and clarify universal structures of the world, while engineering sciences want to find solutions for practical problems (Poser 2012, pp. 317–324). In this sense, technological progress and progress in laboratory research means progress of problem solving, a process where scientific discoveries, technological and economic innovations go hand in hand (Irrgang 2003, p. 58). While classical natural sciences find their objects predefined in the natural environment, engineering-oriented laboratory sciences fundamentally generate and constitute their research objects (phenomena, visualizations, material objects, organisms, procedures, models, know how) actively in a technological situation (Irrgang 2003, pp. 75–77, 2004, pp. 287–292; Poser 2004, p. 178, 2012, p. 315).

Those technological situations in which research objects are constructed are shaped by specific conditions. Each condition needs to be interpreted adequately from a teleological point of view in order to fulfill its pre-defined technical aims and functions (Irrgang 2003, p. 158; Poser 2012, pp. 319–321). The focus of synthetic biology is not an experimental procedure, but the construction of organisms—the objects of the research practice itself. This leads to a significant methodological consequence: synthetic biology cannot be understood in terms of a strong dualism between nature and culture. Boundaries between natural objects and technological objects become blurry (Irrgang 2003, pp. 219–222, 2004, p. 290, 2012, p. 354). The reason for the blurry boundary is not a new ontology of things, but the research practice of technological laboratory research—in this case synthetic biology.

Understanding and interpreting technological situations in terms of technological constitution of biological objects leads to a more complex methodological heuristic and hermeneutics of interpreting specific levels of organic materials and structures (Irrgang 2003, p. 161, 2004, p. 291). A pragmatic framework is needed in order to understand concrete organic situations in their technological meanings. In the next section, I am going to re-frame fundamentals of such a hermeneutic heuristics. But before that, I want to close this section with a short summary of some general characteristics of engineering sciences:

1. engineering is more than applied natural science, it is an epistemologically specific *form of technological knowledge,*
2. engineering is based on *rules of procedure, efficiency, functionality,* and the *criteria of pragmatic truth: successful means-end oriented actions,* and *technical tests,*
3. there is *no dualism between technology and nature* as engineering-oriented laboratory-sciences *construct their own research objects:*

   (a) phenomena
   (b) visualizations
   (c) material objects

(d) organisms
(e) procedures
(f) models
(g) know how

4. engineering intents solutions for *concrete problems*, and
5. therefore, engineers must be able to *understand concrete technological situations* in their specific singularity.

4 *Technikhermeneutik* (Hermeneutics of Technologies)

Tasks of understanding specific situations in their functional singularity are characteristic for engineering sciences. Those tasks call for hermeneutical skills, which have traditionally been associated with social sciences (*Geisteswissenschaften*). But equivalent heuristics and competences can be found in natural sciences and engineering as well. The idea of *Technikhermeneutik* (hermeneutics of technologies)—which is more than social scientific methodology—has been discussed by Ihde (1998) and Irrgang (1996). In his book *Expanding Hermeneutics* (Ihde 1998), Ihde combined a post-phenomenological approach with a trans-classical form of so called "material hermeneutics". His main interests have been shaped by embodied knowledge and multistable practices (dual use, etc.) of using scientific tools and interpreting scientific visualizations.

According to Ihde, scientific observations are activities which need to be interpreted in order to constitute a technological and scientific result. Pure facts do not tell anything about their relations to scientific theories or rules of procedure. Also, visualizations achieved by laboratory instruments (microscopes, gel electrophoresis apparatus, quantitative polymerase chain reaction instruments etc.) must be interpreted in relation to intended aims, technical functions or material organic phenomena. Technique is described as practice and handling of technical tools in concrete bodily and cultural situations. Technological interpretations are not determined, not linear, and include aspects of instability and multistability (dual use etc.). They are shaped by social aspects (values, expectations etc.), institutional aspects (funding, law etc.), research traditions and local research cultures, methodologies and technological processes, and personal skills and competence of the actors (Ihde 1979, 1990). I want to emphasize epistemological issues of *Technikhermeneutik* related to the last two points: methodology and personal skills.

A hermeneutic heuristic of synthetic biology needs to include methodology and skills related to concrete *levels of organic phenomena*. It makes a difference whether a biologist works with isolated molecules (e.g., pieces of the DNA) or whole organisms (such as fruit flies or zebrafishes). Each level of organic structure demands specific skills of interpretation. A hermeneutic heuristic of synthetic biology needs to consider this circumstance by differentiating several levels with genuine conditions for growth and development. The levels are in permanent autopoietic feedback relations. Their interrelations are not causally determined and do not follow linear cause and effect chains. Within processes of self-organization and self-structuring in quasi-teleological development step by step and level by level, internal genetic codes and environmental factors constitute concrete structures. Every currently realized structure can be understood as a starting point for subsequent organic and epigenetic developments (Irrgang 2003, pp. 214–232, 2004, p. 291, 2012, pp. 354–357). In a heuristic of five levels the genuine levels of interpretation can be used as templates for describing and interpreting processes of organic development:

1. genetic code and genomics,
2. proteins and proteomics,
3. cells, cytology and organs,
4. organism, and
5. environment and ecosystems (Irrgang 2003, p. 130, 2004, p. 291).

From an epigenetic and autopoietic point of view, this template involves methodological perspectives. We cannot synthesize all five levels into one causal model. Every level generates a specific perspective for interpretation and understanding of organic development. At the same time, the methodological and technological construction of phenomena differs from level to level. Especially the molecular perspective—which is significant for synthetic biology today—is technologically mediated, because we cannot see "pure and natural" molecules without visualization technologies. As we can learn from Eugene S. Ferguson, technologies, visual thinking and tacit skills of imagination shape bodily forms of engineering knowledge (Ferguson 1992). According to Ihde, the analysis that uses the framework of technological hermeneutics should start with a phenomenology of perception and interpretation of bodily-sensory (*leibliche*) phenomena (Ihde 1998). So it is not the intellectual, theory-oriented understanding that is the first step, but rather, technical interpretation starts with bodily actions in laboratory situations; and these are the situations which can be divided into the five hermeneutic levels listed above. Each level carries a perspective of instrumental realism (Ihde 1991), and all levels together constitute an instrumental hermeneutics of life and its molecular basis (Irrgang 2003, p. 161).

Technical objects, including synthetic organisms, become results and realizations of epistemic hermeneutic actions. Consequently, current theories of engineering and laboratory practice need to be technological multi-level-theories (Münk 2011, p. 117; Poser 2004, p. 191, 2012, p. 330). One of those level-tablets has already been introduced (1. genomics, 2. proteomics, 3. cytology and organs, 4. organism, 5. environment) and is the basis for the visualization of the hermeneutic heuristic of the methodology of synthetic biology in Fig. 1. *Concrete situations* need to be interpreted adequately at every level. Thereby *the actual state of development and the intended aim of development* need to be understood in every singular situation. Following Poser, *Technikhermeneutik* therefore includes three aspects of understanding specific situations: First, *facts* need to be related with *values*. Second, *causes* are related with *aims*. Especially in the latter, rules of

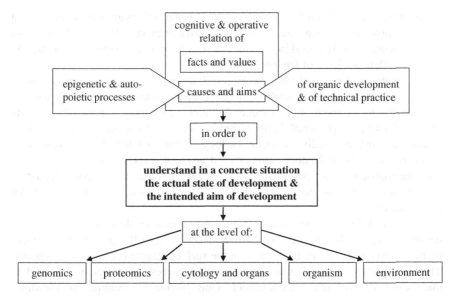

Fig. 1 Scheme of hermeneutic methodology of synthetic biology

procedure, means, functions and concrete conditions are interwoven. Third, both areas of tension—facts versus values, causes versus aims—are brought together in specific situations as *cognitive and operative relations* (Poser 2004, pp. 188–190, 2012, pp. 326–331).

The relation between *facts and values* includes a normative element which also calls for responsibility of engineers and ethical assessment. These are important points; however, the focus of this chapter is on the methodology and epistemology. The relation between *causes and aims* includes, for example, rules of procedure, means, functionality, and efficiency. As these components are general aspects of engineering methodology, I have included two wings in the visualization in order to illustrate the specific characteristics of biotechnology, genetic engineering and synthetic biology. The left wing is directed to the word *causes* and shows the genetic forms of non-linear causality: *epigenetic and autopoietic processes*. On the right side, *aims of organic development and of technical practice* are specified in the second wing. Both aspects need to be interrelated by engineers. There is a pragmatic means-end oriented technical aim—e.g. the construction of a bacteria with a specific detox function—as well as an organic aim—the teleological goal(s) of the development of the organism (for detailed analysis of teleological thinking see Toepfer 2004). Organisms are not static, they are always included in genetic, proteomic etc. processes, and so the engineer does not only need to understand the pragmatic aim of his actions, but also the organic aim(s) of the biological material.

Figure 1 could be understood (with some power of imagination) as a coffee mill: At each step on the engineering-timeline the whole coffee bean (the complex situation) is poured in at the top, shredded by a grinder (following the arrows, grinded by the wings) and cut into five pieces (levels). Insofar, this is an analytic concept in which complex phenomena are dissected into smaller pieces. The coffee mill is not only working once. It's working as often and as long as necessary for solving the whole problem—a kind of feedback-loop.

This is the reason why *Technikhermeneutik* is more than classical hermeneutics: it is multi-perspective interpretation in spiral feedback-loops (classical hermeneutics) related to technologically mediated engineering products and their genuine forms of knowing (material hermeneutics). Genuine for gene technology, biotechnology and especially synthetic biology is the fact that those engineering products cannot be understood at one causal physical level. All five levels and the combination of related perspectives are necessary in order to interpret adequately the relations between genes, proteins, cells and organs, organisms, and the environment.

Personal skills and competences play a major role in these hermeneutic processes and cannot be replaced by theories. According to Borgmann (1984), technological processes can be understood at the surface texture (*Oberflächenstruktur*). This means that we can use and handle a tool successfully without the need to understand its theoretical background. One prominent example is breeding. Breeding is an old skill which has been deeply embedded in a variety of human cultures for more than 10,000 years, and of course it is much older than synthetic biology. Surface texture knowledge shaped the practice of breeding exclusively before the first theoretical rules of heredity were formulated by Gregor Mendel in the 19th century and the DNA-model was developed by Watson, Crick and Franklin in the 1950s. For 10,000 years, before those scientific milestones, breeding was a successful cultural competence (*kulturelle Kompetenz*) on the basis of trial and error, and implicit knowledge (Irrgang 2003, pp. 227–231).

The concept of implicit knowledge—or tacit knowledge—was introduced by Michael Polanyi in 1958 and describes the circumstance that we often know (how to do) things without being able to put it into words (Polanyi 1958, 2009). Body movement or social gestures belong to the tacit dimension as well as perceptual or technical skills. Explicit knowledge, in contrast, describes knowledge which can be explained in strict and true words (logics, mathematical formulas, or natural scientific laws).

Construction of synthetic organisms is the result of implicit knowledge:

1. sensorimotor and perceptual skills, and
2. cultural and social competences,

explicit knowledge which is close to the tacit knowledge-base:

3. rough-and-ready rules, and
4. technical drawings and heuristics, and

explicit knowledge:

5. methodological rules, and
6. experimental rules and technical instructions (Irrgang 2003, p. 133).

These forms of knowledge are interrelated in feedback loops of synthetic biology laboratory research practice. They are what protagonists of synthetic biology need to know in order to be able—metaphorically speaking—to crush the coffee beans into the five pieces (levels) with the heuristic grinder. Implicit as well as explicit knowledge is needed, but starting with biotechnology and genetic engineering, and now with synthetic biology, implicit knowledge seems to become more dominant over explicit knowledge. The reason is that in laboratories of synthetic biology an increasing number of—with Albert Borgmann's words—surface structures are generated. Practice has its own rules. Even if we cannot know all rules theoretically, we can generate an implicit understanding of pragmatic success. And again with imaging technologies or computer-simulations more and more surface structures are generated, which also need to be adequately related to concrete practical situations.

Computers can replace explicit knowledge much better than implicit knowledge. Computers are the better calculators, statisticians or simulators. But they don't tell anything about a successful interpretation of their screen-results in a concrete technological situation. Here, the implicit knowledge of human practitioners comes into play. And the more explicit capacities are replaced by computers, the more important becomes the role of implicit knowledge.

I want to close with a little outlook. By developing a methodological heuristic I have tried to shed some light on the epistemology of synthetic biology as a hybrid of engineering and natural science. In future steps this hermeneutic heuristic could be enhanced by integrating more aspects, e.g.:

1. the similarities and differences between several forms of technical objects and artifacts such as synthetic organisms, robots, machines, or handcraft tools,
2. interrelations between synthetic biology and systems biology: bottom-up and top-down approaches (Münk 2011, pp. 106–110),
3. the impact of standardized components (bio-bricks etc.) in the growing biotechnological industry,
4. the circumstance that products of synthetic biology (and, although to a lesser extent, of genetic engineering)—in contrast with classical breeding—do not stay within the limits of species boundaries and natural evolution; how to interpret artificial organic behavior?,
5. a differentiation of values and ethics, and
6. possible future links between synthetic biology and robotics research or nanotechnologies.

5 Conclusion

In this chapter, I have been developing a hermeneutic methodology of synthetic biology, primarily based on the works of Don Ihde, Bernhard Irrgang, Hans J. Münk and Hans Poser, in order to answer the following questions: (1) How can we understand the practice and generation of knowledge in synthetic biology? (2) Is synthetic biology an engineering science or a natural science? (3) What is the relation between engineering and natural sciences? (4) What is the methodology of synthetic biology? One conclusion is the visualization of a hermeneutic heuristic based on five levels of organic development and their epigenetic and autopietic non-linear causality (see Sect. 4). It is a try to answer the first question on a basis of three turns: from physics to biology, from scientific experiments to engineering and laboratory research, and from theory to practice (see Sect. 2). The second question has been emphasized at the beginning of my chapter and can be easily answered: synthetic biology is engineering. But easy answers are often complicated as well. Yes, synthetic biology is shaped by an engineering-paradigm, but it is another form of engineering than the classical physics oriented forms—such as mechanical engineering. Therefore, characteristics of engineering knowledge and methodology have been described (see Sect. 3) and compared with natural sciences. Engineering, and especially synthetic biology, is much more than applied natural sciences. In synthetic biology, a construction oriented engineering-paradigm and the natural scientific approach of biology form a technoscientific laboratory research hybrid. So, the answer to the third question is: In synthetic biology there is no relation between engineering and natural sciences. Rather, engineering and natural sciences are *the same*—but on a new technological level, which is different to the relations between classical engineering and physics. Answering the fourth question I have suggested a hermeneutic heuristic methodology of synthetic biology. This hermeneutic scheme could be applicable in different concrete situations in order to understand the concrete state of organic development and aim of technical practice at five levels (genomics, proteomics, cytology and organs, organism, environment).

Acknowledgements I'd like to thank the participants of the summer school for the very fruitful and multifaceted presentations and discussions which have contributed to my elaborations in this paper.

References

Banse G, Grunwald A, König W, Ropohl G (eds) (2006) Erkennen und Gestalten. Eine Theorie der Technikwissenschaften, Sigma, Berlin

Borgmann A (1984) Technology and the character of contemporary life. A philosophical inquiry. University of Chicago Press, Chicago

Ferguson ES (1992) Engineering and the mind's eye. MIT Press, Cambridge

Ihde D (1979) Technics and praxis. A philosophy of technology. Reidel, Dordrecht

Ihde D (1990) Technology and the lifeworld. From garden to earth. IUP, Bloomington

Ihde D (1991) Instrumental realism. The interface between philosophy of science and philosophy of technology. IUP, Bloomington

Ihde D (1998) Expanding hermeneutics. Visualism in science. NUP, Evanston

Irrgang B (1996) Von der Technologiefolgenabschätzung zur Technologiegestaltung. Plädoyer für eine Technikhermeneutik. Jahrbuch für christliche Sozialwissenschaften 37:51–66

Irrgang B (2003) Von der Mendelgenetik zur synthetischen Biologie. Epistemologie der Laboratoriumspraxis Biotechnologie. Thelem, Dresden

Irrgang B (2004) Epistemologie der Bio- und Gentechnologie. In: Kornwachs K (ed) Technik – System – Verantwortung. Lit, Münster, pp 285–297

Irrgang B (2010) Von der technischen Konstruktion zum technologischen Design. Philosophische Versuche zur Theorie der Ingenieurpraxis. Lit, Berlin

Irrgang B (2012) Synthetische Biologie und künstliche Organismen. ETHICA 20 (2012) 4:345–361

Janich P, Weingarten M (1999) Wissenschaftstheorie der Biologie. Methodische Wissenschaftstheorie und die Begründung der Wissenschaften. Fink, München

Kornwachs K (2012) Strukturen technologischen Wissens. Analytische Studien zu einer Wissenschaftstheorie der Technik. Sigma, Berlin

Münk HJ (2011) Stellt uns die Synthetische Biologie (SB) vor neue Fragen? ETHICA 19 (2011) 2:99–122

Polanyi M (1958) Personal knowledge. Towards a post-critical philosophy. University of Chicago Press, Chicago/London

Polanyi M (2009) The tacit dimension. With a new foreword by Amartya Sen. The University of Chicago Press, Chicago/London

Poser H (2004) Technikwissenschaften im Kontext der Wissenschaften. In: Banse G, Ropohl G (eds) Wissenskonzepte für die Ingenieurpraxis. Technikwissenschaften zwischen Erkennen und Gestalten. VDI-Report 35, Düsseldorf, pp 175–193

Poser H (2012) Wissenschaftstheorie. Eine philosophische Einführung. 2. Auflage. Reclam, Stuttgart

Toepfer G (2004) Zweckbegriff und Organismus. Über die teleologische Beurteilung biologischer Systeme. Königshausen & Neumann, Würzburg

Epistemological Implications of Synthetic Biology: A Heideggerian Perspective

Martin G. Weiss

1 From Genitals to Laptops

Craig Venter's announcement of the successful replacement of a bacterial genome with a synthetic one in 2010 prompted a huge media reaction. *The Economist* titled "And man made life. Artificial life, the stuff of dreams and nightmares, has arrived", the German weekly *Die Zeit* asked "What is Life?" and declared "now Man can play God." Asked at a press conference about the meaning of his experiment, Craig Venter himself—after a dramatic pause—guessed: "Perhaps it is a giant philosophical change in how we view life."[1]

What synthetic biology is about is best illustrated by the cover of *The Economist*. What we see is Michelangelo's Adam from the famous fresco at the Sistine Chapel depicting the creation of man. But whereas in the original it is the hand of God, which ensouls men, here it is man that generates new life—admittedly not in the old-fashioned way of sexual reproduction, but using a laptop that has literally replaced his genitals.

The illustration captures the alleged essence of synthetic biology as the informatization of life is, at least according to leading figures in this field, its core element: Material DNA is translated into mere information which is then processed, i.e. retranslated into material DNA by computerized gene synthesis machines, and inserted into existing cells, which then replicate. In the words of Drew Endy: "We go from abstract information to physical living design" (Endy 2010, p. 5).

---

[1] Venter's talk is available at http://www.ted.com/talks/craig_venter_unveils_synthetic_life.html. Accessed 11 Nov 2014.

M.G. Weiss (✉)
Department of Philosophy, University of Klagenfurt, Universitaetsstr. 65, 9020 Klagenfurt, Austria
e-mail: Martin.Weiss@aau.at

© Springer International Publishing Switzerland 2016
K. Hagen et al. (eds.), *Ambivalences of Creating Life*, Ethics of Science and Technology Assessment 45, DOI 10.1007/978-3-319-21088-9_18

The possible applications of synthetic biology are amazing:

Applications include (*inter alia*) renewable biofuels, pharmaceuticals, technical materials, nuclear waste disposal, chemical detoxification, superefficient agriculture, energy harvesting and conversion, and geoengineering. Creating a so-called 'minimal microbe' template could streamline the design of biological systems or their components, which could then be engineered for a host of specialized applications, such as those described above. (Buchanan and Powell 2010, p. 1)

2 Playing God to Understand Nature

The importance the public, at least decision-makers, have given to synthetic biology is illustrated by the fact that on the very same day Venter made his announcement, Barack Obama, the President of the United States, ordered his newly appointed *Commission for the Study of Bioethical Issues* to take action and to deliver a report on the pros and cons of synthetic biology within six weeks:

In its study, the Commission should consider the potential medical, environmental, security, and other benefits of this field of research, as well as any potential health, security or other risks. Further, the Commission should develop recommendations about any actions the Federal government should take to ensure that America reaps the benefits of this developing field of science while identifying appropriate ethical boundaries and minimizing identified risks. (Obama 2010, p. 1)

Two months later the first hearings before the commission took place. The cited bioethicists not only mentioned biosafety implications—by stressing the difficulty to foresee all possible consequences of this technology—but also addressed biosecurity issues:

A government-funded antibioterrorist initiative may stimulate research that the government then uses for offensive purposes. Notoriously, the line between defensive and offensive bioweapons development is hard to draw and accountability for keeping within the limits of defense is difficult. (Buchanan and Powell 2010, p. 2)

Surprisingly, Allen Buchanan and Russel Powell explicitly dismissed the argument, used by former Bioethics Commissions against technological innovations, that synthetic biology was problematic because it would violate the intrinsic essence of nature, or amount to "playing God", as such an allegation would probably presuppose a normative notion of nature on grounds of which one would be able to distinguish between allowed (i.e. therapeutic) and forbidden (i.e. manipulative) interventions in nature:

Because intervention in nature in furtherance of human good is widely accepted in religious and secular circles alike (consider the eradication and treatment of disease, for example), one would need to come up with a principled basis by which to distinguish hubristic interventions that amount to playing God from those that are laudably foresighted, realistic, and desirable. (Buchanan and Powell 2010, p. 5)

But even if Buchanan and Powell were wrong and the distinction between therapeutic and manipulative interventions beyond question, it still would remain questionable if deliberate manipulations of nature really amount to "playing God", as in the history of the three great monotheistic traditions (Judaism, Christianity and Islam) the relationship of man to nature has taken at least three very different forms: domination, stewardship, and co-creation (Coady 2009), of which the latter is of particular interest in the context of the debate on synthetic biology.

Thus for Eric Parens,

> [...] according to Genesis, and it seems to me much of Judaism, our responsibility is not merely to be grateful and remember that we are not the creator of the whole. It is also our responsibility to use our creativity to mend and transform ourselves and the world. As far as I can tell, Genesis and Judaism do not exhort us to choose between gratitude and creativity. (Parens 2009, p. 189)

The idea of human co-creation, well known also in Christianity (Rahner 1970), is especially important in Judaism. Here, man is seen as being as natural as everything else and therefore a fundamental part of creation. And if man is part of created nature, his products are as well. Rabbi Barry Freundel, a consultant with the former United States Presidential Commission on Cloning, states:

> If God has built the capacity for gene redesign into nature, then He chose for it to be available to us, and our test remains whether we will use that power wisely or poorly. (Freundel 2000, p. 119)

Here the idea that man is the image of God results in the conception that man, as poor as he is, has to be creative himself, and use technology to help God creating the world (Prainsack 2006). In this perspective synthetic biology would not amount to blasphemous hubris, but be the ultimate expression of the idea that man is created in Gods image and shares the ability to create with its creator. Thus the capacity to create life is nothing that distances us from God but rather a way to experience our proximity to God. It is noteworthy that the capacity to create life makes man similar to God not only with regard to his "manufacturing abilities", but also with regard to his knowledge, as the ability to construct life is widely identified with understanding life, as we will see in the next section.

3 The Epistemology of Synthetic Biology

Apart from the prospective practical applications of synthetic biology, the strongest argument put forward to promote this technology is the assertion, that the ability to create new life-forms would increase our understanding of how life-forms work and ultimately lead to a better, God-like, understanding of life itself:

> The capacity to engineer and reverse-engineer organisms may vastly increase our knowledge of the complex causal relations between genomes and the functional properties of living things, leading to unanticipated benefits. (Buchanan and Powell 2010, p. 1)

That Venter and his team also aimed primarily at increasing our knowledge about life is evident from the "Watermarks" engraved into their bacteria, as the three quotations encrypted into the non-coding DNA of their synthetic genome are: "To live, to err, to fall, to triumph, and to recreate life out of life" (James Joyce), "See things not as they are, but as they might be" (Robert Oppenheimer) and finally "What I cannot build, I cannot understand" (Cf. Angier 2010; Dillow 2010). Especially the last (wrong) quotation from the physicist Richard Feynman (which correctly reads: "What I cannot create, I do not understand"),[2] can be taken as the hidden epistemological principle on which the entire endeavor of synthetic biology is based: We can only understand what we can construct.

The association of knowledge with the notion of building is commonly identified as the specific novelty of synthetic biology (Irrgang 2008, p. 303). The venerable epistemological principle that identified knowledge with observation, i.e. *theoria*, has allegedly been replaced by the idea that knowledge consists in construction. The idea that we know something when we know *what* it is, has allegedly been replaced be the notion that we know something when we are able to *construct* it. Science has finally become engineering.

However, the equation of knowledge with construction is nothing new at all. On the contrary, it is the leading principle of modern science. In his *De Corpore* Thomas Hobbes wrote "Ubi ergo generatio nulla [...] ibi nulla philosophia intelligitur"[3] and Immanuel Kant asserted: "Finite beings cannot [...] understand other things, because they are not their creator" (Kant 1882, p. 229). What at first glance may appear as the "hypermodern" (Irrgang 2008, p. 303) replacement of the old-fashioned epistemological paradigm of description with the new paradigm of construction, at a second glance dates back at least to the 17th century.

Leaving aside the question when this change of paradigm occurred exactly, the problem remains to explain how and why this transition from description to construction took place? Or has this transition never happened at all? Has construction perhaps always been the hidden paradigm of knowledge? And even more pressing: Do we really understand something when we can build it? What kind of knowledge do we gain in constructing something? What do we learn about a thing when we know how to construct it? By constructing something, do we learn what it is, or only how we can use it? What it is in itself, or what it is for us?

3.1 Martin Heidegger on Synthetic Biology: The Constructivist Paradigm of Western Philosophy

That the occidental philosophical tradition, at least since Plato, has conceived the real as produced and identified knowledge with construction, is the thesis put

---

[2]A photo of a blackboard with Feynman's phrase can be found on the Homepage of Caltech: http://archives.caltech.edu/pictures/1.10-29.jpg. Accessed 11 Nov 2014.

[3]https://archive.org/stream/thomhobbesmalme03molegoog#page/n120/mode/2up. Accessed 20 June 2015.

forward by Martin Heidegger in his lectures on the *Basic Problems of Phenomenology* held in 1927. According to Heidegger the basic concepts of Western philosophy, *existentia et essentia*, being and essence, have always been conceived from a constructivist perspective. Heidegger writes: "We must interpret them with a view to production" (Heidegger 1975, p. 106). What does that mean?

In regard to the notion of existence Heidegger explicates that we understand reality, i.e. "actuality" (*Wirklichkeit*) as that which has the power of agency (*Wirkung*). Real is what generates effects. But this model implies that before something can be real, i.e. generate effects, it must itself have been generated, i.e. produced. Thus to be real in the first place means to be produced and only secondarily to have the power to produce.

According to Heidegger, existence has therefore always been conceived as something realized, generated, constructed, produced. Also the notion of essence, which the ancient Greeks called *idea* or *morphe*, makes sense for Heidegger only within a constructivist framework:

> The potter forms a vase out of clay. All forming of shaped products is effected by using an image, in the sense of a model, as guide and standard. [...] It is this anticipated look of the thing, sighted beforehand, that the Greeks mean ontologically by *eidos, idea*. (Heidegger 1975, p. 106)

Thus according to Heidegger, the reigning paradigm of Western ontology has always been production. Since the origins of philosophy, *to be* meant *to be the product of intentional construction*. Reality has always been understood as some sort of artefact, a concept palpable also in the Christian idea of creation: To exist means to be created.

According to Heidegger, the fact that production—thought of as formation—needs some sort of matter does not contradict the general constructivist paradigm. On the contrary:

> The concepts of matter and material have their origin in an understanding of being that is oriented to production. Otherwise, the idea of material as that from which something is produced would remain hidden. The concepts of matter and material, *hule*, that is, the counter-concepts of *morphe*, form, play a fundamental role in ancient philosophy not because the Greeks were materialists but because matter is a basic ontological concept that arises necessarily when a being ... is interpreted in the horizon of the understanding of being which lies as such in productive comportment [*herstellendes Verhalten*]. (Heidegger 1975, p. 116)

The paradigm of "productive comportment" identifies reality with formed matter, i.e. with a product of construction. But has reality not always also been identified with objectivity, meaning independence from any subjectivity? How does the notion of "productive comportment" relate to objectivity?

Even if we accept Heidegger's explanation of traditional ontology as an effect of an underlying implicit "productive comportment", it still seems difficult to make sense of the notion of objectivity within this paradigm. If reality has always been implicitly understood as the product of construction, then how comes that reality has also always been described as independent from any producing subjectivity?

Heidegger argues that objectivity, the being-in-itself, which we traditionally assign to reality, is also the effect of the paradigm of construction and correlates with the notion of a producing subjectivity:

> Productive comportment's understanding of the being [...] takes [...] being beforehand as one that is to be released for its own self so as to stand independently on its own account. The being that is understood in productive comportment is exactly the being-in-itself of the product. (Heidegger 1975, p. 113)

In constructivist ontology, to produce means to release into independence. But the conception of reality as an independent product does not mean that reality is granted autonomy. On the contrary, reality, understood as an independent product of subjectivity, is always already implicitly identified with the disposable, the instrument that we can use for our purpose:

> The thing to be produced is not understood in productive action as something which, as product in general, is supposed to be existent in itself. Rather, in accordance with the productive intention implicit in it, it is already apprehended as something that, qua finished, is available at any time for use. (Heidegger 1975, p. 114)

On the ground of Heidegger's analysis of Western ontology as the result of a constructivist paradigm, it not only becomes clear where the fundamental concepts of Western philosophy, i.e. existence and essence, come from, but also why today our understanding of something can be defined as the capacity to construct it:

> We heard earlier that the concept of *eidos* also had grown from the horizon of production. [...] What constitutes the being of the being is already anticipated in the *eidos*. [...] The anticipation of the prototypical pattern which takes place in production is the true knowledge of what the product is. It is for this reason that only the producer of something, its originator, perceives a being in the light of what it is. Because the creator and producer imagines the model beforehand, he is therefore also the one who really knows the product. (Heidegger 1975, p. 151)

Heidegger's analysis of Western ontology convincingly shows that "productive comportment" has always been the leading attitude towards being, which explains how synthetic biology can claim that in order to understand living beings we must be able to construct them.

My chapter could end here, if Heidegger would not insist on the dubiousness of this paradigm. Actually, Heidegger stresses that the traditional interpretation of being within the paradigm of construction misses the true essence of being or nature, if it reduces nature to that which is produced. To grasp the essence of nature, according to Heidegger, we have to go back to pre-Socratic philosophy, which described nature as that which generates itself. Thus for Heidegger the term *physis*, nature, indicates that which appears by itself, or the event of appearance, and is exactly not a product, because there is no producer. According to Heidegger, Nature is not something produced, not *natura naturata*, but the power and process of pro-duction itself (*Hervorbringung*), i.e. *natura naturans*.

If nature is that which appears by itself, or the event of appearance, and artifacts are the products of human activity, i.e. technology, how then are nature and technology related? Are nature and technology in sharp contrast, or do they have something in common? What happens to nature, i.e. to that which appears by itself, when it becomes the object of technology—as in synthetic biology?

For Heidegger, technology is not opposed to nature. On the contrary: technology as a specific phenomenon is itself something that appears by itself and therefore a manifestation of nature. The singular technological products may be the result of human intentions, but technology, i.e. the fact that man manipulates matter to produce usable artifacts, is nothing man could deliberately start or stop. If nature is that which appears by itself, then the technical production of artifacts is itself a specific mode of this appearance. Thus for Heidegger technology, which at first glance seems to be the exact opposite of nature—as technology is the production of usable artifacts and nature that which is beyond human disposition—at second glance turns out to be the best example of that which appears by itself. Technology, commonly envisioned as a means to dominate nature, becomes itself a natural force in which the true nature of nature becomes visible. In this sense synthetic biology, the technology as well as its products, may reveal the essence of technology, i.e. the untamable and uncontrollable nature of nature.

4 Bioart: Back to Life

If Heidegger is right in claiming that the original meaning of nature is *physis*—i.e. that which has the principle of its movement, including becoming and dying, in itself (as Aristotle puts it), so that to be natural means to posses some kind of autonomy, to posses something one could call "selfhood"—then natural phenomena, i.e. living beings, can't be grasped neither by describing them as mere physical cause-effect relations nor by ascribing them appropriateness for human purposes. Living beings are neither biochemical mechanism nor mere machines, i.e. instruments to achieve human aims, but irreducible things in themselves, i.e. subjects.[4] In this respect, however, living beings resemble works of art, which are also irreducible neither to their physical makeup, nor to a fixed purpose, but autonomous entities. In this sense artworks and living beings are analogous, in this sense both may be called organisms.

This venerable analogy between artworks and living beings has gained new relevance with the rise of some lines of Bioart (Kac 2007a; Stocker 2005; Wray in this book) which use the same methods and technologies as synthetic biology, but not to construct dead biochemical machines as means to predefined human ends, but to produce life in the proper sense of the word, i.e. to produce nature in the

---

[4] „Das Kunstwerk ist eine Art Analogon des ‚von Natur Seienden', weil es selbst ein Zentrum von Bedeutsamkeit ist, das wir wahrnehmen können, und das sich für uns nicht erschöpft in dem, was es in *unserem* Lebenszusammenhang bedeutet. So wie jedes Lebendige, so stiftet das schöne Ding einen eigenen Horizont von Bedeutsamkeit. Und so erst ist es im vollen Sinne des Wortes wirklich und, wie bei allem Wirklichen, ist seine Bestimmtheit unendlich" (Spaemann 2007, p. 258).

sense of *physis*. Paradoxically in a situation where nature and living beings are commonly perceived as dead biochemical machines, one possible way back to an original experience of nature, i.e. *physis*, seems to be its technological manipulation. Resembling the Hegelian "double negation", the technological manipulation (i.e. negation) of biological material, always already predefined as mere cause-effect relations (i.e. as negation of *physis*), leads to a reaffirmation of *physis*.

The work that best illustrates this return to *physis* trough the negation of its negation is Eduardo Kac's *Genesis* exhibited for the first time 1999 at the *Ars Electronica* festival in Linz, Austria:

> *Genesis* is a transgenic artwork [...]. The key element [...] is [...] a synthetic gene that was created by translating a sentence from the biblical Book of Genesis into Morse code, and converting the Morse code into DNA base pairs [...]. The sentence reads: 'Let man have dominion over the fish of the sea, and over the fowl of the air and over every living thing that moves upon the earth.' [...] The Genesis gene was incorporated into bacteria, which were shown in the gallery. Participants on the Web could turn on an ultraviolet light in the gallery, causing real, biological mutations in the bacteria. This changed the biblical sentence in the bacteria. After the show, the DNA of the bacteria was translated back into Morse code and than back into English. The mutation that took place in the DNA had changed the original sentence from the Bible (Kac 2007b, p. 164).

Thus Kac's *Genesis* shows how the attempt to negate the autonomy of the living material by manipulating it technologically turns into the experience of the uncontrollability of *physis*.

But if Bioart uses the same methods as synthetic biology and if Heidegger is right in claiming that ontology and epistemology have followed a constructivist paradigm at least since Plato (leading us to forget the original meaning *physis*), what then marks the difference between synthetic biology and Bioart, respectively between the genetically engineered synthetic life-forms and the artistically created artworks of Bioart?

In the constructivist perspective reality is (at least implicitly) conceived as an intentional product, which as such possess an intelligible essence (i.e. its model), which is furthermore always linked to a specific human aim. Therefore, in the constructivist ontological and epistemological perspective of Western philosophy, reality is always already conceived as something "at hand", as an instrument for human purposes without any autonomous meaning.

The work created in Bioart in contrast has no pre-visioned idea at its origin, no fixed essence. The work of Bioart is not constructed with reference to an already known model, but an open process of production, therefore a form of *natura naturans* and not of *natura naturata*. As the work of art can not be traced back to a model of which it would represent the realization, but must be conceived as mere becoming without an identifiable end in the sense of aim, the work of Bioart resists all attempts to be reduced to a mere instrument. Whereas in the Western constructivist perspective all reality is always already instrumentalised, the works of Bioart and the living beings resist reification and thus represent last remnants of *physis* in an otherwise completely instrumentalised world.

But if the production of *physis*, i.e. of that which has the principle of its movement in itself, is the aim of Bioart, then from a Heideggerian perspective, Bioart makes explicitly visible what is the essence of all technology. Or put the other way round: Technology is, by unintentionally producing *physis* (or revealing to be *physis* itself), a sort of disguised Bioart.

5 Conclusion

By identifying technology as a natural phenomenon which allows to (re)experience the (forgotten) nature of nature, i.e. its spontaneity, Heidegger seems to sustain not only some sort of technological determinism, but to evade the whole issue of technology assessment, as even the greatest technical failure would have to be considered a welcomed hint towards the true essence of reality, i.e. its unavailability (*Unverfügbarkeit*). On the one side this concept may permit to connect the philosophy of technology to the philosophy of nature, thus avoiding unfruitful dualisms, but on the other side it may make impossible any kind of critique of technological developments, let alone the formulation of strategies to control technology.[5]

It is certainly true that Heidegger is quite pessimistic about the possibility of mankind to willingly stop or even turn back technological evolution:

> No single man, no group of men, no commission of prominent statesmen, scientists, and technicians, no conference of leaders of commerce and industry, can brake or direct the progress of history in the atomic age. No merely human organization is capable of gaining dominion over it. Is man, then, a defenseless and perplexed victim at the mercy of the irresistible superior power of technology? He would be if man today abandons any intention to pit meditative thinking decisively against merely calculative thinking. (Heidegger 1966, p. 55)

Actually, Heidegger's suggestion how man could possibly react to technology is quite sobering. As opposition is not feasible—neither in the form of a naïve turning back technological evolution, nor in the self-contradicting form of fixing problems created by technology by technological means—Heidegger proposes a change in attitudes towards technology. As, like it or not, we live in a technological world, the only thing we can do to oppose the reduction of nature to mere manipulable objects, is to think technology differently. Technology must be freed from the conceptual framework of "calculating thinking", i.e. reductionism, and be put in the framework of what Heidegger calls "meditative thinking", i.e. a thinking that accepts that technology is essentially unavailable (*unverfügbar*). In this perspective it becomes evident that technological developments, especially synthetic biology, are an appearance of nature, i.e. a brute force that, according to Heidegger, has the power to eliminate true, i.e. free, human life, as he noted already in 1955:

---

[5]I owe this insight to Georg Toepfer.

The international meeting of Nobel Prize winners took place again in the summer of this year of 1955 in Lindau. There the American chemist, Stanley, had this to say: ‚The hour is near when life will be placed in the hands of the chemist who will be able to synthesize, split and change living substance at will.' We take notice of such a statement. We even marvel at the daring of scientific research, without thinking about it. We do not stop to consider that an attack with technological means is being prepared upon the life and nature of man compared with which the explosion of the hydrogen bomb means little. For precisely if the hydrogen bombs do not explode and human life on earth is preserved, an uncanny change in the world moves upon us. (Heidegger 1966, p. 52)

Thus Heidegger saw clearly the dangers of the capacity "to synthesize, split and change living substance at will", i.e. the dangers of synthetic biology, but as for him the problem is not a particular technology but the underlying technological worldview—one that reduces reality, including the human being, to mere causal mechanism—the adequate response to this problem cannot consist in particular guidelines for particular technologies, which have no effect on the underlying technological ideology, but only in a complete change of the way in which we conceive technology, which then may also lead to a different kind of technology. Bioart may be a precursor of this natural technology to come.

References

Angier N (2010) Peering over the fortress that is the mighty cell, New York Times, 31 May 2010. http://www.nytimes.com/2010/06/01/science/01angi.html. Accessed 11 Nov 2014

Buchanan A, Powell R (2010) The ethics of synthetic biology. Suggestions for a comprehensive approach. https://bioethics.gov/sites/default/files/The-Ethics-of-Synthetic-Biology-Suggestions-for-a-Comprehensive-Approach.pdf Accessed 28 Apr 2015

Coady C (2009) Playing God. In: Bostrom N (ed) Human enhancement. Oxford University Press, Oxford, pp 155–181

Dillow C (2010) Venter Institute's synthetic cell genome contains hidden messages, Popular Science, 25 May 2010 http://www.popsci.com/science/article/2010-05/venter-institutes-synthetic-cell-genome-contains-hidden-messages-watermarks#. Accessed 11 Nov 2014

Endy D (2010) In: Presidential commission for the study of bioethical issues (ed. and transcript): Transcript from meeting one: 8–9 July, 2010, Washington, D.C., Overview and Context of the Science and Technology of Synthetic Biology, Session 1. http://bioethics.gov/node/164. Accessed 05 Mar 2015

Freundel B (2000) Gene modification technology. In: Stock G (ed) Engineering the human germline: an exploration of the science and ethics of altering the genes we pass to our children. Oxford University Press, Oxford, pp 119–122

Heidegger M (1966) Discourse on thinking. Harper & Row, New York

Heidegger M (1975) The basic problems of phenomenology. Indiana University Press, Bloomington

Irrgang B (2008) Philosophie der Technik. WBG, Darmstadt

Kac E (ed) (2007a) Signs of life. Bio art and beyond. MIT, Cambridge, MA

Kac E (2007b) Life transformation—Art mutation. In: Kac E (ed) Signs of life. Bio art and beyond. MIT, Cambridge, MA, pp 164–184

Kant I (1882) Reflexionen Kants zur kritischen Philosophie. Aus Kants handschriftlichen Aufzeichnungen. Ed. by Benno Erdmann, I/1, Reflexionen zur Anthropologie, Fues's Verlag (R. Reisland), Leipzig

Obama B (2010) Letter from president Obama. https://bioethics.gov/sites/default/files/news/Letter-from-President-Obama-05.20.10.pdf. Accessed 28 Apr 2015

Parens E (2009) Toward a more fruitful debate about enhancement. In: Bostrom N (ed) Human enhancement. Oxford University Press, Oxford, pp 181–198

Prainsack B (2006) Negotiating life: the regulation of human cloning and embryonic stem cell research in Israel. Soc Stud Sci 36(2):173–205

Rahner K (1970) Zum Problem der genetischen Manipulation aus der Sicht des Theologen. In: Friedrich Wagner F (ed) Menschenzüchtung: Das Problem der genetischen Manipulierung des Menschen. C. H. Beck, München, pp 135–166

Spaemann R (2007) Was heißt ‚Die Kunst ahmt die Natur nach'? Philosophisches Jahrbuch 2007 (2):247–264

Stocker G (ed) (2005) Hybrid. Living in paradox. Ars Electronica 2005, Hatje Cantz, Ostfildern